Biometric Technologies and Verification Systems

Biometric Technologies and Verification Systems

John R. Vacca

AMSTERDAM • BOSTON • HEIDELBERG • LONDON
NEW YORK • OXFORD • PARIS • SAN DIEGO
SAN FRANCISCO • SINGAPORE • SYDNEY • TOKYO

Butterworth-Heinemann is an imprint of Elsevier

Acquisitions Editor: Pamela Chester
Assistant Editor: Kelly Weaver
Senior Marketing Manager: Phyllis Cerys
Project Manager: Jeff Freeland
Cover Designer: Stewart Larking
Compositor: Cepha Imaging Private Limited
Cover Printer: Phoenix Color Corp.

Butterworth-Heinemann is an imprint of Elsevier
30 Corporate Drive, Suite 400, Burlington, MA 01803, USA
Linacre House, Jordan Hill, Oxford OX2 8DP, UK

Recognizing the importance of preserving what has been written, Elsevier prints its books on
acid-free paper whenever possible.

Library of Congress Cataloging-in-Publication Data
Vacca, John R.
  Biometric technologies and verification systems / by John Vacca.
       p. cm.
  Includes bibliographical references and index.
  ISBN-13: 978-0-7506-7967-1 (alk. paper)
  ISBN-10: 0-7506-7967-0 (alk. paper)
  1. Biometric identification. I. Title.

  TK7882.B56V33 2007
  006.4–dc22

                                                                    2006051915

British Library Cataloguing-in-Publication Data
A catalogue record for this book is available from the British Library.

ISBN: 978-0-7506-7967-1

For information on all Butterworth-Heinemann publications
visit our Web site at www.books.elsevier.com

Transferred to Digital Printing in 2010

This book is dedicated to David Lee.

Contents

Foreword

From the movie screen to the office building—biometric verification systems that were once the fancy of moviemakers and science fiction writers are quickly becoming commonplace. Just a few years ago, most people were skeptical that biometric technology would ever be used as widely as it is now. The consensus that biometrics would take decades to find its way into common use was a gross error. Publishers, editors, writers, and forecasters missed the mark by at least a decade.

Finally, a book that explains and illustrates what individuals and organizations can do with biometric technologies and verification systems has arrived. I know that John Vacca wanted to write this book ten years ago, but editors thought that biometric technologies and verification systems were not going to make it out of the lab.

This book provides comprehensive coverage of biometric technologies and verification systems, and provides a solid education for any student or professional in a world where concerns about security have become the norm.

It covers biometric technologies and verification systems from top to bottom, and also provides explanations of the most important aspects of the technology and how to best use that technology to improve security.

I highly recommend this book for all IT or security professionals as well as those entering the field of security. I also highly recommend it to curriculum planners and instructors for use in the classroom.

Michael Erbschloe
Security Consultant and Author
St. Louis, Missouri

Acknowledgments

There are many people whose efforts on this book have contributed to its successful completion. I owe each a debt of gratitude and want to take this opportunity to offer my sincere thanks.

A very special thanks to my Senior Acquisitions Editor, Mark A. Listewnik, without whose continued interest and support this book would not have been possible, and Assistant Editor Kelly Weaver, who provided staunch support and encouragement when it was most needed. Thanks to my Project Manager, Jeff Freeland, and my copyeditor, Janet Parkinson, whose fine editorial work has been invaluable. Thanks also to my marketing manager, Chris Nolin, whose efforts on this book have been greatly appreciated. Finally, thanks to all of the other people at Academic Press/Butterworth-Heinemann and Elsevier Science and Technology Books, whose many talents and skills are essential to a finished book.

Thanks to my wife, Bee Vacca, for her love, her help, and her understanding of my long working hours. Also, a very, very special thanks to Michael Erbschloe for writing the foreword. Finally, I wish to thank all the organizations and individuals who granted me permission to use the research material and information necessary for the completion of this book.

Introduction

Biometric technologies are crucial components of secure personal identification and verification systems, which control access to valuable information, to economic assets, and to parts of the national infrastructure. Biometric-based identification and verification systems support the information-based economy by enabling secure financial transactions and online sales, and by facilitating many law enforcement, health, and social service activities. Since September 11, 2001, the national requirements to strengthen homeland security have fallen short, hindering government and industry interest in attempting to apply biometric technologies to the automated verification of the identity of individuals.

As you know, biometric technologies are automated methods for identifying a person or verifying a person's identity based on the person's physiological or behavioral characteristics. Physiological characteristics include fingerprints, hand geometry, and facial, voice, iris, and retinal features; behavioral characteristics include the dynamics of signatures and keystrokes. Biometric technologies capture and process a person's unique characteristics, and then verify that person's identity based on comparison of the record of captured characteristics with a biometric sample presented by the person to be verified. After many years of research and development, biometric technologies have become reliable and cost-effective, and acceptable to users. However, new applications of biometrics are being somewhat successfully implemented in more secure travel documents, visas, and personal identity verification cards. These applications help to safeguard valuable assets and information and contribute to the safety and security of automated transactions, but have fallen short of strengthening homeland security.

Both public and private sectors are looking for reliable, accurate, and practical methods for the automated verification of identity. And they are using biometric technologies in a wide variety of applications, including health and social service programs, passport programs, driver licenses, electronic banking, investing, retail sales, and law enforcement (such as it is). Verification systems are usually characterized by three factors:

- Something that you know, such as a password;

- Something that you have, such as an ID badge;

- Something that you are, such as your fingerprints or your face.

Systems that incorporate all three factors are stronger than those that use only one or two factors. Verification using biometric factors can help to reduce identity theft and the need to remember passwords or to carry documents, which can be counterfeited. When biometric factors are used with one or two other factors, it is possible to achieve new and highly secure identity applications. For example, a biometric factor can be stored on a physical device, such as a smart card that is used to verify the identification of an individual. Today, the identification cards that are issued to employees for access to buildings and to information, and the cards that are used for financial transactions, often include biometric information.

Biometric factors can also be used with encryption keys and digital signatures to enhance secure verification. For example, biometric information could use public key infrastructure (PKI) systems that incorporate encryption (such as Federal Information Processing Standard [FIPS] 197, Advanced Encryption Standard). Encrypting the biometric information helps to make the system more tamper-resistant.

What's So Special About This Book?

Knowing when and how to weave biometrics into the security fabric of a customer's enterprise requires a comprehensive understanding of:

- The magnitude of the end user's unique security needs/desires;

- The size of the end user's budget;

- The environment in which the technologies will be used;

- What technologies the customer is already using;

- Which specific biometric technology and verification systems best address the end user's unique needs within the available budget.

Different types of businesses require different levels of security. Biometrics have been particularly popular as a physical access strategy with data centers and network co-location facilities. For example, this book will show how customers in this industry are using a combination of biometrics, CCTV, and mantraps to control access into main entry points and biometric readers to restrict access to network equipment cages. Common denominators in these kinds of applications are:

- Mission-critical servers, storage devices and miles of CAT-5, 6, and 7 cable reside throughout their facilities;

- The data residing and circulating through the facility is extremely sensitive;

- The locations are remote and unmarked and access is tightly restricted, so throughput is not as critical an issue;

- Robust budgets that accommodate the maximum levels of security.

However, most businesses do not share these characteristics. Networking equipment and data storage devices usually are stored centrally, creating a localized security hot spot. In contrast, most enterprise or campus environments have to provide access to a large number of employees, partners, vendors, and customers, all with varying levels of access privileges. In this situation, throughput, convenience, and transparency are priority issues. Proximity card access currently offers the best method of addressing these issues and also provides the basis for the photo identification requirement most organizations have. The best practice here would be to harden security as traffic approaches the organization's hot spots with the use of biometric readers, most of which are compatible with and are designed to easily replace card readers.

In addition, biometric technology and verification systems suppliers have made radical improvements in the costs of their products. For example, a finger scan reader that may have cost $500 only two years ago is now available for under $100, with many other readers available for under $200 per unit. More sophisticated iris scan readers have moved from the $7,000 range into the $4,000 range, and some manufacturers are predicting sub-$1,000 units soon. That being said, biometric technology and verification systems are still substantially more expensive to purchase than most card technologies, which are also dropping in cost. So while end users may express interest in deploying biometrics in their facilities, corporate budgets will often determine whether that will actually happen.

Also, current biometric product design necessitates that units be deployed indoors, as most have not been made rugged enough for prolonged exposure to outdoor conditions or vandalism. The amount and kind of traffic may also affect the selection of biometrics or cards. For instance, in parking structure applications or near main entrances, wireless card technologies like proximity are more convenient than biometrics.

End users will be more inclined to buy off on biometric value propositions if they can leverage rather than replace their current systems. This leveraging can be accomplished in a number of ways. A pure biometric system would function almost exactly like a card access system. Individuals attempting to gain access present their finger, hand, eye, or face, or speak into a microphone, in the same way they would present their card. The difference is that the typical proximity cardholder identification number requires 26 to 85 bits of memory. The typical fingerprint template used by a biometric system requires 250 to 1,000 bytes or, if you recalculate those numbers into bits for comparison, 2,000 to 8,000 bits. Obviously, it takes substantially more processing time and power to verify the identity of an individual biometric scan against a database of hundreds or thousands of others versus a cardholder number.

There are a few ways to use a customer's existing card-based system to solve this problem. One way is to associate each individual cardholder number with that person's biometric template. This can be done easily during the enrollment process, and requires that individuals present their existing card to a card reader either installed next to a biometric reader or actually built into it. The cardholder number tells the biometric system where to look on the template database for the individual's stored template, greatly reducing the amount of processing required to verify the authenticity of the biometric scan. Another way to simplify processing is to store the biometric template on a smart card. This eliminates the need for a separate biometric template database and the infrastructure needed to support it, because the smart card provides all of the storage and security needed. This is an especially popular method for government agency customers who are already using smart card technology for both physical and logical access. The third way to get around the processing problem is to store the biometric template on the controller panel.

After settling these issues, you still have to determine which kind of biometric technology and verification system best matches your customer's situation. The three technologies that this book will show to be the most practical currently are finger scan, hand scan (or hand geometry), and eye scan (either retina or iris). This book will also show you how to use voice or facial scan technologies to provide a practical solution for most commercial physical security applications. In addition, this book will show you how biometric technologies and verification systems offer the user the ability to adjust sensitivity or tolerance levels to balance false-accept and false-reject rates.

There is usually an indirect correlation between accuracy, as measured in the number of unique characteristics the technology can discern, and cost.

The level of intrusiveness is also an important consideration, because customers who deploy intrusive procedures into the organization could become the target of enterprise-wide hostility. Eye scan technology is probably the most accurate technology of the group, but it is also the most expensive and perceived to be the most intrusive. Retina scan products require that users position their eye within half an inch of the reader while over 400 unique features are scanned from the back of the eye. Iris scan technology offers a similar level of accuracy (around 260 unique features) and similar price, but is less intrusive. Individuals need only get within three feet for a reliable scan. Because either eye scanning process requires the individual to get into position and hold their eyes steady (usually for around two seconds), only the most security-conscious employees will be able to truly appreciate the reliability of eye scan technology.

Finger scan technology is probably the most popular of the biometric technologies and verification systems for a wide range of applications including logical access, Internet security, banking, and point-of-purchase. It offers a good balance between accuracy and cost and generally has managed to shake the criminal identification stigma. Traditional optical finger scan technology will most likely be replaced with newer silicon technology that requires less surface scanning area and less maintenance than optical scanning.

Given the current state of development among the various biometric technology and verification systems alternatives, hand scan, also known as hand geometry, integrates best with physical access systems and is the preferred choice for combining accuracy (up to 90 unique features or measurements) and cost, with a minimal perceived amount of intrusion. Hand geometry templates are the smallest available from current biometric technology and verification systems at around 9 bytes (72 bits), which translates into reduced processing and storage requirements. Hand geometry readers are designed to correctly position the individual's hand and ensure quick, efficient reads.

Once the decision has been made about where biometric technology and verification systems will be used in your customer's organization, which kind of technology will be used, and how it will be integrated with existing systems, the final step is to train customer security personnel. Not only will they need to know how to adjust the tolerances of the readers to balance false-accept and false-reject rates, they also will need to know how to calm employees' fears that their identities may be stolen. Additionally, the security director should expect some level of animosity toward the biometric readers when some employees are unable to access areas to which they are authorized due to improper use or narrow tolerance settings. Thoroughly preparing the security personnel can go a long way toward smoothing the path to acceptance of the new technology.

So, with the preceding in mind, the three most important selling points of this book are:

1.   Positive identification technology and systems

2.   Physical access control technology and systems

3.   Biometric engineering design techniques

Furthermore, biometric technology and verification systems offer a number of benefits to both businesses and consumers. It is these benefits, in addition to the factors noted earlier, that are driving their increased usage and acceptance:

- Combating credit card fraud
- Preventing identity theft
- Restoring identity
- Enhanced security
- Data verification/authentication

Any situation that allows for an interaction between man and machine is capable of incorporating biometrics. The benefits of biometrics will make the technology's use, and consequently its acceptance, inevitable.

As discussed in this book, the public acceptance of biometrics is not necessarily inevitable. It will only come if the privacy concerns associated with the technology are effectively addressed.

Whether biometrics are privacy's friend or foe is entirely dependent upon how the systems are designed and how the information is managed. While the biometric industry has made some positive initial steps, without private sector data protection legislation, companies are still free to use biometric data without restriction.

It must be recognized that the use of biometrics needs to conform to the standards and expectations of a privacy-minded society. The responsibility to ensure that this new technology does not knowingly or unknowingly compromise consumer privacy lies not only with businesses, but also with consumers.

Businesses must acknowledge and accept their obligation to protect their customers' privacy. Prior to introducing any biometric system, the impact that such an application may have on consumer privacy should be fully assessed. To appropriately and effectively balance the use of biometric information for legitimate business purposes with the consumer's right to privacy, companies should adopt and implement the fair information practices and requirements discussed in this book. Voluntary adoption of such practices is essential if there is to be meaningful privacy protection of consumers' biometric data in the private sector.

Finally, consumers need to advocate for their own privacy rights. They can make a difference by only doing business with companies that follow fair information practices and that make use of the privacy-enhancing aspects of biometrics in the design of their information management systems protection techniques. Consumer preferences will be key in defining the appropriate uses and protection of biometrics. Consumers have the power—they need to use it wisely.

Purpose

With the preceding in mind, the purpose of this book is to show experienced (intermediate to advanced) industry, government, and law enforcement professionals how to analyze and conduct biometric security, and how to report the findings leading to incarceration of the perpetrators. This book also provides the fundamental knowledge you need to analyze risks to your system and to implement a workable biometric security policy that protects your information assets from potential intrusion, damage, or theft. Through extensive hands-on examples (field and trial experiments) and case studies, you will gain the knowledge and skills required to master the deployment of biometric security systems to thwart potential attacks.

Scope

This book discusses the current state of the art in biometric verification/authentication, identification, and system design principles. The book also provides a step-by-step discussion of how biometrics works; how biometric data in human beings can be collected and analyzed in a number of ways; how biometrics are currently being used as a method of personal identification in which people are recognized by their own unique corporal or behavioral characteristics; and how to create detailed menus for designing a biometric verification system. Furthermore, the book will also discuss how human traits and behaviors can be used in biometrics, including fingerprints, voice, face, retina, iris, handwriting, and hand geometry. Essentially, biometrics is the same system the human brain uses to recognize and distinguish the man in the mirror from the man across the street. Using biometrics for identifying and verifying/authenticating human beings offers some unique advantages over more traditional methods. Only biometric verification/authentication is based on the identification of an intrinsic part of a human being. Tokens, such as smart cards, magnetic stripe cards, and physical keys, can be lost, stolen, or duplicated. Passwords can be forgotten, shared, or unintentionally observed by a third party. Forgotten passwords and lost smart cards are a nuisance for users and an expensive time-waster for system administrators. In addition, this book will show how biometrics can be integrated into any application that requires security, access control, and identification or verification of users. With biometric security, the key, the password, the PIN code can be dispensed with; the access-enabler is you—not something you know, or something you have.

Finally, this book leaves little doubt that the field of biometric security is about to evolve even further. This area of knowledge is now being researched, organized, and taught. No question, this book will benefit organizations and governments, as well as their biometric security professionals.

Target Audience

This book is primarily targeted at those in industry, government, and law enforcement who require the fundamental skills to develop and implement security schemes designed to protect their organizations' information from attacks, including managers, network and systems administrators, technical staff, and support personnel. This list of personnel also includes, but is not limited to, security engineers, security engineering designers, bioinformatics engineers, computer security engineers, molecular biologists, computer security officers, computational biologists, security managers, university-level professors, short course instructors, security R&D personnel, security consultants, and marketing staff.

Organization of This Book

The book is organized into nine parts composed of 30 chapters and an extensive glossary of biometric terms and acronyms at the end.

Part 1: Overview of Biometric Technology and Verification Systems

Part 1 discusses what biometrics are, types of biometrics technology and verification systems, and biometrics technology and verification systems standards.

Chapter 1, "What Is Biometrics?," sets the stage for the rest of the book by showing the importance of biometrics as a method of protection for enterprises, government, and law enforcement.

Chapter 2, "Types of Biometric Technology and Verification Systems," provides an overview of biometric technologies that are currently available and being developed, current uses of these technologies, and issues and challenges associated with the implementation of biometrics.

Chapter 3, "Biometric Technology and Verification Systems Standards," discusses related biometric standards development programs and business plans.

Part 2: How Biometric Eye Analysis Technology Works

Part 2 discusses how iris pattern recognition and retina pattern recognition works.

Chapter 4, "How Iris Pattern Recognition Works," discusses how iris-based personal identification (PI) or recognition uses the unique visible characteristics of the human iris (the tinted annular portion of the eye bounded by the black pupil and the white sclera) as its biometric.

Chapter 5, "How Retina Pattern Recognition Works," examines the anatomy and uniqueness of the retina, and forms the foundation for the following: the technology behind retinal pattern recognition, sources of problems (errors) and biometric performance standards, strengths and weaknesses of retinal pattern recognition, and the applications of retinal pattern recognition.

Part 3: How Biometric Facial Recognition Technology Works

Part 3 discusses how video face recognition and facial thermal imaging works.

Chapter 6, "How Video Face Recognition Works," shows how computers are turning your face into computer code so it can be compared to thousands, if not millions, of other faces.

Chapter 7, "How Facial Thermal Imaging in the Infrared Spectrum Works," proposes a method that enhances and complements Srivastava's approach.

Part 4: How Biometric Fingerscanning Analysis Technology Works

Part 4 discusses how finger image capture and finger scanning verification and recognition works.

Chapter 8, "How Finger Image Capture Works," thoroughly discusses finger image capture technology, which is also called fingerprint scanning.

Chapter 9, "How Fingerscanning Verification and Recognition Works," discusses how fingerprint sensors solve the size, cost, and reliability problems that have limited the widespread application of fingerscanning verification.

Part 5: How Biometric Geometry Analysis Technology Works

Part 5 discusses how hand geometry image technology and finger geometry technology works.

Chapter 10, "How Hand Geometry Image Technology Works," discusses how handprint recognition scans the outline or the shape of a shadow, and not the handprint.

Chapter 11, "How Finger Geometry Technology Works," discusses how a few biometric vendors use finger geometry or finger shape to determine identity.

Part 6: How Biometric Verification Technology Works

Part 6 discusses how dynamic signature verification technology, voice recognition technology, keystroke dynamics technology, palm print pattern recognition technology, vein pattern analysis recognition technology, ear shape analysis technology, body odor analysis technology, and DNA measurement technology works.

Chapter 12, "How Dynamic Signature Verification Technology Works," explores what new dynamic signature verification technology is doing to solve problems.

Chapter 13, "How Voice Recognition Technology Works," discusses how voice recognition technology is a viable solution to securely and inexpensively authenticate users both at a physical location and remotely.

Chapter 14, "How Keystroke Dynamics Technology Works," discusses how keystroke dynamics, a behavioral measurement, is a pattern exhibited by an individual using an input device in a consistent manner.

Chapter 15, "How Palm Print Pattern Recognition Technology Works," provides a brief overview of the historical progress of and future implications for palm print biometric recognition.

Chapter 16, "How Vein Pattern Analysis Recognition Technology Works," discusses why vein pattern recognition has gained sponsorship from companies that have developed reputations for developing products that compete successfully in global markets.

Chapter 17, "How Ear-Shape Analysis Technology Works," proposes a simple ear shape model-based technique for locating human ears in side face range images.

Chapter 18, "How Body Odor and/or Scent Analysis Technology Works," discusses how research laboratories envision tools that could identify and track just about every person, anywhere—and sound alarms when the systems encounter hazardous objects or chemical compounds.

Chapter 19, "How DNA Measurement Technology Works," discusses how an interesting application of the DNA "ink" would be to use it for the authentication of passports or visas.

Part 7: How Privacy-Enhanced Biometric-Based Verification/Authentication Works

Part 7 discusses how fingerprint verification/authentication technology, vulnerable points of a biometric verification system, brute force attacks, data hiding technology, image-based challenges/response methods, and cancelable biometrics works.

Chapter 20, "How Fingerprint Verification/Authentication Technology Works," contains an overview of fingerprint verification methods and related issues.

Chapter 21, "Vulnerable Points of a Biometric Verification System," outlines the inherent vulnerability of biometric-based verification, identifies the weak links in systems employing biometric-based verification, and presents new solutions for eliminating some of these weak links.

Chapter 22, "How Brute Force Attacks Work," proposes a technique for generating keys for symmetric cipher algorithms (such as the widely used Data Encryption Standard (DES) and 3-DES), to show how brute force attacks work and how they can be prevented

Chapter 23, "How Data-Hiding Technology Works," introduces two applications of an amplitude modulation-based watermarking method, in which the researchers hid a user's biometric data in a variety of images.

Chapter 24, "Image-Based Challenges/Response Methods," covers the inherent strengths of an image-based biometric user verification scheme and also describes the security holes in such systems.

Chapter 25, "How Cancelable Biometrics Work," discusses handwriting, voiceprints, and face recognition.

Part 8: Large-Scale Implementation/Deployment of Biometric Technologies and Verification Systems

Part 8 discusses specialized biometric enterprise deployment and how to implement biometric technology and verification systems.

Chapter 26, "Specialized Biometric Enterprise Deployment," provides an overview of the main types of device "form factors" that are available for practical use today.

Chapter 27, "How to Implement Biometric Technology and Verification Systems," deals with the implementation of social, economic, legal, and technological aspects of biometric and verification systems.

Part 9: Biometric Solutions and Future Directions

Part 9 discusses how mapping the body technology works, selecting biometric solutions, biometric benefits, and a glossary consisting of biometric security-related terms and acronyms.

Chapter 28, "How Mapping-the-Body Technology Works," presents a continuous human movement recognition (CHMR) framework, which forms a basis for the general biometric analysis of the continuous mapping of the human body in motion as demonstrated through tracking and recognition of hundreds of skills, from gait to twisting saltos.

Chapter 29, "Selecting Biometric Solutions," briefly describes some emerging biometric technologies to help guide your decision making.

Chapter 30, "Biometric Benefits," shows you the benefits of using biometric systems that use handwriting, hand geometry, voiceprints, and iris and vein structures.

And, finally, the "Glossary" consists of biometric security–related terms and acronyms.

John R. Vacca
Author and IT Consultant
visit us at http://www.johnvacca.com/

Part 1: Overview of Biometric Technology and Verification Systems

What Is Biometrics?

Once a tool primarily used by law enforcement, biometric technologies increasingly are being used by government agencies and private industry to verify a person's identity, secure the nation's borders (as possible), and to restrict access to secure sites including buildings and computer networks. Biometric systems recognize a person based on physiological characteristics, such as fingerprints, hand and facial features, and iris patterns, or behavioral characteristics that are learned or acquired, such as how a person signs his name, types, or even walks (see sidebar "Definition of Biometrics") [1].

Definition of Biometrics

Biometrics are automated methods of recognizing a person based on a physiological or behavioral characteristic. Biometric technologies are becoming the foundation of an extensive array of highly secure identification and personal verification solutions. Examples of physiological characteristics include hand or finger images, facial characteristics, and iris recognition. Behavioral characteristics are traits that are learned or acquired. Dynamic signature verification, speaker verification, and keystroke dynamics are examples of behavioral characteristics.

Biometrics is expected to be incorporated in solutions to provide for increased homeland security, including applications for improving airport security, strengthening our national borders, in travel documents and visas, and preventing ID theft. Now, more than ever, there is a wide range of interest in biometrics across federal, state, and local governments. Congressional offices and a large number of organizations involved in many markets are addressing the important role that biometrics will play in identifying and verifying the identity of individuals and protecting national assets.

There are many needs for biometrics beyond homeland security. Enterprise-wide network security infrastructures, secure electronic banking, investing and other financial transactions, retail sales, law enforcement, and health and social services are already benefiting from these technologies. A range of new applications can be found in such diverse environments as amusement parks, banks, credit unions, and other financial organizations, enterprise and government networks, passport programs and driver licenses, colleges, physical access to multiple facilities (nightclubs), and school lunch programs.

Biometric-based verification applications include workstation, network, and domain access, single sign-on, application logon, data protection, remote access to resources, transaction security, and Web security. Trust in these electronic transactions is essential to the healthy growth of the global economy—especially in the area of outsourced American jobs. Utilized alone or integrated with other technologies such as smart cards, encryption keys [9], and digital signatures, biometrics are set to pervade nearly all aspects of the economy and our daily lives. Utilizing biometrics for personal verification is becoming convenient and considerably more accurate than current methods (such as the utilization of passwords or PINs). This is because biometrics links the event to a particular individual (a password or token may be used by someone other than the authorized user); is convenient (nothing to carry or remember); accurate (it provides for positive verification); can provide an audit trail; and is becoming socially acceptable and inexpensive [2].

The successful use of the classic biometric, fingerprints, owes much to government and private industry research and development. For more than 30 years, computer scientists have helped the Federal Bureau of Investigation (FBI) improve the automation process for matching rolled fingerprints taken by law enforcement agencies or latent prints found at crime scenes against the FBI's master file of fingerprints. Test data have been used to develop automated systems that can correctly match fingerprints by the minutiae, or tiny details, that investigators previously had to read by hand. In cooperation with the American National Standards Institute (ANSI), the Commerce Department's National Institute of Standards and Technology (NIST) also developed a uniform way for fingerprint, facial, scar, mark, and tattoo data to be exchanged between different jurisdictions and between dissimilar systems made by different manufacturers [1].

In conjunction with the FBI, NIST has developed several databases, including one consisting of 858 latent fingerprints and their matching rolled file prints. This database can be used by researchers and commercial developers to create and test new fingerprint identification algorithms, test commercial and research systems that conform to the NIST/ANSI standard, and assist in training latent fingerprint examiners. The increasing use of specialized live fingerprint scanners will help ensure that a high-quality fingerprint can be captured quickly and added to the FBI's current files. Use of these scanners also should speed up the matching of fingerprints against the FBI database of more than 80 million prints [1].

Improved Biometrics Is Critical to Security!
But Is It?

Under the unpopular Patriot Act and the Enhanced Border Security and Visa Entry Reform Act (such as it is), the U.S. government is evaluating the ability of biometrics to enhance border security. But, that's all it is doing: still evaluating, with no promise of actual implementation in this present political climate of insecure borders and nonenforcement of deportation of illegal aliens. Nevertheless, these acts, when legally enforced, call for developing and certifying a technology standard for verifying the identity of individuals and determining the accuracy of biometric technologies, including fingerprints, facial recognition, and iris recognition [1].

For example, NIST recently tested both face and fingerprint recognition technologies using large realistic samples of biometric images obtained from several federal, state, and county agencies. Testing showed that fingerprints provide higher accuracy than facial recognition systems [1].

This program is producing standard measurements of accuracy for biometric systems, standard scoring software, and accuracy measurements for specific biometrics required for the system scenarios mandated under the Border Security Act. This work will have wide impact beyond the mandated systems when the present political climate changes, and border security is enforced. Standard test methods are likely to be accepted as international standards. Presently, discussions are still under way concerning the use of these same standards for airport security [1].

In November 2003, NIST submitted its report on this work to the State and Justice Departments for transmittal to the U.S. Congress in February 2004. The report recommended a dual approach that employs both fingerprint and facial recognition technology for a biometrics system to make the nation's borders more secure. Additional NIST studies evaluated the effectiveness and reliability of computerized facial recognition and fingerprint matching systems [1].

The Department of Homeland Security announced in July 2005 that to ensure the highest levels of accuracy in identifying people entering and exiting the United States, the United States Visitor and Immigrant Status Indicator Technology (US-VISIT) program will require a one-time 10-fingerscan capture for all first-time visitors. Subsequent entries will require two-print verification [1].

In addition to fingerprint systems, computer scientists at NIST have extensive experience working with systems that match facial images. While facial recognition systems employ different algorithms than fingerprint systems, many of the underlying methods for testing the accuracy of these systems are the same. Researchers have designed tests to measure the accuracy and reliability of software programs in matching facial patterns, using both still and video images [1].

Iris recognition is another potentially valuable biometric, but before its use is widespread, more testing is needed to determine its accuracy in operation. Researchers recently began the first large-scale evaluation to measure the accuracy of the underlying technology that makes iris recognition possible [1].

Different Biometric Standards

Open consensus standards, and associated testing, are critical for providing higher levels of security through biometric identification systems. For decades, NIST has been involved with the law enforcement community in biometric testing and standardization, and NIST has intensified its work in biometric standardization over the past nine years. For example, following the terrorist attacks of Sept. 11, 2001, NIST championed the establishment of formal national and international biometric standards development bodies to support deployment of standards-based solutions and to accelerate the development of voluntary consensus standards. These standards bodies are the Technical Committee M1 on Biometrics (established in November 2001 by the executive board of the International Committee for Information Technology Standards (INCITS)) as shown in sidebar "INCITS"; and the International Organization for Standardization (ISO)/International Electrotechnical Commission (IEC) Joint Technical Committee 1 Subcommittee on biometrics (known as JTC 1 SC 37-Biometrics, created in June 2002 (http://www.iso.org/iso/en/stdsdevelopment/tc/tclist/TechnicalCommitteeDetailPage.TechnicalCommitteeDetail?COMMID=5537)). NIST chairs both the INCITS committee and the JTC 1 SC 37-Biometrics and contributes to the work of these standard development bodies with technical expertise. INCITS has approved seven standards for the exchange of biometric data: two biometric application profiles, two biometric interface standards, and the Common Biometric Exchange Formats Framework (discussed later in this chapter). In 2005, ISO approved four biometric data interchange standards developed by the JTC 1 SC 37-Biometrics. These standards are being adopted both in the United States and abroad. Also, NIST has been charged with developing a Personal Identity Verification standard for secure and reliable forms of identification issued by the federal government to its employees and contractors [1].

Warning: URLs are subject to change without notice.

INCITS

The Executive Board of INCITS established Technical Committee M1, Biometrics, in November 2001 to ensure a high-priority, focused, and comprehensive approach in the United States for the rapid development and approval of formal national and international generic biometric standards. The M1 program of work includes biometric standards for data interchange formats, common file formats, application program interfaces, profiles, and performance testing and reporting. The goal of M1's work is to accelerate the deployment of significantly better, standards-based security solutions for purposes such as homeland defense and the prevention of identity theft [5], as well as other government and commercial applications based on biometric personal authentication.

M1 serves as the U.S. Technical Advisory Group (U.S. TAG) for the international organization ISO/IEC JTC 1/SC 37 on Biometrics, which was established in June 2002. As the U.S. TAG to SC 37, M1 is responsible for establishing U.S. positions and contributions to SC 37, as well as representing the U.S. at SC 37 meetings.

M1 Ad-Hoc Group:

This is the Ad-Hoc Group on Evaluating Multi-Biometric Systems (AHGEMS). The Ad-Hoc Group is responsible for a Study Project on the concepts of operation and methods of performance evaluation for multi-biometric systems. The Ad-Hoc Group concluded its work at its October 2005 meeting. The Final Report developed by AHGEMS can be found at: http://www.incits.org/tc_home/m1htm/docs/m1050676.pdf

M1 has created five new Task Groups to handle increased activity in biometrics. The purview of the five Task Groups is as follows:

- M1.2
- M1.3
- M1.4
- M1.5
- M1.6

M1.2

M1.2, the Task Group on Biometric Technical Interfaces, covers the standardization of all necessary interfaces and interactions between biometric components and subsystems, including the possible use of security mechanisms to protect stored data [6] and data transferred between systems. M1.2 will also consider the need for a reference model for the architecture and operation of biometric systems in order to identify the standards that are needed to support multivendor systems and their applications.

M1.3

M1.3, the Task Group on Biometric Data Interchange Formats, focuses on the standardization of the content, meaning, and representation of biometric data interchange formats. Currently, assigned projects are:

- Finger Pattern Based Interchange Format
- Finger Minutiae Format for Data Interchange
- Face Recognition Format for Data Interchange
- Iris Interchange Format
- Finger Image Based Interchange Format
- Signature/Sign Image Based Interchange Format
- Hand Geometry Interchange Format

M1.3 Ad-Hoc Group:

This is the Ad-Hoc Group on Data Quality. The Ad-Hoc is addressing means of quality and ways of expressing and interpreting the quality of a biometric sample.

M1.4

M1.4, the Task Group on Biometric Profiles, covers the standardization of Application Profile projects. Currently, assigned projects are:

- Application Profile for Interoperability and Data Interchange: Biometric Based Verification and Identification of Transportation Workers
- Application Profile for Interoperability, Data Interchange and Data Integrity: Biometric Based Personal Identification for Border Management
- Application Profile for Point-of-Sale Biometric Verification/Identification

M1.4 Ad-Hoc Group:

The M1.4 Ad-Hoc Group on Biometrics and E-Authentication (AHGBEA) is responsible for developing a technical report describing suitability of biometric architectures, security requirements, and recommendations for the use of biometrics for e-authentication. AHGBEA is also responsible for examining related biometrics and security issues related to the topics addressed in the Ad-Hoc Group's Terms of Reference.

M1.5

M1.5 is the Task Group on Biometric Performance Testing and Reporting. It handles the standardization of biometric performance metric definitions and calculations. These are approaches to test performance and requirements for reporting the results of these tests. M1.5 is responsible for the development of a Multi-Part Standard on Biometric Performance Testing and Reporting.

M1.6

M1.6, the Task Group on Cross Jurisdictional and Societal Issues, addresses study and standardization of technical solutions to societal aspects of biometric implementations. Excluded from the TG's scope is the specification of policies, the limitation of usage, or imposition of nontechnical requirements on the implementations of biometric technologies, applications, or systems. M1.6 is responsible for U.S. technical contributions to JTC1 SC 37 WG 6 on Cross-Jurisdictional and Societal Issues.

Membership on M1 and its Task Groups:

Membership on M1 and its Task Groups is open to all materially affected parties. There are two current Ad-Hoc Groups of M1 and its Task Groups.

First, is the Ad-Hoc Group on Evaluating Multi-Biometric Systems. This Ad-Hoc Group is responsible for a Study Project on the concepts of operation and methods of performance evaluation for multi-biometric systems.

Second, is the Ad-Hoc Group on Issues for Harmonizing Conformity Assessment to Biometric Standards. This Ad-Hoc Group is developing a taxonomy that identifies and defines the possible types of activities that may occur under Conformity Assessment schemes. The Terms of Reference for this Ad-Hoc Group includes identifying the standards that M1 could develop for use in biometric standards–based conformance testing programs [3].

Consortium Helps Advance Biometric Technologies

The Biometric Consortium serves as a focal point for the federal government's research, development, testing, evaluation, and application of biometric-based personal identification and verification technology (see sidebar, "Biometric Consortium"). The consortium now has more than 1,500 members, including 60 government agencies. NIST and the National Security Agency co-chair the consortium (no big surprise there, with regards to NSA). NIST has collaborated with the consortium, the biometric industry, and other biometric organizations to create a Common Biometric Exchange Formats Framework (CBEFF). The format already is part of government requirements for data interchange and is being adopted by the biometric industry. The specification defines biometric data structures that allow for exchange of many types of biometric data files, including data on fingerprints, faces, palm prints, retinas, and iris and voice patterns. NIST co-chaired the CBEFF Technical Development Team [1].

Biometric Consortium

As previously mentioned, the Biometric Consortium [4] serves as a focal point for research, development, testing, evaluation, and application of biometric-based personal identification/verification systems. The Biometric Consortium now has over 1,500 members from government, industry, and academia. Over 60 different federal agencies and members from 140 other organizations participate in the Biometric Consortium. Approximately 60% of the members are from industry. An electronic discussion list is maintained for Biometric Consortium members. This electronic discussion list provides an on-line environment for technical discussions among the members on all things biometric. The National Institute of Standards and Technology (NIST) [4] and the National Security Agency (NSA) [4] co-chair the Biometric Consortium (BC) and co-sponsor most of the BC activities. Recently NIST and NSA have co-sponsored and spearheaded a number of biometric-related activities, including the development of a Common Biometric Exchange File Format (CBEFF) [4], NIST Biometric Interoperability, Performance, and Assurance Working Group [4], a BioAPI Users' and Developers' Seminar [4], and the NIST BioAPI Interoperability Test Bed.

CBEFF describes a set of data elements necessary to support biometric technologies in a common way independently of the application and the domain of use (mobile devices, smart cards, protection of digital data, biometric data storage). CBEFF facilitates biometric data interchange between different system components or between systems; promotes interoperability of biometric-based application programs and systems; provides forward compatibility for technology improvements; and simplifies the software and hardware integration process. CBEFF was developed by a Technical Development Team comprised of members from industry, NIST, and NSA, and in coordination with industry consortiums (BioAPI Consortium [4] and TeleTrusT [4]) and a standards development group (ANSI/ASC X9F4 Working Group [4]).

The International Biometric Industry Association (IBIA) [4] is the Registration Authority for CBEFF format owner and format type values for organizations and vendors that require them. The NIST Biometric Interoperability, Performance and Assurance Working Group supports advancement of technically efficient and compatible biometric technology solutions on a national and international basis. It promotes and encourages exchange of information and collaborative efforts between users and private industry in all things biometric. The Working Group consists of 105 organizations representing biometric universities, government agencies, national labs, and industry organizations. The Working Group is currently addressing development of a simple testing methodology for biometric systems as well as addressing issues of biometric assurance. In addition, the Working Group is addressing the utilization of biometric data in smart card applications by developing a smart card format compliant with the Common Biometric Exchange File Format (CBEFF).

NIST and NSA also provide advice to other government agencies such as the General Services Administration (GSA) Office of Smart Cards Initiatives and DoD's Biometric Management Office. The Biometric Consortium (BC) holds annual conferences for its members and the general public.

The BC website is http://www.biometrics.org. It contains a variety of information on biometric technology, research results, federal and state applications, and other topics. With over 780,000 hits per month, it is one of the most used reference sources on biometrics. There is no cost to join the Biometric Consortium [4].

How Biometric Verification Systems Work

A door silently opens, activated by a video camera and a face recognition system. Computer access is granted by checking a fingerprint. Access to a security vault is allowed after an iris check. Are these scenes from the TV shows *24* or *Alias*, or the latest spy thriller movie? Perhaps, but soon this scenario could be in your office or on your desktop. Biometric verification technologies such as face, finger, hand, iris, and speaker recognition are commercially available today and are already coming into wide use. Recent advances in reliability and performance and declines in cost make these technologies attractive solutions for many computer and network access, protection of digital content, and physical access control problems [4].

What Is Biometric Verification?

Biometric verification requires comparing a registered or enrolled biometric sample (biometric template or identifier) against a newly captured biometric sample (for example, a fingerprint captured during a login). During enrollment, as shown in Figure 1-1, a sample of the biometric trait is captured, processed by a computer, and stored for later comparison [4].

Biometric recognition can be used in identification mode, where the biometric system identifies a person from the entire enrolled population by searching a database for a match based solely on the biometric. For example, an entire database can be searched to verify a person has not applied for entitlement benefits under two different names. This is sometimes called one-to-many matching. A system can also be used in verification mode in which the biometric

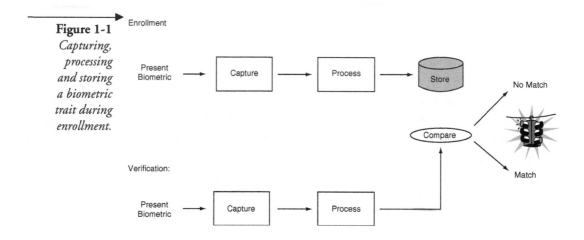

Figure 1-1
Capturing,
processing
and storing
a biometric
trait during
enrollment.

system verifies a person's claimed identity from their previously enrolled pattern. This is also called one-to-one matching. In most computer access or network access environments, verification mode would be used. A user enters an account number user name, or inserts a token such as a smart card, but instead of entering a password, a simple touch with a finger or a glance at a camera is enough to authenticate the user [4].

Uses for Biometrics

Biometric-based verification applications include workstation and network access, single sign-on, application logon, data protection, remote access to resources, transaction security, and Web security. The promises of e-commerce and e-government can be achieved through the utilization of strong personal verification procedures. Secure electronic banking, investing, and other financial transactions, retail sales, law enforcement, and health and social services are already benefiting from these technologies. Biometric technologies are expected to play a key role in personal verification for large-scale enterprise network verification environments, for point-of-sale and for the protection of all types of digital content such as in digital rights management and healthcare applications. Utilized alone or integrated with other technologies such as smart cards, encryption keys, and digital signatures, biometrics are anticipated to pervade nearly all aspects of the economy and our daily lives. For example, biometrics is used in various schools, such as in lunch programs in Pennsylvania [4] and a school library in Minnesota [4]. Examples of other current applications include verification of annual pass holders in an amusement park, speaker verification for television home shopping, Internet banking, and users' verification in a variety of social services [4].

Types of Biometrics

There are many types of biometrics currently in use, and many more types to come in the very near future (DNA, holograms [8], etc. ...). Today, some of the most common ones in use are:

- Fingerprints
- Face recognition
- Speaker recognition
- Iris recognition
- Hand and finger geometry
- Signature verification

Fingerprints

The patterns of friction ridges and valleys on an individual's fingertips are unique to that individual. For decades, law enforcement has been classifying and determining identity by matching key points of ridge endings and bifurcations. Fingerprints are unique for each finger of every person, including identical twins. One of the most commercially available biometric technologies, fingerprint recognition devices for desktop and laptop access are now widely available from many different vendors at a low cost. With these devices, users no longer need to type passwords—instead, a touch provides instant access. Fingerprint systems can also be used in identification mode. Several states check fingerprints for new applicants to social services benefits to ensure recipients do not fraudulently obtain benefits under fake names. New York state has over 1,500,000 people enrolled in such a system [4].

Face Recognition

The identification of a person by their facial image can be done in a number of different ways, such as by capturing an image of the face in the visible spectrum using an inexpensive camera or by using the infrared patterns of facial heat emission. Facial recognition in visible light typically models key features from the central portion of a facial image. Using a wide assortment of cameras, the visible light systems extract features from the captured image(s) that do not change over time, while avoiding superficial features such as facial expressions or hair. Several approaches to modeling facial images in the visible spectrum are principal component analysis, local feature analysis, neural networks, elastic graph theory, and multi-resolution analysis [4].

Some of the challenges of facial recognition in the visual spectrum include reducing the impact of variable lighting and detecting a mask or photograph. Some facial recognition systems may require a stationary or posed user in order to capture the image, though many systems use a real-time process to detect a person's head and locate the face automatically. Major benefits of facial recognition are that it is nonintrusive, hands-free, continuous, and accepted by most users [4].

Speaker Recognition

Speaker recognition has a history dating back some four decades, where the output of several analog filters were averaged over time for matching. Speaker recognition uses the acoustic features of speech that have been found to differ between individuals. These acoustic patterns reflect both anatomy (size and shape of the throat and mouth) and learned behavioral patterns (voice pitch, speaking style). This incorporation of learned patterns into the voice templates (the latter called voiceprints) has earned speaker recognition its classification as a behavioral biometric. Speaker recognition systems employ three styles of spoken input: text-dependent, text-prompted, and text-independent. Most speaker verification applications use text-dependent input, which involves selection and enrollment of one or more voice passwords. Text-prompted input is used whenever there is concern of imposters. The various technologies used to process and store voiceprints includes hidden Markov models, pattern-matching algorithms, neural networks, matrix representation, and decision trees. Some systems also use anti-speaker techniques, such as cohort models, and world models.

Ambient noise levels can impede both collection of the initial and subsequent voice samples. Performance degradation can result from changes in behavioral attributes of the voice and from enrollment using one telephone and verification on another telephone. Voice changes due to aging also need to be addressed by recognition systems. Many enterprises market speaker recognition engines, often as part of large voice processing, control, and switching systems. Capture of the biometric is seen as noninvasive. The technology needs little additional hardware by using existing microphones and voice-transmission technology, allowing recognition over long distances via ordinary telephones (wire line or wireless [7]) [4].

Iris Recognition

This recognition method uses the iris of the eye, which is the colored area that surrounds the pupil. Iris patterns are thought to be unique. The iris patterns are obtained through a video-based image acquisition system. Iris scanning devices have been used in personal authentication applications for several years. Systems based on iris recognition have substantially decreased in price, and this trend is expected to continue. The technology works well in both verification and identification modes (in systems performing one-to-many searches in a database). Current systems can be used even in the presence of eyeglasses and contact lenses. The technology is not intrusive. It does not require physical contact with a scanner. Iris recognition has been demonstrated to work with individuals from different ethnic groups and nationalities [4].

Hand and Finger Geometry

These methods of personal verification are well established. Hand recognition has been available for over 30 years. To achieve personal verification, a system may measure physical characteristics of either the fingers or the hands. These include length, width, thickness, and surface area of the hand. One interesting characteristic is that some systems require a small biometric sample (a few bytes). Hand geometry has gained acceptance in a range of applications. It can frequently be found in physical access control in commercial and residential applications, in time and attendance systems, and in general personal authentication applications [4].

Signature Verification

This technology uses the dynamic analysis of a signature to verify a person. The technology is based on measuring speed, pressure, and angle used by the person when a signature is produced. One focus for this technology has been e-business applications and other applications where signature is an accepted method of personal verification [4].

Why Use Biometrics?

Using biometrics for identifying human beings offers some unique advantages. Biometrics can be used to identify you as you. Tokens, such as smart cards, magnetic stripe cards, photo ID cards, physical keys and so forth, can be lost, stolen, duplicated, or left at home. Passwords can be forgotten, shared, or observed. Moreover, today's fast-paced electronic world means people are asked to remember a multitude of passwords and personal identification numbers for computer accounts, bank ATMs, e-mail accounts, wireless phones [7], websites and so forth. Biometrics holds the promise of fast, easy-to-use, accurate, reliable, and less expensive authentication for a variety of applications [4].

There is no one perfect biometric that fits all needs. Each biometric system has its own advantages and disadvantages. There are, however, some common characteristics needed to make a biometric system usable. First, the biometric must be based upon a distinguishable trait. For example, for over a century, law enforcement has used fingerprints to identify people. There is a great deal of scientific data supporting the idea that no two fingerprints are alike. Technologies such as hand geometry have been used for many years, and technologies such as face or iris recognition have come into widespread use. Some newer biometric methods may be just as accurate, but may require more research to establish their uniqueness [4].

Another key aspect is how user-friendly a system is. The process should be quick and easy, such as having a picture taken by a video camera, speaking into a microphone, or touching a fingerprint scanner. Low cost is important, but most implementers understand that it is not only the initial cost of the sensor or the matching software that is involved. Often, the life cycle support cost of providing system administration and an enrollment operator can overtake the initial cost of the biometric hardware [4].

Finally, the advantage that biometric verification provides is the ability to require more instances of verification in such a quick and easy manner that users are not bothered by the additional requirements. As biometric technologies mature and come into wide commercial use, dealing with multiple levels of verification or multiple instances of verification will become less of a burden for users.

Summary/Conclusion

Recent advances in biometric technology have resulted in increased accuracy at reduced cost. Biometric technologies are positioning themselves as the foundation for many highly secure identification and personal verification solutions. Today's biometric solutions provide a means to achieve fast, user-friendly verification with a high level of accuracy and cost savings. Many areas will benefit from biometric technologies. Highly secure and trustworthy electronic commerce, for example, will be essential to the healthy growth of the global Internet economy. Many biometric technology providers are already delivering biometric verification for a variety of Web-based and client/server-based applications to meet these and other needs. Continued improvements in technology will bring increased performance at a lower cost [4].

Finally, interest in biometrics is growing substantially. Evidence of the growing acceptance of biometrics is the availability in the marketplace of biometric-based verification solutions that are becoming more accurate, less expensive, faster, and easier to use. The Biometric Consortium, NIST, and

NSA are supporting this growth. While biometric verification is not a magical solution that solves all authentication concerns, it will make it easier and cheaper for you to use a variety of automated information systems—even if you're not a secret agent [4].

References

1. "NIST and Biometrics," NIST, 100 Bureau Drive, Stop 1070, Gaithersburg, MD 20899-1070 [US Department of Commerce, 1401 Constitution Avenue, NW, Washington, DC 20230], 2006.

2. "About Biometrics," NIST, 100 Bureau Drive, Stop 1070, Gaithersburg, MD 20899-1070 [US Department of Commerce, 1401 Constitution Avenue, NW, Washington, DC 20230], 2006.

3. "M1-Biometrics," INCITS Secretariat, c/o Information Technology Industry Council, 1250 Eye Street NW, Suite 200, Washington, DC 20005, 2006. Copyright © 2004 Information Technology Industry Council.

4. Fernando L. Podio and Jeffrey S. Dunn. "Biometric Authentication Technology: From the Movies to Your Desktop," NIST, 100 Bureau Drive, Stop 1070, Gaithersburg, MD 20899-1070 [US Department of Commerce, 1401 Constitution Avenue, NW, Washington, DC 20230], 2005.

5. John R. Vacca, *Identity Theft*, Prentice Hall (2002).

6. John R. Vacca, *The Essentials Guide to Storage Area Networks*, Prentice Hall, Professional Technical Reference, Pearson Education (2001).

7. John R. Vacca, *Guide to Wireless Network Security*, Springer (2006).

8. John R. Vacca, *Holograms: Design, Techniques, and Commercial Applications*, Charles River Media (2001).

9. John R. Vacca, *Public Key Infrastructure: Building Trusted Applications and Web Services*, CRC Press (2005).

2

Types of Biometric Technology and Verification Systems

Biometric technologies are available today that can be used in security systems to help protect assets. Biometric technologies vary in complexity, capabilities, and performance and can be used to verify or establish a person's identity. Leading biometric technologies include facial recognition, fingerprint recognition, hand geometry, iris recognition, retina recognition, signature recognition/verification, RFID chip implant, and speaker recognition/voice verification. Biometric technologies under development include palm print, vein patterns, DNA, ear shape, body odor, holography [10], and body scan [1].

Biometric technologies have been used in federal applications such as access control, criminal identification, surveillance, aviation/airports, and border security. Other applications include benefit-payment systems, shopping networks, drivers' licenses, prison visitor systems, voting systems, and so forth (see sidebar, "Applications for Biometric Technologies") [1].

Applications for Biometric Technologies

Biometrics has been used for a long time in areas not publicly advertised, so most of society is probably unaware of its use. These covert applications include surveillance systems that make use of facial recognition, voice verification, and signature verification. Governments have always demonstrated a keen interest in biometric devices. They have been the first to test and purchase the different types of biometric technologies and integrate them into their systems and facilities. But the overt terrorism of the 21st century has heightened businesses' awareness of vulnerabilities as well, prompting the growth of biometrics in nongovernment sectors. Use of biometrics in the commercial environment is expanding worldwide and is spilling over into the consumer marketplace.

Benefit-payment systems are beginning to implement more biometric verification systems in order to authenticate individuals applying for and claiming benefits. Such systems are now adding a fingerprint or hand-geometry biometric and sometimes a voice-verification biometric, as well as a photograph, to authenticate the claimant. As an example, the state of New York found that when a biometric was implemented in the welfare system, making it difficult for applicants to impersonate others, trade IDs, and have multiple identities, fraud was dramatically reduced. This resulted in savings of millions of dollars annually.

Television shopping networks such as QVC and financial brokerage houses like Charles Schwab are now implementing and using voice verification in order to authenticate customers and carry out transactions. Both of these companies use Nuance, a well-established voice-verification company. Customers can place orders and check on accounts without waiting for intervention from a human operator.

Drivers' licenses generally feature a picture that matches a state DMV database. Some states are now adding a fingerprint in order to prevent swapping of licenses and to make it more difficult to create fraudulent licenses.

Prison visitor systems have implemented facial recognition, fingerprint readers, and hand-geometry readers in an effort to authenticate visitors and to avoid identity swapping during visits. In some cases, biometrics are also being used to authenticate prisoners so that identities are not switched and only the correct individuals are paroled.

Airports have adopted facial recognition systems in an effort to pick known criminals and terrorists out of a crowd, even if the targets are wearing a disguise. In the United States, a trial of the INSPASS system used biometrics to allow travelers to bypass lengthy immigration lines.

In the United Kingdom and the Netherlands, facial recognition systems are being used in conjunction with cameras that are located on prominent streets and thoroughfares, continually scanning pedestrians to find fugitives and terrorists.

Border-control agencies are regularly using facial recognition and fingerprint biometrics to authenticate individuals passing through the borders. Many Mexican workers who travel to California daily for work have been issued identity cards containing a biometric fingerprint.

Voting systems in some areas of the world require politicians to verify their identity during the voting process in order to avoid "proxy" voting.

ATM machines in some cities are now using facial recognition with or instead of PINs in order to verify the identity of customers and allow them to conduct transactions at the machines. Additionally, in an attempt to combat fraud and theft, banks are looking at implementing additional biometrics such as fingerprints into bank cards and processes such as access to safe-deposit boxes and check cashing.

Many vendors, such as Compaq, KeyTronics, Samsung, and Sony, have integrated fingerprint-reading technology into their keyboards and workstations in order to authenticate individuals and limit access to machines and networks. Other companies, such as Digital Persona and Veridian, sell fingerprint readers commercially that can be used for individual computers or networks.

Several credit card companies, like American Express, Visa, and MasterCard, are discussing incorporating fingerprint technology into their cards to work in place of or in conjunction with photographs and smart chips, which are already being used for identification purposes.

Disneyland and other private companies are implementing hand-geometry biometrics to permit access to customers and workers. These readers not only identify and authenticate users; they also speed up the process and can be used to log time of entry and exit.

Facial recognition systems have been used successfully in Germany to provide customers with 24-hour access to their safe-deposit boxes. In the United States and Canada some gas stations, stores, and banks are using facial recognition in order to identify and record check-cashing transactions.

Since September 11, 2001, numerous government agencies have been discussing a multipurpose, federally issued ID card containing a biometric, such as a fingerprint, that would be used by American citizens to verify their identity whenever necessary (for purchasing power, credit applications, or travel validation, for example). However, these discussions have incurred some arguments among legislators and individuals that feel such a card would jeopardize individual privacy.

The federal government, followed by some state governments and the military, has implemented iris- and retinal-scanning biometrics, hand geometry, fingerprint scanning, facial recognition, and voice verification in a variety of instances where tight security is required. Generally, until some official statement is made to the public, government-sponsored trials of these technologies are kept under wraps, making it difficult to get exact figures and accurate information on how much biometric technology is actually in use.

Biometrics can offer a number of significant benefits to governments, industry, and individual consumers. For governments, biometrics provides a method for maintaining tight security and preventing impersonation and data theft and manipulation. For industry, biometric access-control systems can limit access by person, by location, and by time while keeping accurate audit logs. For the consumer, biometric systems are fast and easy to use and can provide customized service and unlimited access.

Currently, biometric systems do not guarantee 100% accuracy, but no security technology does. The degree of uniqueness varies among the different types of biometric characteristics, and there may also be a significant variation in the performance figures quoted by vendors. Generally, biometrics is thought to offer enhanced security over knowledge-based and token-based identification methods. However, there are strong and weak biometrics from a security perspective. While biometrics are considered secure, some biometric techniques can be spoofed or circumvented by fraudulent means. There are also privacy concerns that have been raised concerning retaining unique personal identifiers in a template over which individuals have no control.

As a result of these ongoing difficulties, inaccuracies, and fears, public acceptance of biometrics is not guaranteed. Future events will probably have significant effect on the growth, acceptance, and profitability of biometric technology [6].

Nevertheless, it is important to bear in mind that effective security cannot be achieved by relying on technology alone. Technology and people must work together as part of an overall security process. Weaknesses in either area diminishes the effectiveness of the security process. The security process needs to account for limitations in biometric technology. For example, some people cannot enroll in a biometric system. Similarly, errors sometimes occur during matching operations. Procedures need to be developed to handle these situations. Exception processing that is not as good as biometric-based primary processing could also be exploited as a security hole. Thus, three key considerations need to be addressed before a decision is made to design, develop, and implement biometrics into a security system:

1.   Decisions must be made on how the technology will be used;

2.   A detailed cost-benefit analysis must be conducted to determine that the benefits gained from a system outweigh the costs;

3.   A trade-off analysis must be conducted between the increased security that the use of biometrics would provide, and the effect on areas such as privacy and convenience [1].

Security concerns need to be balanced with practical cost and operational considerations as well as political and economic interests. A risk management approach can help federal agencies identify and address security concerns. As federal agencies consider the development of security systems with biometrics, they need to define what the high-level goals of their systems will be and develop the concept of operations that will embody the people, process, and technologies required to achieve these goals. With these answers, the proper role of biometric technologies in security can be determined. If these details are not resolved, the estimated cost and performance of the resulting system will be at risk [1].

Keeping the preceding in mind, this chapter provides an overview of biometric technologies that are currently available and being developed; current uses of these technologies; and issues and challenges associated with the implementation of biometrics. In addition, this chapter discusses the current maturity of several biometric technologies; the possible implementation of these technologies in current border control processes; and the policy implications and key considerations for using these technologies [1].

Biometric Technologies for Personal Identification

When used for personal identification, biometric technologies measure and analyze human physiological and behavioral characteristics. Identifying a person's physiological characteristics is based on direct measurement of a part of the body—fingertips, hand geometry, facial geometry, and eye retinas and irises. The corresponding biometric technologies are fingerprint recognition, hand geometry, and facial, retina, and iris recognition. Identifying behavioral characteristics is based on data derived from actions, such as speech and signature, the corresponding biometrics being speaker recognition and signature recognition [1].

Biometrics can theoretically be very effective personal identifiers because the characteristics they measure are thought to be distinct to each person. Unlike conventional identification methods that use something you have, such as an identification card to gain access to a building, or something you know, such as a password to log on to a computer system, biometric characteristics are integral to something you are. Because they are tightly bound to an individual, they are more reliable, cannot be forgotten, and are less easily lost, stolen, or guessed [1].

How Biometric Technologies Work

Biometric technologies vary in complexity, capabilities, and performance, but all share several elements. Biometric identification systems are essentially pattern recognition systems. They use acquisition devices such as cameras and scanning devices to capture images, recordings, or measurements of an individual's characteristics, and use computer hardware and software to extract, encode, store, and compare these characteristics. Because the process is automated, biometric decision making is generally very fast, in most cases taking only a few seconds in real time [1].

Depending on the application, biometric systems can be used in one of two modes: verification or identification. Verification (also called authentication) is used to verify a person's identity—that is, to authenticate that individuals are who they say they are. Identification is used to establish a person's identity—that is, to determine who a person is. Although biometric technologies measure different characteristics in substantially different ways, all biometric systems involve similar processes that can be divided into two distinct stages: enrollment and verification or identification [1].

Enrollment

In enrollment, as briefly discussed in Chapter 1, a biometric system is trained to identify a specific person. The person first provides an identifier, such as an identity card. The biometric is linked to the identity specified on the identification document. He or she then presents the biometric (fingertips, hand, or iris) to an acquisition device. The distinctive features are located and one or more samples are extracted, encoded, and stored as a reference template for future comparisons. Depending on the technology, the biometric sample may be collected as an image, a recording, or a record of related dynamic measurements. How biometric systems extract features and encode and store information in the template is based on the system vendor's proprietary algorithms. Template size varies depending on the vendor and the technology. Templates can be stored remotely in a central database or within a biometric reader device itself; their small size also allows for storage [9] on smart cards or tokens [1].

Minute changes in positioning, distance, pressure, environment, and other factors influence the generation of a template, making each template likely to be unique as an individual's biometric data are captured and a new template is generated. Consequently, depending on the biometric system, a person may need to present biometric data several times in order to enroll. The reference template may then represent an amalgam of the captured data, or several enrollment templates may be stored. The quality of the template or templates is critical in the overall success of the biometric application. Because biometric features can change over time, people may have to re-enroll to update their reference template. Some technologies can update the reference template during matching operations [1].

The enrollment process also depends on the quality of the identifier the enrollee presents. The reference template is linked to the identity specified on the identification document. If the identification document does not specify the individual's true identity, the reference template will be linked to a false identity [1].

Thus, verification is a one-to-one comparison of the biometric sample with the reference template on file. A reference template is the enrolled and encoded biometric sample of record for a user. Identification makes a one-to-many comparison to determine a user's identity. It checks a biometric sample against all the reference templates on file. If any of the templates on file match the biometric sample, there is a good probability the individual has been identified [2].

Verification

In verification systems, the step after enrollment is to verify that a person is who he or she claims to be (the person who enrolled). After the individual provides whatever identifier he or she enrolled with, the biometric is presented, and the biometric system captures it, generating a trial template that is based on the vendor's algorithm. The system then compares the trial biometric template with this person's reference template, which was stored in the system during

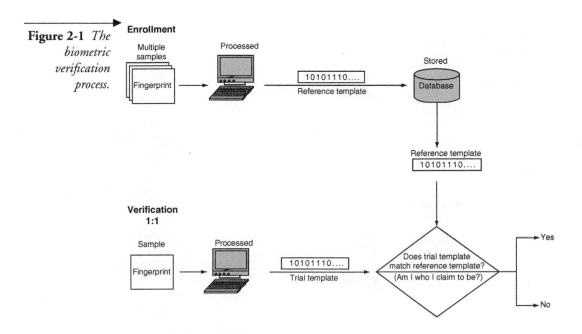

Figure 2-1 *The biometric verification process.*

enrollment, to determine whether the individual's trial and stored templates match (see Figure 2-1) [1].

Verification is often referred to as 1:1 (one-to-one) matching. Verification systems can contain databases ranging from dozens to millions of enrolled templates but are always predicated on matching an individual's presented biometric against his or her reference template. Nearly all verification systems can render a match/no-match decision in less than a second. A system that requires employees to authenticate their claimed identities before granting them access to secure buildings or to computers is a verification application [1].

Identification

In identification systems, the step after enrollment is to identify who the person is. Unlike verification systems, no identifier need be provided. To find a match, instead of locating and comparing the person's reference template against his or her presented biometric, the trial template is compared against the stored reference templates of all individuals enrolled in the system (see Figure 2-2) [1]. Identification systems are referred to as 1: N (one-to-N, or one-to-many) matching because an individual's biometric is compared against multiple biometric templates in the system's database [1].

There are two types of identification systems: positive and negative. Positive identification systems are designed to ensure that an individual's biometric is

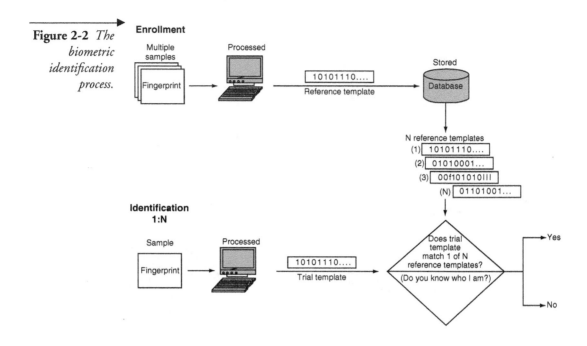

Figure 2-2 *The biometric identification process.*

enrolled in the database. The anticipated result of a search is a match. A typical positive identification system controls access to a secure building or secure computer by checking anyone who seeks access against a database of enrolled employees. The goal is to determine whether a person seeking access can be identified as having been enrolled in the system [1].

Negative identification systems are designed to ensure that a person's biometric information is not present in a database. The anticipated result of a search is a nonmatch. Comparing a person's biometric information against a database of all who are registered in a public benefits program, for example, can ensure that this person is not "double-dipping" by using fraudulent documentation to register under multiple identities [1].

Another type of negative identification system is a surveillance system that uses a watch list. Such systems are designed to identify people on the watch list and alert authorities for appropriate action. For all other people, the system is to check that they are not on the watch list and allow them normal passage. The people whose biometrics are in the database in these systems may not have provided them voluntarily. For instance, for a surveillance system, the biometrics may be faces captured from mug shots provided by a law enforcement agency [1].

No match is ever perfect in either a verification or an identification system, because every time a biometric is captured, the template is likely to be unique. Therefore, biometric systems can be configured to make a match or no-match decision based on a predefined number, referred to as a threshold, that establishes the acceptable degree of similarity between the trial template and the enrolled reference template. After the comparison, a score representing the degree of similarity is generated, and this score is compared to the threshold to make a match or no-match decision. Depending on the setting of the threshold in identification systems, sometimes several reference templates can be considered matches to the trial template, with the better scores corresponding to better matches [1].

Leading Biometric Technologies

A growing number of biometric technologies have been proposed over the past several years, but only in the past nine years have the leading ones become more widely deployed. Some technologies are better suited to specific applications than others, and some are more acceptable to users [1].

Many different types of unique physiological or behavioral characteristics exist for humans. Some of the more traditional uses of these biometric methods for identification or verification include (a detailed discussion of each of these leading biometric technologies follows this chapter):

- **Facial recognition:** Facial recognition attempts to identify a subject based on facial characteristics such as eye socket position, space between cheekbones, etc.

- **Fingerprint recognition:** Fingerprint recognition systems rely on the biometric device's ability to distinguish the unique impressions of ridges and valleys made by an individual's finger.

- **Hand geometry:** Hand geometry solutions take more than 90 dimensional measurements to record an accurate spatial representation of an individual's hand.

- **Iris scanning/recognition:** Iris scanning uses a camera mounted between three and 10 feet away from the person to take a high-definition photograph of the individual's eyes. It then analyzes 266 different points of data from the trabecular meshwork of the iris.

- **Retina scanning/recognition:** Retinal scanning involves an electronic scan of the retina, the innermost layer of the wall of the eyeball.

- **Signature dynamics/recognition:** Dynamic signature verification not only compares the signature itself, but also marks changes in speed, pressure and timing that occur during signing.

- **Keystroke dynamics:** Keystroke dynamics technology measures dwell time (the length of time a person holds down each key) as well as flight time (the time it takes to move between keys). Taken over the course of several login sessions, these two metrics produce a measurement of rhythm unique to each user.

- **Voice/speaker recognition:** Voice recognition biometrics digitizes a profile of a person's speech into a template voiceprint and stores it as a table of binary numbers. During authentication, the spoken passphrase is compared to the previously stored template.

- **RFID chip implant:** RFID chips contain a unique identification number and can carry other personal data about the implantee. When radio-frequency energy passes from a scanner, it energizes the chip, which is passive (not independently powered), that then emits a radio-frequency signal transmitting the chip's information to the reader, which in turn links with a database [2].

Facial Recognition

Facial recognition technology identifies people by analyzing features of the face not easily altered—the upper outlines of the eye sockets, the areas around the cheekbones, and the sides of the mouth. The technology is typically used to compare a live facial scan to a stored template, but it can also be used in comparing static images such as digitized passport photographs. Facial recognition can be used in both verification and identification systems. In addition, because facial images can be captured from video cameras, facial recognition is the only biometric that can be used for surveillance purposes [1].

Fingerprint Recognition

Fingerprint recognition is one of the best known and most widely used biometric technologies. Automated systems have been commercially available since the early 1970s. There are more than 121 fingerprint recognition technology companies today. Until recently, fingerprint recognition was used primarily in law enforcement applications [1].

Fingerprint recognition technology extracts features from impressions made by the distinct ridges on the fingertips. The fingerprints can be either flat or rolled. A flat print captures only an impression of the central area between the fingertip and the first knuckle; a rolled print captures ridges on both sides of the finger [1].

An image of the fingerprint is captured by a scanner, enhanced, and converted into a template. Scanner technologies can be optical, silicon, or ultrasound technologies. Ultrasound, while potentially the most accurate, has not been demonstrated in widespread use. Today, optical scanners are the most commonly used. During enhancement, "noise" caused by such things as dirt, cuts, scars, and creases or dry, wet, or worn fingerprints is reduced, and the definition of the ridges is enhanced. According to industry analysts, approximately 84% of vendors base their algorithms on the extraction of minute points relating to breaks in the ridges of the fingertips. Other algorithms are based on extracting ridge patterns [1].

Hand Geometry

Hand geometry systems have been in use for almost 34 years for access control to facilities ranging from nuclear power plants to daycare centers. Hand geometry technology takes 96 measurements of the hand, including the width, height, and length of the fingers; distances between joints; and shapes of the knuckles [1].

Hand geometry systems use an optical camera and light-emitting diodes with mirrors and reflectors to capture two orthogonal two-dimensional images of the back and sides of the hand [3]. Although the basic shape of an individual's hand remains relatively stable over his or her lifetime, natural and environmental factors can cause slight changes [1].

Iris Recognition

Iris recognition technology is based on the distinctly colored ring surrounding the pupil of the eye. Made from elastic connective tissue, the iris is a very rich source of biometric data, having approximately 266 distinctive characteristics. These include the trabecular meshwork, a tissue that gives the appearance of dividing the iris radially, with striations, rings, furrows, a corona, and freckles. Iris recognition technology uses about 173 of these distinctive characteristics. Formed during the eighth month of gestation, these characteristics reportedly remain stable throughout a person's lifetime, except in cases of injury. Iris recognition can be used in both verification and identification systems [1].

Iris recognition systems use a small, high-quality camera to capture a black and white, high-resolution image of the iris. The systems then define the boundaries of the iris, establish a coordinate system over the iris, and define the zones for analysis within the coordinate system [1].

Retina Recognition

Retina recognition technology captures and analyzes the patterns of blood vessels on the thin nerve on the back of the eyeball that processes light entering through the pupil. Retinal patterns are highly distinctive traits. Every eye has its own totally unique pattern of blood vessels; even the eyes of identical twins are distinct. Although each pattern normally remains stable over a person's lifetime, they can be affected by disease such as glaucoma, diabetes, high blood pressure, and autoimmune deficiency syndrome [1].

The fact that the retina is small, internal, and difficult to measure makes capturing its image more difficult than most biometric technologies. An individual must position the eye very close to the lens of the retina-scan device, gaze directly into the lens, and remain perfectly still while focusing on a revolving light while a small camera scans the retina through the pupil. Any movement can interfere with the process and can require restarting. Enrollment can easily take more than a minute [1].

Signature Dynamics/Recognition

Signature recognition authenticates identity by measuring handwritten signatures. The signature is treated as a series of movements that contain unique biometric data, such as personal rhythm, acceleration, and pressure flow. Unlike electronic signature capture, which treats the signature as a graphic image, signature recognition technology measures how the signature is signed [1].

In a signature recognition system, a person signs his or her name on a digitized graphics tablet or personal digital assistant. The system analyzes signature dynamics such as speed, relative speed, stroke order, stroke count, and pressure. The technology can also track each person's natural signature fluctuations over time. The signature dynamics information is encrypted and compressed into a template [1].

Keystroke Dynamics

Keystroke dynamics, being a behavioral measurement, is a pattern exhibited by an individual using an input device in a consistent manner. Raw measurements already available via the standard keyboard can be manipulated to determine dwell time (the time one keeps a key pressed) and flight time (the time it takes a person to jump from one key to another). Variations of algorithms differentiate between absolute versus relative timing. The captured data is analyzed to determine aggregate factors such as cadence, content, spatial corrections, and consistency. This is then fed through a signature processing routine, which deduces the primary (and supplementary) patterns for later verification. Signature processing is not unique to biometrics; in fact, many of these algorithms are present in actuarial sciences, from economic trending to quantum mechanics [4].

Speaker/Voice Recognition

Differences in how different people's voices sound result from a combination of physiological differences in the shape of vocal tracts and learned speaking habits. Speaker recognition technology uses these differences to discriminate between speakers [1].

During enrollment, speaker recognition systems capture samples of a person's speech by having him or her speak some predetermined information into a microphone a number of times. This information, known as a passphrase, can be a piece of information such as a name, birth month, birth city, favorite color, or a sequence of numbers. Text-independent systems are also available that recognize a speaker without using a predefined phrase. This phrase is converted from analog to digital format, and the distinctive vocal characteristics, such as pitch, cadence, and tone, are extracted, and a speaker model is established. A template is then generated and stored for future comparisons [1].

Speaker recognition can be used to verify a person's claimed identity or to identify a particular person. It is often used where voice is the only available biometric identifier, such as telephone and call centers [1].

RFID Chip Implant

The RFID technology process starts with a tag, which is made up of a microchip with an antenna, and a reader with an antenna. The reader sends out radio-frequency waves that form a magnetic field when they join with the antenna on the RFID tag. A passive RFID tag creates power from this magnetic field and uses it to energize the circuits of the RFID chip. The chip in the radio-frequency identification tag sends information back to the reader in the form of radio-frequency waves. The RFID reader converts the new waves into digital information. Semi-passive RFID tags use a battery to run the circuits of the chip, but communicate by drawing power from the RFID reader [5].

Harnessing technology to enhance human operations opens the way to a new future in which those with physical difficulties or impairments could benefit from cybernetics to gain improved quality of life. Ultimately, RFID chip transplants will provide a viable route for people to regain the use of their limbs. Soon it will be possible for people to control their external environment. Using microchip implants, it becomes possible for a disabled person to undertake basic lifestyle operations, such as directing an electric wheelchair or controlling room temperature and lighting. In essence, cybernetics seeks to establish how humans and technology can operate together. RFID chip implants open the way to exciting new applications in the fields of medical science, bionics, and human biometrics [5].

Biometric Technologies Under Development

Other technologies that are emerging or that are being studied and developed include blood pulse, nailbed identification, body salinity (salt) identification, palm print, vein patterns, facial thermography, DNA, sweat pores, hand grip, fingernail bed, body odor, ear shape, gait, skin luminescence, brain wave pattern, electronic nose identification, footprint recognition, and foot dynamics. A detailed discussion of each of these leading biometric technologies under development follows this chapter.

Blood Pulse

Blood pulse biometrics measure the blood pulse on a finger with infrared sensors. This technology is still experimental and has a high false match rate, making it impractical for personal identification.

The exact composition of all the skin elements is distinctive to each person. For example, skin layers differ in thickness, the interfaces between the layers have different undulations, pigmentation differs, collagen fibers and other proteins differ in density, and the capillary beds have distinct densities and locations beneath the skin. Skin pattern recognition technology measures the characteristic spectrum of an individual's skin. A light sensor illuminates a small patch of skin with a beam of visible and near-infrared light. The light is measured with a spectroscope after being scattered by the skin. The measurements are analyzed, and a distinct optical pattern can be extracted [7].

Nailbed Identification

Nailbed identification technology is based on the distinct longitudinal, tongue-in-groove spatial arrangement of the epidermal structure directly beneath the fingernail. This structure is mimicked in the ridges on the outer surface of the nail. When an interferometer is used to detect phase changes in back-scattered light shone on the fingernail, the distinct dimensions of the nailbed can be reconstructed and a one-dimensional map can be generated [7].

Body Salinity (Salt) Identification

Development in this area has been conducted by IBM and the Massachusetts Institute of Technology (MIT). Their joint product (The Personal Area Network [PAN]) works by exploiting the natural level of salinity, or salt, in the human body. This is accomplished by an electric field that passes a tiny electrical current through the body (salt is an effective conductor of electricity), on which data can be carried. The electrical current that is used is in the order of a nanoamp (one-billionth of an amp), which is less than the natural currents already present in the body. Speeds equivalent to a 2,400-baud modem have been claimed, giving 400,000 bits per second data transfer.

Applications of this kind of biometric technology could include the interaction (data transfer) between communication devices carried on the body including watches, mobile phones, and pagers. Other applications could include "waking up" household appliances or devices as one enters a room.

Palm Print

This physical biometric analyzes the unique patterns on the palm of a user's hand, unlike hand geometry, which concentrates on shapes and relativity. In other words, this system uses the lines on one's palm to identify an individual. Like fingerprint identification systems, palm print systems measure ridges and minute points found on the palm [6].

Vein Patterns

This method analyzes the pattern of veins in the back of a person's hand. The underlying vein structure or "vein tree" is captured using a camera and infrared light.

In other words, much like retinal identification, it uses infrared light to produce an image of one's vein pattern in one's face, wrist, or hand. The advantage of this type of biometric technology is that veins are relatively stable through one's life and cannot be erased or tampered with.

Facial Thermography

Facial thermography detects heat patterns created by the branching of blood vessels and that are emitted from the skin. These patterns, called thermograms, are highly distinctive. Even identical twins have different thermograms. Developed in the mid-1990s, thermography works much like facial recognition, except that an infrared camera is used to capture the images. The advantages of facial thermography over other biometric technologies are that it is not intrusive (no physical contact is required), every living person presents a usable image, and the image can be collected on the fly. Also, unlike visible light systems, infrared systems work accurately even in dim light or total darkness. Although identification systems using facial thermograms were undertaken in 1997, the effort was suspended because of the cost of manufacturing the system [7].

DNA

Humans have 23 pairs of chromosomes containing their DNA blueprint. One member of each chromosomal pair comes from their mother; the other comes from their father. Every cell in a human body contains a copy of this DNA. The large majority of DNA does not differ from person to person, but 0.10% of a person's entire genome is unique to each indiviual. This represents 3 million base pairs of DNA.

Genes make up 5% of the human genome. The other 95% are noncoding sequences (which used to be called junk DNA). In noncoding regions there are identical repeat sequences of DNA, which can be repeated anywhere from one to 30 times in a row. These regions are called variable number tandem repeats (VNTRs). The number of tandem repeats at specific places (called loci) on chromosomes varies between individuals. For any given VNTR loci in an individual's DNA, there will be a certain number of repeats. The higher number of loci are analyzed, the smaller the probability of finding two unrelated individuals with the same DNA profile.

DNA profiling determines the number of VNTR repeats at a number of distinctive loci, and uses it to create an individual's DNA profile. The main steps in creating a DNA profile are isolate the DNA (from a sample such as blood, saliva, hair, semen, or tissue); cut the DNA up into shorter fragments containing known VNTR areas; sort the DNA fragments by size; and compare the DNA fragments in different samples.

Sweat Pores Analysis

The distribution of the pores in the area of the finger is distinct for each individual. Based on this observation, sweat pores analyzers have been developed that analyze the sweat pores on the tip of the finger. When the finger is placed on the sensor, the software records the pores as stars and stores their position relative to the area of the finger.

Grip Recognition

A new technology dubbed "Grip" is significantly different from other commercially available systems, according to the developer. Grip technology analyzes highly unique internal features of the human hand such as veins, arteries, and fatty tissues. Grip uses infrared light to scan and read the patterns of tissue and blood vessels under the skin of the hand presented in the gripped pose. The technology completely maps the substructure of the person's hand, and 16 scans are then taken. This is a physiological biometric.

Body Odor

Body odor can be digitally recorded for identification. For example, a British company, Mastiff Electronic Systems Ltd., is working on a system that uses your hand to identify your body odor. The British company has developed a sensor named Scentinel that is used to "capture" your body odor. The product is still three years away from commercial release and is still very expensive ($48,600), but there is interest in its implementation from the British embassy in Buenos Aires, Saudi Arabia's National Guard, and private Indian and Japanese companies.

Ear Shape

Ear shape recognition is still a research topic. It is based on the distinctive shape of each person's ears and the structure of the largely cartilaginous, projecting portion of the outer ear. Although ear biometrics appears to be promising, no commercial systems are available.

However, police and other law enforcement agents have found that criminals listening at windows and doors leave earprints. Working on that premise, several different nations began to use ear-shape biometrics. In the Netherlands, police have used this biometric to obtain criminal convictions, and at least one French company plans to market the Octophone, a telephone-like biometric device that captures images of the ear [7].

Gait

Gait recognition, recognizing individuals by their distinctive walk, captures a sequence of images to derive and analyze motion characteristics. A person's gait can be hard to disguise because a person's musculature essentially limits the variation of motion, and measuring it requires no contact with the person. However, gait can be obscured or disguised if the individual, for example, is wearing loose-fitting clothes. Preliminary results have confirmed its potential, but further development is necessary before its performance, limitations, and advantages can be fully assessed [7].

Skin Luminescence

Skin luminescence is light from the skin not generated by high temperatures alone. It is different from incandescence, in that it usually occurs at low temperatures. Examples include fluorescence, bioluminescence, and phosphorescence.

Skin luminescence can be caused by chemical or biochemical changes, electrical energy, subatomic motions, reactions in crystals, or stimulation of an atomic system. The following kinds of skin luminescence are known to exist:

- Chemoluminescence (including bioluminescence)
- Crystalloluminescence
- Electroluminescence
  - Cathodoluminescence
- Photoluminescence
  - Phosphorescence
  - Fluorescence
- Radioluminescence
- Sonoluminescence
- Thermoluminescence
- Triboluminescence

Historically, radioactivity was first thought of as a form of "radioluminescence," although it is today considered to be separate since it involves more than electromagnetic radiation.

Brain-Wave Pattern

While it is true that a person has the ability to alter most of their own brain-wave patterns, they cannot alter what is referred to as their baseline brain-wave pattern. So, an individual's baseline brain-wave pattern has the ability to be recognized as the newest undiscovered biometric solution. This is a solution that is referred to as an "EEG fingerprint."

Another type of technology that can assist individuals is known by many names: electroencephalogram interface (EEGI), brain-computer interface (BCI), human-computer interface (HCI), neural human-computer interface (NHCI), and neural interface (NI). However, a more accurate description of this type of interface technology is the neural wave analysis interface (NWAI).

The neural waves can either emanate from a subject's brain (in the form of brain waves) or muscles (in the form of bioelectrical impulses).

Electronic Nose Identification

Global interoperability of equipment needs to be put in place, as does a coordination of national practices. Some nations may adopt algorithms that compare the geometry of the electronic nose bridge between the live person and the stored ID image, while others may compare the larger, facial triangle.

Footprint Recognition

Biometrics, however, is not solely concerned with who's who in the human world. Environmental groups are using it as a noninvasive way to track and monitor endangered species. International animal protection group Wild Watch is using a biometric footprint recognition system as part of its efforts to save the black rhino, Bengal tiger, and other threatened species. In the case of the rhino, the system is designed to distinguish subtle differences in the footprints of black and white species while allowing observers to monitor their behavior.

Foot Dynamics

Like hand dominance (right/left handedness), foot dynamics (right/left leggedness) also exists. While matching, therefore, you can assume that improperly aligned (right/left leg forward) reference and test sequences affects the performance. This is an issue because it is not possible to distinguish between the left/right limbs from the 2-D binarized silhouettes. Suppose there are five (half-) cycles in both the gallery and probe sequences for a particular subject. To account for foot dynamics, you match the first four half-cycles of the two sequences and generate a matrix of similarity scores. Then you match the gallery sequence with a phase-shifted probe sequence to generate another matrix of similarity scores. Of the two phase-shifted test sequences, only one can provide a match that is in-phase unless the subject does not exhibit foot dominance.

Without loss of generality, you may assume that foot dynamics exists in all subjects. Then one of the two test sequences is a better match unless corrupted by noise. Therefore, the two similarity scores are combined using the MIN rule.

Accuracy of Biometric Technology

Biometrics is a very young technology, having only recently reached the point at which basic matching performance can be acceptably deployed. It is necessary to analyze several metrics to determine the strengths and weaknesses of each technology and vendor for a given application [1].

The three key performance metrics are false match rate (FMR), false non-match rate (FNMR), and failure to enroll rate (FTER). A false match occurs when a system incorrectly matches an identity; FMR is the probability of individuals being wrongly matched. In verification and positive identification systems, unauthorized people can be granted access to facilities or resources as the result of incorrect matches. In a negative identification system, the result of a false match may be to deny access. For example, if a new applicant to a public benefits program is falsely matched with a person previously enrolled in that program under another identity, the applicant may be denied access to benefits [1].

A false nonmatch occurs when a system rejects a valid identity; FNMR is the probability of valid individuals being wrongly not matched. In verification and positive identification systems, people can be denied access to some facility or resource as the result of a system's failure to make a correct match. In negative identification systems, the result of a false nonmatch may be that a person is granted access to resources to which she or he should be denied. For example, if a person who has enrolled in a public benefits program under another identity is not correctly matched, she or he will succeed in gaining fraudulent access to benefits [1].

False matches may occur because there is a high degree of similarity between two individuals' characteristics. False nonmatches occur because there is not a sufficiently strong similarity between an individual's enrollment and trial templates, which could be caused by any number of conditions. For example, an individual's biometric data may have changed as a result of aging or injury. If biometric systems were perfect, both error rates would be zero. However, because biometric systems cannot identify individuals with 100% accuracy, a trade-off exists between the two [1].

False match and nonmatch rates are inversely related; they must therefore always be assessed in tandem, and acceptable risk levels must be balanced with the disadvantages of inconvenience. For example, in access control, perfect security would require denying access to everyone. Conversely, granting access to everyone would result in denying access to no one. Obviously, neither extreme is reasonable, and biometric systems must operate somewhere between the two [1].

For most applications, how much risk one is willing to tolerate is the overriding factor, and this translates into determining the acceptable FMR. The greater the risk entailed by a false match, the lower the tolerable FMR. For example, an application that controlled access to a secure area would require that the FMR be set low, which would result in a high FNMR. However, an application that controlled access to a bank's ATM might have to sacrifice some degree of security and set a higher FMR (and hence a lower FNMR) to avoid the risk of irritating legitimate customers by wrongly rejecting them. Selecting a lower FMR increases the FNMR. Perfect security would require setting the FMR to 0, in which case the FNMR would be 1. At the other extreme, setting the FNMR to 0 would result in an FMR of 1 [1].

Vendors often use equal error rate (EER), an additional metric derived from FMR and FNMR, to describe the accuracy of their biometric systems. EER refers to the point at which FMR equals FNMR. Setting a system's threshold at its EER will result in the probability that a person is falsely matched equaling the probability that a person is falsely not matched. However, this statistic tends to oversimplify the balance between FMR and FNMR, because in few real-world applications is the need for security identical to the need for convenience [1].

Note: Equal error rate is the point at which FMR equals FNMR.

FTER is a biometric system's third critical accuracy metric. It measures the probability that a person will be unable to enroll. Failure to enroll (FTE) may stem from an insufficiently distinctive biometric sample or from a system design that makes it difficult to provide consistent biometric data. The fingerprints of people who work extensively at manual labor are often too worn to be captured. A high percentage of people are unable to enroll in retina recognition systems because of the precision such systems require. People who are mute cannot use voice systems, and people lacking fingers or hands from congenital disease, surgery, or injury cannot use fingerprint or hand geometry systems. Although between 1% and 3% of the general public does not have the body part required for using any one biometric system, they are normally not counted in a system's FTER [1].

Using Multiple Biometrics

Because biometric systems based solely on a single biometric may not always meet performance requirements, the development of systems that integrate two or more biometrics is emerging as a trend. Multiple biometrics could be two types of biometrics, such as combining facial and iris recognition.

Multiple biometrics could also involve multiple instances of a single biometric, such as one, two, or 10 fingerprints, two hands, and two eyes. One prototype system integrates fingerprint and facial recognition technologies to improve identification. A commercially available system combines face, lip movement, and speaker recognition to control access to physical structures and small office computer networks. Depending on the application, both systems can operate for either verification or identification. Experimental results have demonstrated that the identities established by systems that use more than one biometric could be more reliable, be applied to large target populations, and improve response time [1].

It's a given that biometric technologies can be combined to provide enhanced security. This combined use of two or more biometric technologies in one application is called a multimodal biometric system. A multimodal system allows for an even greater level of assurance of a proper match in verification and identification systems. Multimodal systems help overcome limitations of single biometric solutions, such as when a user does not have a quality biometric sample to present to the system (an individual with a cold attempts to authenticate to a voice recognition system), and reduces the ability of the system to be tricked fraudulently [2].

Understanding Biometric Systems' Performance Measures

Performance measures are used to create baselines to help organizations evaluate products. The performance of a biometric system is based on measures such as false rejection rate, false acceptance rate, crossover rate, verification time, and failure to enroll rate. The following is a brief description of these performance measures [2].

False rejection rate (FRR), also commonly referred to as a type I error, measures the percentage of times an individual who should be positively accepted is rejected—in other words, how many times the "good guys" cannot gain access. If users who should be granted access are repeatedly rejected, they will not have access to the protected application or location to perform their assigned duties. Biometric vendors strive to have a low FRR [2].

False acceptance rate (FAR), also commonly referred to as a type II error, measures the percentage of times an individual who should be rejected is positively matched by the biometric system—how many times the "bad guys" beat the system. If an attacker gains access to a protected application or location, the security of the system has been breached. Biometric vendors strive to have a low FAR [2].

Crossover rate, also referred to as the equal error rate (EER), is the point on a graph where the lines representing the FAR and FRR intersect. A lower crossover rate indicates a system with a good level of sensitivity and generally means the system will perform well [2].

Note: Verification time is the average time taken for the actual matching process to occur.

Failure to enroll rate (FTER) is used to determine the rate of failed enrollment attempts. Factors such as quality of the enrollment equipment, ease of enrollment, environment surrounding enrollment, and quality of the user's biometric influence the FTER [2].

It should be noted that vendors typically market products using measures based on laboratory tests in ideal situations. However, practical applications of these products show different statistical results and change the actual performance baseline. These differences are caused by factors such as user familiarity, network speeds, environmental effects, and product design. Organizations and standards groups, such as the National Biometric Test Center, INCITS M1, and the ISO SC37 Biometrics group, are working to provide real-world statistics on biometric systems so consumers have a better guide to a biometric solution's true performance. As more effective standards become available, published performance measures will become more reliable, but organizations should still consider performing independent testing. These independent tests should be executed within the organization's own environment and user population guidelines to provide the best understanding of actual performance in the installed system [2].

Business and Federal Applications of Biometric Technologies

Biometrics have been used in several federal applications, including access control to facilities and computers, criminal identification, and border security. In the last six years, laws have been passed that will require a more extensive use of biometric technologies in the federal government [1].

Business Drivers of Biometrics: Increased Security and Convenience

Biometric technology is designed to provide a greater degree of security than traditional authentication techniques since biometric credentials are difficult to steal, lose, forget, or compromise. Biometrics may be leveraged as a complementary form of authentication to increase security for a critical resource. In addition, biometric systems are designed to improve the verifiability of IT audit trails and user accountability because the technology provides a higher level of confidence in the identity verification process [2].

Convenience is another goal. Unlike traditional authentication methods, a biometric is based on a user characteristic that is not easily lost or forgotten. For that reason, users would not have to remember as many passwords or worry about misplacing authentication tokens [2].

Enterprise Applicability

Biometric systems can be applied to areas across the enterprise requiring logical or physical access solutions. Biometric authentication readers for workstations can be integrated with desktop applications for logical authentication to provide a stronger alternative to a username and password. Biometric devices can also be used to control physical access to buildings, safes, or rooms [2].

Biometric authentication integration efforts are becoming easier with the vendor adoption of industry standards, such as Biometric Application Programming Interface (BioAPI) and the Common Biometric Exchange File Format (CBEFF). The BioAPI is designed to provide a cross-platform interface that simplifies development and standardizes programmatic interaction with biometric devices. The CBEFF was developed to facilitate improved interoperability between biometric systems and simplify hardware and software integration [2].

Federal Access Control

Biometric systems have long been used to complement or replace badges and keys in controlling access to entire facilities or specific areas within a facility. The entrances to more than half the nuclear power plants in the United States employ biometric hand geometry systems [1].

Recent reductions in the price of biometric hardware have spurred logical access control applications. Fingerprint, iris, and speaker recognition are replacing passwords to authenticate individuals accessing computers and networks. The Office of Legislative Counsel of the U.S. House of Representatives, for example, is using an iris recognition system to protect confidential files and working documents. Other federal agencies, including the Department of Defense, Department of Energy, and Department of Justice, as well as the intelligence community, are adopting similar technologies [1].

The Department of Homeland Security's Transportation Security Administration (TSA) is working to establish a systemwide common credential to be used across all transportation modes for all personnel requiring unescorted physical and/or logical access to secure areas of the national transportation system, such as airports, seaports, and railroad terminals. Called the Transportation Worker Identification Credential (TWIC), the program was developed in response to recent laws and will include the use of smart cards and biometrics to provide a positive match of a credential to a person for 14–19 million transportation workers across the United States [1].

Criminal Identification

Fingerprint identification has been used in law enforcement over the past 104 years and has become the de facto international standard for positively identifying individuals. The FBI has been using fingerprint identification since 1928. The first fingerprint recognition systems were used in law enforcement about 44 years ago [1].

The FBI's Integrated Automated Fingerprint Identification System (IAFIS) is an automated 10-fingerprint matching system that stores rolled fingerprints. The over 84 million records in its criminal master file are connected electronically with all 50 states and some federal agencies like the CIA and NSA [1].

IAFIS was designed to handle a large volume of fingerprint checks against a large database of fingerprints. IAFIS processes, on average, approximately 82,000 fingerprints per day and has processed as many as 126,000 in a single day. IAFIS's target response time for criminal fingerprints submitted electronically is two hours; for civilian fingerprint background checks, 24 hours [1].

The Immigration and Naturalization Service (INS) began developing the Automated Biometric Fingerprint Identification System (IDENT) around 1990 to identify illegal aliens who were repeatedly apprehended trying to enter the United States illegally. INS's goal was to enroll virtually all apprehended aliens; however, the agency never did, because the Bush administration gave orders not to. IDENT can also identify aliens who have outstanding warrants or who have been deported. When such aliens are apprehended, a photograph and two index fingerprints are captured electronically and queried against three databases. IDENT has over 8.9 million entries. A fingerprint query of IDENT normally takes about two minutes. IDENT is also being used as a part of the National Security Entry-Exit Registration System (NSEERS) that was recently implemented [1].

Note: Under NSEERS, certain nonimmigrants who may pose a national security risk are being registered, fingerprinted, and photographed when they arrive in the United States. These nonimmigrants are required to periodically report and update (but never do) when changes occur to their registration information, and record their departure from the country (which never happens).

Border Insecurity

INS Passenger Accelerated Service System (INSPASS), a pilot program in place since 1993, has more than 89,000 frequent fliers enrolled at nine airports, and has admitted more than 744,000 travelers. It is open to citizens of the United States, Canada, Bermuda, and visa waiver program countries who travel to the United States on business three or more times a year. INSPASS permits frequent travelers to circumvent customs procedures and immigration lines. To participate, users undergo a background screening and registration. Once enrolled, they can present their biometric at an airport kiosk for comparison against a template stored in a central database [1].

In a misguided joint INS and State Department effort to comply with the Illegal Immigration Reform and Immigrant Responsibility Act of 1996, every border-crossing card issued after April 1, 1998 contains a biometric identifier and is machine-readable. The cards, also called laser visas, allow Mexican citizens to enter the United States for the purpose of business or pleasure without being issued further documentation and to stay for 72 hours or less, going no further than 25 miles from the border. Consular staff in Mexico photograph applicants, take prints of the two index fingers, and then electronically forward applicants' data to INS. Both the State Department and INS conduct checks on each applicant, and the fingerprints are compared with prints of previously enrolled individuals to ensure that the applicant is not applying for multiple cards under different names. (This has not been successful, since illegal aliens have been caught recently with multiple cards under different names.) The cards store a holder's identifying information along with a digital image of his or her picture and the minutiae of the two index finger-prints. As of May 2006, the State Department had issued more than 9 million cards [1].

The State Department has been running pilots of facial recognition technology at 27 overseas consular posts for several years. As a visa applicant's information is entered into the local system at the posts and replicated in the State Department's Consular Consolidated Database (CCD), the applicant's photograph is compared with the photographs of previous applicants stored in CCD to prevent fraudulent attempts to obtain visas. Again, this has not been very successful, since fraudulent attempts to obtain visas continue. Some photographs are also being compared to a watch list [1].

Laws passed in the last six years require a more extensive use of biometrics for border control; but again, this is not being used very effectively or enforced. The Attorney General and the Secretary of State jointly, through the National Institute of Standards and Technology (NIST), are to develop a technology standard (although they have not yet), including biometric identifier standards. When developed, this standard is to be used to verify the identity of persons applying for a U.S. visa for the purpose of conducting a background check, confirming identity, and ensuring that a person has not received a visa under a different name. At the time of this writing, the Departments of State and Justice have still not issued to aliens machine-readable, tamper-resistant visas and other travel and entry documents that use biometric identifiers. At the same time, the Justice Department still has not installed at all ports of entry equipment and software that allow the biometric comparison and authentication of all U.S. visas and other travel and entry documents issued to aliens, including machine-readable passports. The Department of Homeland Security is still developing the United States Visitor and Immigrant Status Indication Technology (US-VISIT) system to address this requirement. Implementation of this equipment will not take place until after 2008, when the current Bush administration is out of office [1].

Challenges and Issues in Using Biometrics

While biometric technology is currently available and is used in a variety of applications, questions remain regarding the technical and operational effectiveness of biometric technologies in large-scale applications. A risk management approach can help define the need and use for biometrics for security. In addition, a decision to use biometrics should consider the costs and benefits of such a system and its potential effect on convenience and privacy [1].

Risk Management Is the Foundation of Effective Strategy

The approach to good security is fundamentally similar, regardless of the assets being protected, whether information systems security, building security, or homeland security. These principles can be reduced to five basic steps that help to determine responses to five essential questions (see Figure 2-3) [1]:

1. **What am I protecting?** The first step in risk management is to identify assets that must be protected and the impact of their potential loss.

2. **Who are my adversaries?** The second step is to identify and characterize the threat to these assets. The intent and capability of an adversary are the principal criteria for establishing the degree of threat to these assets.

3. **How am I vulnerable?** Step three involves identifying and characterizing vulnerabilities that would allow identified threats to be realized. In other words, what weaknesses can allow a security breach?

4. **What are my priorities?** In the fourth step, risk must be assessed and priorities determined for protecting assets. Risk assessment examines the potential for the loss or damage to an asset. Risk levels are established by assessing the impact of the loss or damage, threats to the asset, and vulnerabilities.

5. **What can I do?** The final step is to identify countermeasures to reduce or eliminate risks. In doing so, the advantages and

Figure 2-3 *Five steps in the risk management process.*

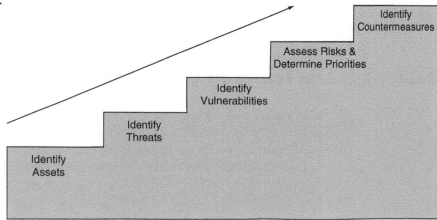

Identify Assets

Identify Threats

Identify Vulnerabilities

Assess Risks & Determine Priorities

Identify Countermeasures

benefits of these countermeasures must also be weighed against their disadvantages and costs [1].

Control Considerations and Management Risks for Biometrics

Biometric technologies present unique risks and need to be managed to allow an organization to achieve an acceptable return on its investment. The organization (management and auditors) should consider the following controls when evaluating, designing, implementing, maintaining, and auditing biometric systems:

- Misuse of biometric data from social and business viewpoints
- False negatives and positives
- Physical and logical controls over access to biometric data
- Security of the computers hosting the application and databases
- Audit trails
- Certification of software and hardware by vendor(s)
- Auditor's role in selecting the system [2]

Misuse of Biometric Data from Social and Business Viewpoints

The adoption of privacy laws throughout the world requires an immediate determination of the applicable laws with regard to biometric data use. In the United States, for example, laws such as the Health Insurance Portability and Accountability Act (HIPAA) contain important privacy restrictions [2].

False Negatives and Positives

Organizations should consider the impact to the organization, from operational and reputational viewpoints, presented by the misuse of biometric controls. False negatives could hinder productivity, because valid individuals would be prevented from accessing the system. False positives can present an opportunity for unauthorized access to the data and systems protected by the biometric control [2].

Physical and Logical Controls Over Access to Biometric Data

The location of biometric storage is a key point in the consideration of controls. The organization should ensure that the underlying digital representation of the biometric is controlled as standing data during transmission, regardless of whether it is stored centrally, in single computers, or on a smart card or other device [2].

Security of the Computers Hosting the Application and Databases

The organization should review access controls and configuration settings of the underlying computers and networks hosting and providing the communication channels for the biometric controls in use. The organization should also ensure that the computers, network lines and equipment, and other equipment used in the authentication process have been secured and are being monitored on an ongoing basis to ensure their security [2].

Audit Trails

Proper audit trails are essential in ensuring proper use, maintenance, and control of biometric systems. Audit trails should exist for all transactions used in the biometric process and should provide a mechanism to trace system users and their activity. Audit trail logs should be backed up and secured offsite to ensure their security and availability [2].

Certification of Software and Hardware by Vendor(s)

The vendor should properly test the software and hardware used by the biometric authentication process to ensure that it meets required standards. The organization should determine if steps have been taken to ensure that the vendor has supplied evidence and/or a certification of the software and hardware abilities [2].

Auditor's Role in Selecting the System

The organization must determine that the system has been thoroughly reviewed to ensure compatibility with the existing network and legacy applications. The auditor can help by understanding the intended use of the system to determine that the biometric system chosen will comply with standards required for external systems with which it may interface [2].

Protection, Detection, and Reaction Are Integral Security Concepts

Countermeasures identified through the risk management process support the three integral concepts of a holistic security program: protection, detection, and reaction. Protection provides countermeasures such as policies, procedures, and technical controls to defend against attacks on the assets being protected. Detection monitors for potential breakdowns in protective mechanisms that could result in security breaches. Reaction, which requires human involvement, responds to detected breaches to thwart attacks before damage can be done.

Because absolute protection is impossible to achieve, a security program that does not incorporate detection and reaction is incomplete [1].

Biometrics can support the protection component of a security program. It is important to realize that deploying them will not automatically eliminate all security risks. Technology is not a solution in isolation. Effective security also entails having a well-trained staff to follow and enforce policies and procedures. Weaknesses in the security process or failures by people to operate the technology or implement the security process can diminish the effectiveness of technology [1].

Furthermore, there is a need for the security process to account for limitations in technology. Biometrics can help ensure that people can only enroll into a security system once and that a person presenting himself or herself before the security system is the same person who enrolled into the system. However, biometrics cannot necessarily link a person to his or her true identity. While biometrics would make it more difficult for people to establish multiple identities, if the one identity a person claimed were not his or her true identity, then the person would be linked to the false identity in the biometric system. The quality of the identifier presented during the enrollment process is key to the integrity of a biometric system [1].

Procedures for exception processing would also need to be carefully planned. Not all people can enroll in a biometric system. Similarly, false matches and false nonmatches will sometimes occur. Procedures need to be developed to handle these situations. Exception processing that is not as good as biometric-based primary processing could be exploited as a security hole [1].

Deciding to Use Biometric Technology

A decision to use biometrics in a security solution should consider the benefits and costs of the system. This includes the potential effects on convenience and privacy [1].

Weighing Costs and Benefits

Best practices for information technology investment dictate that prior to making any significant project investment, the benefit and cost information of the system should be analyzed and assessed in detail. A business case should be developed that identifies the organizational needs for the project, and a clear statement of high-level system goals should be developed. The high-level goals should address the system's expected outcomes, such as the binding of a biometric feature to an identity or the identification of undesirable individuals on a watch list. Certain performance parameters should also be specified, such as the time required to verify a person's identity or the maximum population that the system must handle [1].

Once the system parameters are developed, a cost estimate can be developed. Not only must the costs of the technology be considered, but also the costs of the effects on people and processes. Both initial costs and recurring costs need to be estimated. Initial costs need to account for the engineering efforts to design, develop, test, and implement the system; training of personnel; hardware and software costs; network infrastructure improvements; and additional facilities required to enroll people into the biometric system. Recurring cost elements include program management costs, hardware and software maintenance, hardware replacement costs, training of personnel, hiring of additional personnel to enroll or verify the identities of people in the biometric system, and possibly the issuance of token cards for the storage of biometrics [1].

Weighed against these costs are the security benefits that accrue from the system. Analyzing this cost-benefit tradeoff is crucial when choosing specific biometric-based solutions. The consequences of performance issues (for example, accuracy problems and their effect on processes and people) are also important in selecting a biometric solution [1].

Effects on Privacy and Convenience

The Privacy Act of 1974 limits federal agencies' collection, use, and disclosure of personal information, such as fingerprints and photographs. Unfortunately, the current Bush administration violated this act when Vice President Cheney ordered the NSA to listen in on the private conversations of U.S. citizens as well as collect (mine data) and use that information to intimidate and manipulate U.S. citizens if the need arises.

Overall, the Privacy Act generally covers federal agency use of personal biometric information. However, the act includes exemptions for law enforcement and national security purposes (which was the cover story used by Cheney to justify the data mining). Representatives of civil liberties groups and privacy experts have expressed concerns regarding (1) the adequacy of protections for security, data sharing, identity theft [8], and other identified uses of biometric data; and (2) secondary uses and "function creep." These concerns relate to the adequacy of protections under current law for large-scale data handling in a biometric system. Besides information security, concern was voiced about an absence of clear criteria for governing data sharing. The broad exemptions of the Privacy Act, for example, provide no guidance on the extent of the appropriate uses that law enforcement may make of biometric information. Because there is no general agreement on the appropriate balance of security and privacy to build into a system using biometrics, further policy decisions are required. The range of unresolved policy issues suggests that questions surrounding the use of biometric technology center as much on management policies as on technical issues [1].

Privacy Impacts of Biometrics

The biometric is a digital representation that could be stolen, lost, or otherwise compromised. Unauthorized access to biometric storage devices could present numerous issues, not the least of which is privacy. Misuse of a biometric is a serious issue, given that the biometric itself cannot be changed and, once compromised, continues to be an issue for the life of the donor. Even when used as intended, the biometric control results in the capture of personal information, such as fingerprints, iris scans, palm geometry, and so forth [2].

Individuals do not always have the choice to opt out of using biometrics because of policy requirements, even if they are aware of the biometric use. They may be required to use biometrics as a job requirement or to gain access to related systems or services. Some may reject its use solely on the basis of the "Big Brother" principle, while others may truly believe that the information may be misused to track their activities, falsify transactions, or for other unauthorized purposes [2].

The adoption of biometric systems is growing and will almost assuredly continue to gain momentum. Organizations must accept biometrics and determine the best approach to ensure that they are used appropriately, that the information stored is adequately secured, and that data collected on the user remain private. Data collection, storage methods, and the consent of the persons from whom the data are being collected are key factors that must be closely examined during an audit or review. Legal considerations also must be clearly reviewed to determine the propriety of the collection process, storage, and use, and the possible contingencies posed by the use of biometrics within an organization [2].

Keeping the preceding in mind, consideration must be given to the convenience and ease of using biometrics and their effect on the ability of the agency to complete its mission. For example, some people find biometric technologies difficult, if not impossible, to use. Still others resist biometrics because they believe them to be intrusive, inherently offensive, or just uncomfortable to use. Lack of cooperation or even resistance to using biometrics can affect a system's performance and widespread adoption [1].

Furthermore, if the processes to use biometrics are lengthy or erroneous, they could negatively affect the ability of the assets being protected to operate and fulfill its mission. For example, there are significant challenges in using biometrics for border security—none of which are being met today. The use of biometric technologies could potentially impact the length of the inspection process. Any lengthening in the process of obtaining travel documents or entering the United States could affect travelers significantly. Delays inconvenience travelers and could result in fewer visits to the United States or lost business to the nation. Further studies could help determine whether the increased security from biometrics would result in fewer visits to the United States or lost business to the nation, potentially adversely affecting the American economy and, in particular, the border communities. These communities depend on trade with Canada and Mexico, which totaled $1 trillion in 2005 [1].

Barriers to Future Growth

A successfully implemented biometric application can help organizations address complex authentication issues. While it seems natural to expect that biometrics should be booming, in reality, only a few businesses and government agencies are testing or have deployed biometrics. Skeptics say the technology is still too expensive, is not foolproof, can be hard to integrate with other systems, and requires employees to change the way they work. The following are some of the challenges organizations face trying to incorporate biometrics into their business processes:

- Technology is not foolproof
- Cost of deployment
- Accuracy
- Resistance to change [2]

Technology Is Not Foolproof

Interest probably will not start growing until biometric systems overcome technical problems. These are problems related to the reliability of the biometric application [2].

Cost of Deployment

Deploying biometric readers on every door leading into a building or every PC on a network can be an expensive proposition. Hardware and software costs may not be the only consideration; the organization must bear in mind the associated complexity involved in enrolling new users and administering usage training [2].

Accuracy

Verification and positive identification systems may allow unauthorized users to access facilities or resources as a result of incorrect matches. In a negative identification system, the result of a false match may be to deny access [2].

Resistance to Change

Finally, as with many technologies, some users would rather not change the way they do things. For example, some users have the perception that using a username and password to log onto a system is faster than using a fingerprint scanner. This perception may arise from frustration related to the FRR, a performance measure that tracks the percentage of times an individual who should be positively accepted is rejected [2].

Summary/Conclusion

Biometrics is poised to take off, but before this transformation can occur, obstacles such as overall cost, lack of globally accepted standards, interoperability, reliability, and user perceptions must be overcome. Drivers such as governmental and commercial mandates to improve security and privacy, enterprise application integration, and the ongoing reduction in the cost of hardware will help overcome some of the barriers related to the widespread implementation of biometric technology [2].

References

1. Keith A. Rhodes, "Information Security: Challenges in Using Biometrics" [Testimony before the Subcommittee on Technology, Information Policy, Intergovernmental Relations, and the Census, Committee on Government Reform, House of Representatives (GAO-03-1137T)], United States General Accounting Office (GAO), 441 G Street NW, Room 7149, Washington, DC 20548, September 9, 2003.

2. Michael P. Down and Richard J. Sands, "Biometrics: An Overview of the Technology, Challenges and Control Considerations," volume 4,

3701 Algonquin Road, Suite 1010, Rolling Meadows, IL 60008, 2004. Copyright © 2006 Information Systems Audit and Control Association (ISACA). All Rights Reserved.

3. John R. Vacca, *Optical Networking Best Practices Handbook*, John Wiley & Sons (2006).

4. John C. Checco, "Keystroke Dynamics & Corporate Security," WSTA, 241 Maple Avenue, Red Bank, NJ 07701, 2006.

5. Kevin Warwick, "Are Chip Implants Getting Under Your Skin?" Synopsys, Inc., 700 East Middlefield Road, Mountain View, CA 94043, March 10, 2003. Copyright © 2003 Synopsys, Inc. All Rights Reserved.

6. D. E. Levine, "The Advantages and Applications of Biometric Technology," Security Technology & Design, 100 Colony Park, Drive Ste. 203, Cumming, GA 30040, 2006.

7. "Emerging Biometric Technologies," 300 N. Washington St., Suite B-100, Alexandria, VA 22314, 2006. Copyright © 2000–2006 GlobalSecurity.org. All Rights Reserved.

8. John R. Vacca, *Identity Theft*, Prentice Hall (2002).

9. John R. Vacca, *The Essentials Guide to Storage Area Networks*, Prentice Hall, Professional Technical Reference, Pearson Education (2001).

10. John R. Vacca, *Holograms: Design, Techniques, and Commercial Applications*, Charles River Media (2001).

Biometric Technology and Verification Systems Standards

Deploying new information technology systems for homeland security will require a comprehensive set of both national and international, technically sound standards for biometrics that meet U.S. needs. Over the past four years, NIST has worked in close partnership with other U.S. government agencies and industry to establish formal standards groups for biometric standards development. These groups are actively working on accelerating the development of national and international biometric standards that are of high relevance to the United States. NIST has identified the critical tasks that will contribute to the timely development of these standards. They will support significantly better, open systems standards-based security solutions. To describe in detail NIST's current involvement in formal national and international biometric standards, this chapter discusses related biometric standards development programs and business plans [1].

Developing technically sound consensus biometric standards requires a significant effort made by all the organizations involved. It includes participation by technical experts in the standards development process and working with the other member organizations in the development of technical contributions and positions. This is a labor-intensive work. Although processes for consensus IT standards development have been streamlined in the last few years, the development of a standard can still be inherently time-consuming because of technical and business decisions needed to reach consensus [1].

The timely and adequate support of these efforts serves as the catalyst for ensuring the rapid development of technically sound consensus standards. Bridging from national to international work will not be an easy task. Nevertheless, this effort will be an excellent opportunity to accelerate the deployment of standards-based biometric technologies that will meet U.S. requirements. The national and international community, and especially the United States, need this work to be done. Time is a compelling factor for new homeland security applications critical to this country. Some key standards objectives have already been achieved, and NIST is fully involved in the ongoing development efforts.

The work needs to proceed at the accelerated pace required by the critical needs discussed in this chapter [1].

NIST is involved in many capacities in biometric standard development activities. As a member organization of formal biometric standards development bodies and biometric consortia, NIST provides technical expertise and contributions to the creation of these draft standards and specifications. Technical developments in support of these consensus standards often are required while the standards are developed and also after the standard is approved. These include technical activities that help to implement the standards (reference implementations, conformance tests, evaluation procedures) [1].

Responding to requests from other U.S. government agencies and industry, NIST is also providing the leadership for the national and international bodies that are developing formal biometric standards. These organizations are the InterNational Committee for Information Technology Standards (INCITS) M1-Biometrics Technical Committee (see sidebar, "Technical Committee INCITS M1-Biometrics") and ISO/IEC Joint Technical Committee 1 (JTC 1) Subcommittee SC 37—Biometrics. NIST is also one of the organizations supporting the infrastructure required for successful development of these consensus standards. An efficient infrastructure for consensus standards development includes providing competent leadership (chairpersons, project editors) and providing competent administration (secretariat, websites, ballots). In addition, NIST serves in different capacities in multiple biometric consortia that have developed biometric specifications, as described later in the chapter [1].

Technical Committee INCITS M1-Biometrics

The Technical Committee M1-Biometrics (http://www.ncits.org/tc_home/m1.htm) has been established by the Executive Board of the International Committee for Information Technology Standards (INCITS) (http://www.ncits.org/) to ensure a high-priority, focused, and comprehensive approach in the United States for the rapid development and approval of formal national and international generic biometric standards. Critical generic biometric standards include common file formats and application program interfaces. The M1 Document Register (http://m1.incits.org/m1htm/2006docs/m1docreg_2006.htm) provides information on the current M1 activities, presentations given during the first M1 meeting (January 16–17, 2002), and a summary of the resolutions taken at the meeting or by letter ballots. M1 has 87 members from private industry, government agencies, and academia. A first meeting Convener's Report is available in the M1 Document Register [2].

Background

Measurements, testing, and standards have long been the heart of the mission of NIST. NIST has been working with government users and industry for more than 100 years to develop and apply technology, measurements, testing, and standards to support end-users and industry and to promote U.S. economic growth. In the aftermath of September 11, 2001, NIST, along with all federal agencies, is firmly committed to supporting this nation's new priorities for homeland security. NIST is addressing legislative requirements for biometric standards and conducting biometric testing under new public laws, such as the Patriot Act and the Enhanced Border Security Act. One of the key aspects of these requirements is the acceleration of biometric standards development [1].

Biometric technologies are posed to become the foundation of an extensive array of highly secure identification and personal verification solutions. In addition to supporting homeland security and preventing ID fraud, biometric-based systems are able to provide for confidential financial transactions and personal data privacy. Enterprise-wide network security infrastructures, employee IDs, secure electronic banking, investing and other financial transactions, retail sales, law enforcement, and health and social services are already benefiting from these biometric technologies. The need for standards-based biometric technologies is apparent. To fully realize the benefits of biometric technologies, comprehensive standards are necessary to ensure that information technology systems and applications are interoperable, scalable, usable, reliable, and secure. NIST has made a dramatic impact, and seeks to continue to make an impact, in the development of consensus standards for biometrics and related technical activities such as technology testing and the development of reference implementations and system emulations [1].

For decades, NIST has been involved with end-users and industry, especially within the law enforcement community, in biometric testing and standardization. In the past nine years, NIST has intensified its work in biometric standardization in support of open systems standards-based security solutions. NIST has served as a catalyst in national and international biometric standards developments in order to support the needs of other U.S. government agencies as well as U.S. industry. In 1999, NIST played a significant role in the unification efforts of several industrial groups developing incompatible biometric application programming interfaces (APIs). This work led to the formation of the current BioAPI Consortium, the development of the BioAPI specification, and the approval of this specification as a formal national standard (ANSI INCITS 358-2000, BioAPI v1.1). The development of a single approach specified in the BioAPI standard promotes interoperability among applications and biometric subsystems by defining a generic way of interfacing to a broad range of biometric technologies. NIST also led, in collaboration with the U.S. National Security Agency (NSA), the development of a common biometric exchange file format (CBEFF). The development of a single approach for a biometric data structure assured biometrics companies and their potential customers that different biometric devices and applications could exchange information efficiently. This specification is being incorporated in U.S. government and international requirements, such as the technical specifications drafted by the International Civil Aviation Organization (ICAO) [1].

Based upon a proposal from NIST, the Executive Board of INCITS established in November 2001 a new Technical Committee M1 on Biometrics. The purpose of M1 is to ensure a high-priority, focused, and comprehensive approach in the United States for the rapid development and approval of formal national and international generic biometric standards that are critical to U.S. needs for purposes such as homeland defense and the prevention of identity theft [6]. M1 is developing a portfolio of data interchange and interoperability standards, including biometric data formats for finger, facial, iris, and signature recognition, application profiles for transportation workers, border crossing, and point-of-sale, and a standard specifying biometric performance evaluation and reporting methods. In the international arena, the most critical activity of high relevance to the United States is the recently formed ISO/IEC Joint Technical Committee 1 (JTC 1) Subcommittee 37 on Biometrics. The United States provides the secretariat for this new subcommittee. The formation of JTC 1 SC 37 was initiated and championed by the U.S. National Body to ISO/IEC JTC 1 with strong support from NIST. Other U.S. government agencies and industry have asked NIST to provide leadership for INCITS M1 and JTC 1 SC 37. These new standards groups are providing the needed venues for a focused and comprehensive approach to the rapid development and approval of the required formal national and international biometric standards, and will support the deployment of biometric-based interoperable enterprise systems [1].

In addition to playing a leadership role in formal biometric standardization, associated testing, and related critical technical activities, NIST also partners with the biometric community in other capacities. NIST co-chairs with NSA the Biometric Consortium (BC) and its working groups, the NIST/BC Biometric Interoperability Performance and Assurance Working Group and the Common Biometric Exchange File Format Development Group. The U.S.

Biometric Consortium is an organization that currently consists of over 900 members representing over 60 government agencies, industry, and academia. The Biometric Consortium serves as the U.S. government focal point for research, testing, evaluation, and application of biometric-based personal verification and identification technologies. NIST co-sponsors the Biometric Consortium conferences and biometric technical developments. The NIST/BC Biometric Working Group has been working over the last six years to develop biometric specifications that can now be turned over to formal standards bodies such as INCITS M1 and JTC 1 SC 37. In the last four years, the NIST/BC Biometric Working Group approved and provided to INCITS M1 the following three specifications for consideration as national and international standards:

1.   Biometric Template Protection and Usage

2.   Biometric Application Programming Interface for Java Card^{TM}

3.   An augmented version of the Common Biometric Exchange File Format [1].

NIST is also a member of the BioAPI Consortium, a member of its Steering Committee, and leads the BioAPI Consortium's External Liaisons Working Group. BioAPI Consortium's membership consists of over 500 organizations including biometric vendors, end-users, system developers, and OEMs. NIST has recently developed the Linux version of the BioAPI reference implementation (originally developed as a Windows-compatible implementation) (see sidebar, "NIST Develops BioAPI Linux Reference Implementation") and harmonized the Linux implementation with a Unix implementation developed by another BioAPI member organization (the International Biometric Group) [1].

Standards-Based Biometric Architectures

NIST is currently examining architectures for the utilization of multimodal biometrics in large authentication systems that are BioAPI- and CBEFF-compliant. The focus is on developing and evaluating standards-based biometric systems in response to homeland defense and security requirements. The focus is also on working with government and industry to achieve more biometric-based interoperable open systems [3].

NIST Develops BioAPI Linux Reference Implementation

The Information Technology Laboratory (ITL) of NIST released to the BioAPI Consortium the BioAPI Linux Reference Implementation for testing by BioAPI Consortium member organizations. This effort, undertaken by members of the Convergent Information Systems Division of ITL, entailed "porting" the existing Windows (Win32) implementation to execute in Linux platforms. It was harmonized by NIST with a Unix (Solaris) version developed by another BioAPI Consortium member. The configuration included in the combined Reference Implementation software provided by NIST can handle multiple "flavors" of Unix (currently configurable for Linux and Solaris), and is easily adaptable to handle other Unix systems such as BSD, HPUX, or AIX in the future. The Reference Implementation is the software instantiation of the BioAPI framework. As an "open systems" specification, the BioAPI is intended for use across a broad spectrum of computing environments to insure cross-platform support [3].

NIST's Accomplishments

NIST played a significant role in the development of the BioAPI specification and the approval of this specification as a formal national standard. The development of the BioAPI standard promotes interoperability among applications by defining a generic way of interfacing to a broad range of biometric technologies. NIST also led, in collaboration with the National Security Agency, the development of a common biometric exchange file format (CBEFF). The CBEFF is a "technology-blind" common biometric format that facilitates the exchange and interoperability of biometric data from all types of biometrics, independent of the particular vendor that generates the biometric data. The development of this single approach for a biometric data structure assured biometrics companies and their potential customers that different biometric devices and applications could exchange information efficiently. This specification is being incorporated into U.S. government and international requirements, such as the technical specifications drafted by ICAO [4].

The Biometric Consortium

The U.S. Biometric Consortium (BC) serves as the government focal point for research, development, testing, evaluation, and application of biometric-based personal identification/verification technologies. The membership has substantially grown in the last few years to over 1,300 members from government, industry, and academia. Sixty government agencies have membership in the Consortium. The BC organizes biometric conferences and technical workshops that are open to members and nonmembers. It sponsors a website open to members and an electronic mail discussion list for its members. The BC also sponsors technology workshops and standards activities such as the NIST/BC Biometric Working Group and the CBEFF development. It also sponsors other user/industry activities, as needed, to address required research and other technical issues. The BC identifies technical areas of support to users and industry, such as research and technology evaluation efforts. NIST co-chairs the Biometric Consortium and its Working Groups and is involved in the organization of its outreach activities, including technical workshops and the annual conferences. The BC annual conference has become the largest biometric conference worldwide. The BC is the global clearinghouse for biometrics, is a major biometric standards incubator, and is a catalyst for enterprise integration of biometric technology [1].

Common Biometric Exchange File Format

In February 1999, NIST/ITL and the Biometric Consortium (BC) sponsored a workshop to discuss the potential for reaching industry consensus in common fingerprint template formats. The participants identified the need for a "technology-blind" biometric format that would facilitate the handling of different biometric types, versions, and biometric data structures in a common way. This common format would facilitate exchange and interoperability of biometric data from all modalities of biometrics, independent of the particular vendor that would generate the biometric data. CBEFF's initial conceptual definition was achieved through a series of workshops co-sponsored by NIST and the Biometric Consortium. A technical development team led by NIST and NSA then developed CBEFF. It was published by NIST as NISTIR 6529 in January 2001. This development included efforts focused on harmonizing the data formats among CBEFF, X9.84 (which was later approved as ANSI X9.84-2000), and the BioAPI specification (which was later approved as an ANSI INCITS 358-2002). The International Biometric Industry Association (IBIA) is the Registration Authority for CBEFF biometric data formats. Further CBEFF development has continued under the NIST/BC Biometric Interoperability, Performance, and Assurance Working Group. The result of this further work is an augmented version of CBEFF recently approved by NIST/BC Biometric Working Group. This version is now a candidate for national and international standardization. It includes the specification of a nested structure that accommodates biometric data from multiple biometric types, such as finger, facial, and iris data in the same structure; accommodates multiple samples of a specific biometric type; and includes a format specifying biometric information data objects for use within smart cards or other tokens. This format has been defined with the collaboration of technical experts from ISO/IEC JTC 1 SC 17 WG 4 and INCITS Technical Committee B10—ID Cards and Related Devices.

In addition, CBEFF (http://www.itl.nist.gov/div895/isis/bc/cbeff/) describes a set of data elements necessary to support biometric technologies in a common way independently of the application and the domain of use (mobile devices, smart cards, protection of digital data, biometric data storage) [7]. CBEFF facilitates biometric data interchange between different system components or between systems, promotes interoperability of biometric-based application programs and systems, provides forward compatibility for technology improvements, and simplifies the software and hardware integration process. CBEFF is described in detail in NISTIR 6529, "Common Biometric Exchange File Format (CBEFF)," published January 3, 2000. A copy of NISTIR 6529 can be downloaded from the CBEFF website. CBEFF is being augmented under the NIST/BC Biometric Interoperability, Performance and Assurance Working Group (http://www.itl.nist.gov/div895/isis/bc/bcwg/) to incorporate a compliant smart card format, product ID, and a CBEFF nested structure definition [2].

NIST/BC Biometric Interoperability, Performance and Assurance Working Group

Over six years ago, NIST in cooperation with the U.S. Biometric Consortium established the NIST/BC Biometric Interoperability, Performance and Assurance Working Group to support the advancement of technically efficient and compatible biometric technology solutions on a national and international basis and to promote and encourage exchange of information and collaborative efforts between users and private industry in all things biometric. In the last six years, over 500 organizations (government, industry and academia) contributed to the work of this organization. Recently the NIST/BC WG completed and approved three specifications that were delivered to the InterNational Committee for Information Technology Standards M1-Biometrics for further standardization as national and international standards. They will also be issued as NIST publications. These specifications are:

1.   The augmented version of CBEFF;

2.   A specification that defines biometric template protection and usage techniques;

3. A biometric Application Programming Interface for Java Card (developed in cooperation with the Java Card Forum) [1].

Note: The NIST BC WG is currently working in specifying biometric security techniques of interest to both end-users and the industry.

As previously mentioned, the Working Group consists of 100 organizations representing biometric vendors, system developers, information assurance organizations, commercial end users, universities, government agencies, national labs, and industry organizations. The Working Group has the following Task Groups/Technical Development Teams:

- Testing Ad-Hoc Group: Basic testing methodology

- Assurance Ad-Hoc Group: Biometrics assurance issues, review of protection profiles

- CBEFF Technical Development Team: Augmented CBEFF under development (compliant smart card format, product ID, nested structure)

- Biometric Template Protection and Integrity Task Group: Risk of reinsertion, template transformations

- Biometric Security Task Force: Vulnerability of biometric data to different attacks, nonrepudiation [2]

The BioAPI Consortium

The BioAPI Consortium was formed to develop a widely available and widely accepted application programming interface to serve any type of biometric technology. Harmonization efforts that took place in 1999, sponsored by NIST and the Biometric Consortium, resulted in a single industry standard for biometrics developed by the BioAPI Consortium. Version 1.0 of the specification was approved by the membership and published in March 2000. Version 1.1 of both the specification and the reference implementation were released in March 2001. The implementation of BioAPI-compliant solutions allows for:

1. Easy substitution of biometric technologies;

2. The utilization of biometric technologies across multiple applications;

3.   Easy integration of multiple biometrics using the same interface (the BioAPI interface);

4.   Rapid application development [1].

Utilization of open systems standards such as the BioAPI specification allows for increasing competition (which tends to lower development and implementation costs). Fast-track approval of the BioAPI specification as an ANSI INCITS standard was achieved through INCITS. BioAPI was approved as a standard (ANSI INCITS 358-2002) in February 2002. NIST is a member of the BioAPI Consortium Steering Committee and chairs the External Liaisons Working Group, which was responsible for identifying rapid mechanisms to fast-track the BioAPI specification as an ANSI INCITS standard [1].

National Standards Activities

In response to new requirements for biometric standards after September 11, 2001, the Executive Board of INCITS established Technical Committee M1 on Biometrics in November 2001. INCITS M1 is also the U.S. Technical Advisory Group (TAG) to the international subcommittee in biometrics Joint Technical Committee 1 Subcommittee 37. Therefore, M1 represents the U.S. internationally in biometrics, as well as having maintenance responsibility for the BioAPI standard. INCITS M1 is developing a portfolio of data interchange and interoperability standards and also intends to elevate consortia standards to national and international standards. Two candidates for INCITS' fast-track through M1 are the augmented version of CBEFF and the biometric Application Programming Interface for Java Card developed by the NIST/BC Biometric WG, as discussed previously. M1 has over 90 member organizations. NIST participates in the INCITS Executive Board, and chairs INCITS M1 and its Application Profile Ad-Hoc Group. Current projects under M1 development include:

- Finger Minutiae Format for Data Interchange
- Finger Pattern–Based Interchange Format
- Face Recognition Image Format for Data Interchange
- Finger Image Interchange Format
- Iris Image Format for Data Interchange
- Signature/Sign Image-Based Interchange Format
- Biometric Application Profile: Verification and Identification of Transportation Workers

- Biometric Application Profile: Personal Identification for Border Crossing

- Biometric Application Profile Biometric Verification in Point-of-Sale Systems

- Biometric Performance Testing and Reporting [1]

International Standards Activities of High Relevance to the United States

Post September 11th, many have expressed support for the urgent need for rapid formal international biometric standardization. There was widespread recognition that there was a great deal of work to be done in generic biometric standardization. Within ISO/IEC, Joint Technical Committee 1 (JTC 1) approved the establishment of a new SC 37 for biometrics in June 2002. The establishment of this new subcommittee for biometric standardization provided a venue to exploit the present window of opportunity to accelerate and harmonize international biometric standardization. This harmonization is necessary to support standards-based systems and applications that are interoperable, scalable, reliable, and secure. The formation of JTC 1 SC 37 was initiated and championed by the U.S. National Body to ISO/IEC JTC 1 with strong participation from NIST. Other federal agencies and U.S. industry have asked NIST to provide leadership for JTC 1 SC 37. JTC 1 SC 37 met for the first time on December 11–13, 2002 in Orlando, Florida, and the NIST candidate was endorsed by SC 37 to chair the SC for the next three years. The United States is responsible for providing the SC 37 Secretariat, and NIST has committed to seeing that this critical service is provided. Approximately 70 National Body delegates from 17 countries, prospective liaisons, and other JTC 1 SC chairs attended the inaugural meeting of SC 37. The current scope of work is the standardization of generic biometric technologies pertaining to human beings to support interoperability and data interchange among applications and systems. Generic human biometric standards include common file frameworks; biometric application programming interfaces; biometric data interchange formats; related biometric profiles; application of evaluation criteria to biometric technologies; methodologies for performance testing and reporting; and cross-jurisdictional and societal aspects. SC 37 established a Special Group/Study Group structure to progress its work:

- Special Group on Biometric Data Interchange Formats

- Study Group on Profiles for Biometric Applications

- Special Group on Biometric Technical Interfaces

- Special Group on Biometric Testing and Reporting
- Special Group on Harmonized Biometric Vocabulary and Definitions
- Study Group on Cross-Jurisdictional and Societal Aspects [1]

The U.S. TAG to JTC 1 SC 37 (INCITS M1) submitted multiple contributions to the JTC 1 SC 37 work, including working drafts of its national projects, the BioAPI specification, and the augmented version of CBEFF. NIST chairs the SC and provided the convener for the SC 37 Biometric Profiles Study Group [1].

International Civil Aviation Organization

The International Civil Aviation Organization (ICAO) has been working for several years to establish international biometric standards for machine-readable travel documents. A machine-readable travel document (MRTD) is an international travel document (passport or visa) containing eye- and machine-readable data. In June 2002, ICAO's Technical Advisory Group on Machine Readable Travel Documents endorsed the use of face recognition as the globally interoperable biometric for machine-assisted identity confirmation with machine-readable travel documents. While digital facial image was endorsed as the primary biometric, the Technical Advisory Group stated that ICAO Member States may elect to use fingerprint and/or iris recognition as additional biometric technologies in support of machine-assisted identity confirmation [4].

In March 2003, the Technical Advisory Group provided three key clarifications to the June 2002 resolution. First, digitally stored images (rather than templates) will be used, and these will be on-board (electronically stored in the travel document). Storage of the image, rather than a template created from the biometric feature by a proprietary algorithm, is important to ensure global interoperability. Second, these images are to be standardized. For example, ICAO has issued standards on the degree to which an image may be compressed and/or cropped. Third, high-capacity contactless integrated circuit (IC) chips will be used to store identification information in MRTDs. These chips will provide the additional data storage capacity necessary to incorporate compressed images of one or more biometrics into MRTDs [4].

Note: US VISIT will utilize internationally recognized standards, such as those developed by NIST and ICAO, as provided for in section 303 of the Enhanced Border Security and Visa Entry Reform Act of 2002 (P.L. 107-173).

Summary/Conclusion

An indication of the current substantial growth and interest in biometrics is the emergence of biometric industry standards and related activities. Standards have become a strategic business issues. For any given technology, industry standards assure the availability of multiple sources for comparable products and of competitive products in the marketplace. Standards will support the expansion of the marketplace for biometrics [2].

After the tragic events of September 11, there has been an increased emphasis on biometric standards. ITL is in a unique position to help end-users and the industry in accelerating the deployment of needed, standards-based security solutions in response to critical infrastructure protection and homeland defense/security requirements. ITL is accelerating the development of biometric standards (technology-independent interoperability and data interchange) in collaboration with federal agencies, other end-users, biometric vendors, and the IT industry [2].

ANSI INCITS 358-2002: Information Technology—BioAPI Specification

This specification defines the application programming interface and service provider interface for a standard biometric technology interface. BioAPI defines an open system standard API that allows software applications to communicate with a broad range of biometric technologies in a common way. As an "open systems" specification, the BioAPI is intended for use across a broad spectrum of computing environments to ensure cross-platform support. It is beyond the scope of this specification to define security requirements for biometric applications and service providers, although some related information is included by way of explanation of how the API is intended to support good security practices. BioAPI was developed by the BioAPI Consortium which consists of 120 organizations representing biometric vendors, original equipment manufacturers, major information technology corporations, systems integrators, application developers, and end-users. NIST holds membership in the Consortium and is a member of the Steering Committee. BioAPI specifies standards functions and a biometric data format that is an instantiation of CBEFF [2].

Human Recognition Services Module (HRS) of the Open Group's Common Data Security Architecture

HRS is an extension of the Open Group's Common Data Security Architecture (CDSA). CDSA is a set of layered security services and a cryptographic framework that provides the infrastructure for creating cross-platform, interoperable, security-enabled applications for client-server environments. The CDSA solutions cover all the essential components of security capability to secure electronic commerce and other business applications with services that provide facilities for cryptography, certificate management, trust policy management, and key recovery. The biometric component of the CDSA's HRS is used in conjunction with other security modules (cryptographic, digital certificates, and data libraries) and is compatible with the BioAPI specification and CBEFF [2].

ANSI X9.84-2000 Biometrics Management and Security for the Financial Services Industry

This American National Standards Institute (ANSI) standard was developed by the X9.F4 Working Group of ANSI Accredited Standards Committee X9, an ANSI-accredited standards organization that develops, establishes, publishes, maintains, and promotes standards for the financial services industry. X9.84-2000 specifies the minimum security requirements for effective management of biometrics data for the financial services industry and the security for the collection, distribution, and processing of biometrics data. It specifies:

- The security of the physical hardware used throughout the biometric life cycle;
- The management of the biometric data across its life cycle;
- The utilization of biometric technology for verification/identification of banking customers and employees;
- The application of biometric technology for physical and logical access controls;
- The encapsulation of biometric data;
- Techniques for securely transmitting and storing biometric data: The biometric data object specified in X9.84 is compatible with CBEFF [2].

ANSI/NIST-ITL 1-2000 Fingerprint Standard Revision

On July 27, 2000, ANSI approved ANSI/NIST-ITL 1-2000. This is a revision, re-designation, and consolidation of ANSI/NIST-CSL 1-1993 and ANSI/NIST-ITL 1a-1997. The standard specifies a common format to be used to exchange fingerprint, facial scars, mark, and tattoo identification data effectively across jurisdictional lines or between dissimilar systems made by different manufacturers. NIST has published the document as NIST Special Publication SP 500-245. The revision began with a fingerprint data interchange workshop that was held in September 1998. This revision was performed in accordance with the ANSI procedures for the development of standards using the Canvass method. All federal, state, and local law enforcement data is transmitted using the ANSI-NIST standard. This standard is a key component in allowing interoperability in the justice community [2].

AAMVA Fingerprint Minutiae Format/National Standard for the Driver's License/Identification Card DL/ID-2000

The purpose of the American Association for Motor Vehicle Administration (AAMVA) Driver's License and Identification (DL/ID) Standard is to provide a uniform means to identify issuers and holders of driver's license cards within the United States and Canada. The standard specifies identification information on driver's license and ID card applications. In high-capacity technologies such as bar codes, integrated circuit cards, and optical memory [5], the AAMVA standard employs international standard application coding to make additional applications possible on the same card. The standard specifies minimum requirements for presenting human-readable identification information, including the format and data content of identification in the magnetic stripe, the bar code, integrated circuit cards, optical memories, and digital imaging. It also specifies a format for fingerprint minutiae data that would be readable across state and province boundaries for drivers' licenses. DL/ID-2000 is compatible with the BioAPI specification and CBEFF [2].

Information Technology: Identification Cards

This standard is being developed as Part 11 of the ISO/IEC 7816 standard. The scope is specifying security-related inter-industry commands to be used for personal verification with biometric methods in integrated circuit cards (smart cards). It also defines data elements to be used with biometric methods. This standard is under development in the International Standards Organization (ISO) Subcommittee (SC) 17, Working Group 4 [2].

Finally, to fully realize the benefits of biometric technologies, comprehensive technical and operational standards are necessary to ensure that the systems are interoperable, effective, reliable, and secure. Progress in this area is being made, but more work remains to be done.

References

1. Fernando L. Podio and Michael D. Hogan, "Roles for the National Institute of Standards and Technology (NIST) in Accelerating the Development of Critical Biometric Consensus Standards for US Homeland Security and the Prevention of ID Theft," Convergent Information Systems Division, Information Technology Laboratory, NIST, 100 Bureau Drive, Stop 1070, Gaithersburg, MD 20899-1070 [US Department of Commerce, 1401 Constitution Avenue, NW, Washington, DC 20230], March 11, 2003.

2. "Biometrics Standards and Current Standard-Related Activities," NIST, 100 Bureau Drive, Stop 1070, Gaithersburg, MD 20899-1070 [US Department of Commerce, 1401 Constitution Avenue, NW, Washington, DC 20230], 2006.

3. "Information Technology Laboratory Areas," NIST, 100 Bureau Drive, Stop 1070, Gaithersburg, MD 20899-1070 [US Department of Commerce, 1401 Constitution Avenue, NW, Washington, DC 20230], 2006.

4. "Subcommittee on Aviation: Hearing on the Use of Biometrics to Improve Aviation Security," U.S. House of Representatives, Washington, DC 20515.

5. John R. Vacca, *Optical Networking Best Practices Handbook*, John Wiley & Sons (2006).

6. John R. Vacca, *Identity Theft*, Prentice Hall (2002).

7. John R. Vacca, *The Essentials Guide to Storage Area Networks*, Prentice Hall, Professional Technical Reference, Pearson Education (2001).

Part 2: How Biometric Eye Analysis Technology Works

4

How Iris Pattern Recognition Works

Historically, identity or authentication conventions were based on things one possessed (a key, a passport, or identity credential), or something one knew (a password, the answer to a question, or a PIN). This possession or knowledge was generally all that was required to confirm identity or confer privileges. However, these conventions could be compromised, as possession of a token or the requisite knowledge by the wrong individual could, and still does, lead to security breaches [1].

To bind identity more closely to an individual and appropriate authorization, a new identity convention is becoming more prevalent. Based not on what a person has or knows, but instead on what physical characteristics or personal behavior traits he or she exhibits, these are known as biometrics (measurements of behavioral or physical attributes). In other words, this is how an individual smells, walks, signs their name, or even types on a keyboard, their voice, fingers, facial structure, vein patterns, or patterns in the iris [1].

The iris is the plainly visible ring that surrounds the pupil of one's eye. It is a muscular structure that controls the amount of light entering the eye, with intricate details that can be measured, such as striations, pits, and furrows. The iris is not to be confused with the retina, which lines the inside of the back of the eye (see Figure 4-1) [2]. The iris recognition biometric technology uses the measurable features of the iris to create mathematical algorithms of the iris. The algorithms are then stored and later compared with new algorithms of irises presented to a capturing device for either identification or verification purposes [2].

What Is Iris Pattern Recognition?

Of all the biometric technologies used for human authentication today, iris pattern recognition is generally conceded to be the most accurate. Combining this high-confidence authentication with factors like outlier group size, speed, usage/human factors, and platform versatility and flexibility for use in identification or verification modes (as well as addressing issues like

Figure 4-1 *The* *basic internal structure of the eye. (Source: Reproduced with permission from Ball State University.)*

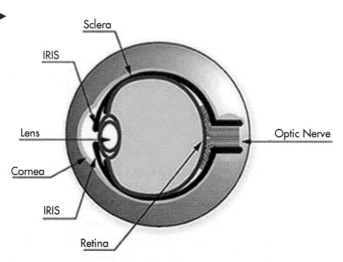

database size/management and privacy concerns), it has also shown itself to be exceedingly versatile and suited for large population applications [1].

Iris recognition technology was developed by Dr. John Daugman, and is patented solely by Iridian Technology Incorporated. There are three basic steps to iris recognition. The first step involves capturing or acquiring an image of the iris. This step is generally fulfilled by a person standing in front of a camera (see Figure 4-2) [2]. The camera then takes a picture of the iris using visible and/or infrared light. The second step is that of converting the image to what is called an Iriscode. In this step, the digital image is filtered, by an algorithm, to map segments of the iris into hundreds of vectors, also known as phasors.

Figure 4-2   *The image on the left shows the visible characteristics of the iris (http://www. cl.cam.ac.uk/users/jgd1000), while the image on the right gives a "camera's eye" view of the subject (http://www.iridiantech.com). (Source: Reproduced with permission from Ball State University.)*

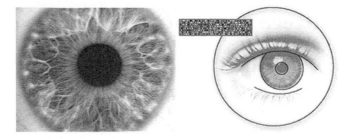

The visible characteristics, including the tribecular meshwork (the appearance of radial divisions in the iris), rings, furrows, freckles, and corona of the iris, are mapped into the different phasors and stored as hexadecimal representations in a computer. The third and final step of iris recognition is to search an already known database of Iriscode information for a match with the Iriscode of a new sample [2].

Iris recognition in an opt-in technology, which means that the user must cooperate with the system for the technology to be used. Since precise measurements must be taken with the image, the subject must hold reasonably still in a specific location, even if momentarily, in order for the image to be taken. Systems in which the user is infrequent may encounter ease-of-use problems. When the biometric is used more frequently, the interaction becomes easier to use. Glasses and colored contact lenses can alter the image of the iris. These items cause glare and color changes in the image, although the algorithms created by Dr. Daugmen recognize and account for most problems when an iris region is obscured by eyelids, contains any eyelash occlusions, specular rejections, and boundary artifacts of hard contact lenses [2].

The converted hexadecimal representation of the iris is stored into a 512-byte template. From the iris's 11 mm diameter, algorithms provide 3.4 bits of data per square mm. This density of information is such that each iris is said to have 266 degrees of freedom instead of the 13–60 for most other biometric technologies. A key difference between iris recognition technologies and others is that its 512-byte templates facilitate extremely fast match speeds. On a 300 MHz CPU, such exhaustive searches are performed at a rate of about 100,000 irises per second. On a 2.2 GHz server, one million Iriscodes can be compared in 1.7 seconds. Iris recognition technology boasts an extremely low false recognition rate (FRR) of 1:1,200,000 [2].

In other words, iris recognition is the best authentication process available today. While many mistake it for retinal scanning, iris recognition simply involves taking a picture of the iris; this picture is used solely for authentication. But what makes iris recognition the authentication system of choice? The following are the reasons:

- **Stable:** The unique pattern in the human iris is formed by 10 months of age, and remains unchanged throughout one's lifetime.

- **Unique:** The probability of two irises producing the same code is nearly impossible.

- **Flexible:** Iris recognition technology easily integrates into existing security systems or operates as a stand-alone.

- **Reliable:** A distinctive iris pattern is not susceptible to theft, loss, or compromise.

- **Non-invasive:** Unlike retinal screening, iris recognition is noncontact and quick, offering unmatched accuracy when compared to any other security alternative, from distances as far as 3″ to 10″ [1].

How Does Iris Pattern Recognition Work?

Iris recognition technology provides accurate identity authentication without PIN numbers, passwords, or cards. Enrollment takes less than two minutes. Authentication takes less than two seconds [1].

> **Tip:** Video-based technology makes it easy to enroll, producing a template that in most cases is good for the life of the subject.

Although the terminology "iris-scanning" is often used when referring to iris pattern recognition technology, there is no scanning involved at all. Iris technology is based on pattern recognition, and the pattern-capturing methodology is based on video camera technology similar to that found in ordinary camcorders. Like these cameras, the image capture process does not require bright illumination or close-up imaging [1].

With a device activated by proximity sensor, a subject positioned 3″ to 10″ from the Enrollment Optional Unit is guided by a mirrored, audio-assisted interactive interface to allow an auto-focus camera to take a digital video of the iris. Individual images from the live video are captured using a frame grabber. The innovative algorithm of the iris recognition process analyzes the patterns in the iris that are visible between the pupil and sclera (white of the eye) and converts them into a 512-byte digital template. This value is stored in a database and communicated to Identification Control Units associated with portals where the subject has access privileges [1].

Recognition takes just two seconds. Upon approaching a portal, proximity sensors activate a Remote Optical Unit (ROU) [4] when the subject nears the operational range of the unit. The same mirror-assisted, audio-prompted interface that the subject became familiar with at enrollment helps ensure proper positioning and speedy recognition. The ROU uses the same video and frame-grabbing methodology to create, select, and digitize an image to be compared against the stored value retained at enrollment [1].

The live presented value is compared against stored values at an Identification Control Unit assigned to the portal. Once the iris is matched, either a direct signal is sent to activate a door, or a Weigand signal sent to a central access panel provides the impetus to open the door to the individual authorized to enter [1].

The Biology Behind the Technology

Like a snowflake, the iris of every human eye is absolutely unique, exhibiting a distinctive pattern that forms randomly in utero in a process called chaotic morphogenesis. In fact, it's estimated the chance of two irises being identical is 1 in 1,078 [1].

As previously mentioned, the iris is not the retina, which is found within the eye itself. Nor should iris recognition be confused with retinal scanning, an older authentication technology based on mapping the vasculature found on this inner part of the eye. Unlike iris recognition, retinal scanning requires the inner eye to be subjected to intense illumination, which adds to a general feeling that this older and waning technology is invasive [1].

Why Iris Recognition Technology?

There are many reasons why iris recognition is a particularly attractive technology for identity management, such as the following:

1. Smallest outlier population
2. Unparalleled stability
3. Unique design facilitates superior management of large databases
4. Unmatched search speed
5. Application mode versatility
6. High-level user acceptance
7. Convenient intuitive user interface [1]

Smallest Outlier Population

There are relatively few people who don't have at least one eye, so there are only a few people who can't use the technology. While blind people can be difficult to enroll, there are instances where blind people have used iris recognition successfully (The technology is pattern-dependent, not sight-dependent.) [1].

Unparalleled Stability

The patterns in the human iris are fixed from about one year of age and remain constant, barring trauma, certain rare diseases, or change that may occasionally ensue from some ophthalmologic surgical procedures. This means that once a subject is enrolled, the need to re-enroll is lower than for other biometric identification options, where changes in voice timbre, weight, hairstyle, finger or hand size, or sustained manual labor or the presence of a superficial cut can require re-enrollment [1].

Unique Design Facilitates Superior Management of Large Databases

Iris recognition is the only biometric authentication technology designed to work in the 1-n or exhaustive search mode. This makes it ideal for handling applications that require management of large user groups, such as a national documentation application might necessitate. The technology is ideally suited to handle large databases, and does so without any degradation in accuracy [1].

Unmatched Search Speed

Unmatched search speed in the one to many search mode is unmatched by any other technology. It is limited not by database size, but by hardware selected for server management [1].

Application Mode Versatility

While the technology was initially designed to work in one-to-many search mode, iris recognition is perfectly suited to applications that require one-to-one matching, or verification mode operation. This makes iris authentication ideal for use in upgrading security systems that have a large base of installed card readers or PIN pads. Compatibility with the Weigand environment means that a high integrity security overlay can be built into such systems. In instances where legislative or strategic considerations demand it, the technology is ideally suited for smart card use, leaving the issue of biometric database management a moot one, as the user retains control of biometric data—in this case a 512-byte template held on the smart card [1].

High-Level User Acceptance

Most people in the developed world are comfortable with the idea of having their picture taken, particularly if there is some benefit to having it done. Iris recognition involves nothing more than taking a digital picture of the iris (from moving video), and recreating an encrypted digital template of that pattern. That 512-byte template cannot be re-engineered or reconstituted to produce any sort of visual image, and provides a high level defense against identity theft [5], a rapidly growing crime. There are no lasers or bright lights involved in iris recognition, and authentication is entirely noncontact. Enrollment is opt-in, and data collected bears no resemblance to that collected for any purpose other than real-time human authentication, so the technology is free of any surveillance-related or criminal/forensic stigmas [1].

Convenient Intuitive User Interface

Using the technology is an almost intuitive experience. Proximity sensors activate the equipment, which includes mirror-assisted alignment functionality. Audio auto-positioning prompts, auto-focus image capture, and visual and audio authentication decision cueing complete the process [1].

How Iris Recognition Compares to Other Biometrics

Few would argue with the generally held view (and evidence) that iris recognition is the most accurate of the commonly used biometric technologies. There are a number of other factors that weigh heavily in iris recognition's favor for applications requiring large databases and real-time authentication.

- Accuracy
- Stability
- Speed
- Scalable
- Noninvasive [1]

Accuracy

As previously mentioned, like a snowflake, every iris is absolutely unique. A subject's left and right iris is as different from each other as they are from any other individual's. It has been calculated that the chance of finding two randomly formed identical irises is on an almost astronomical order of 1 in 1,078 [1].

Another differentiator impacting accuracy is that no human intervention is required to "set" thresholds for false-accept and false-reject performance. The human element plays no role in performance standards for this technology, while an unmatched EER (equal error rate) performance of 1 in 1.2 million is delivered. Other electronic authentication technologies sometimes select a number of templates that represent "possible matches," perpetuating the potential for error, in that final determination of identity relies on a human interpretation [1].

At the root of iris recognition's accuracy is the data-richness of the iris itself. Fingerprints, facial recognition, and hand geometry have far less detailed input in template construction. In fact, it's probably fair to say that one iris template contains more data than is collected in creating templates for a finger, a face, and a hand combined. This is one reason why iris recognition can authenticate with confidence even when significantly less than the whole eye is visible [1].

Stability

Virtually every other biometric template changes significantly over time, detracting from overall system performance and requiring frequent re-enrollment. Voices change. Hands and fingers grow. The type of labor one does, even the weather temperature or one's medical condition, can result in template changes in other technologies. Barring trauma and certain ophthalmologic surgery, the patterns in the iris are constant from infancy to death.

Note: At death, iris tissue is among the most rapidly deteriorating of all body tissues, something that leads to its use by forensic pathologists in estimating time of death.

Speed

No other biometric technology is designed to deliver 1-n searching of large databases in real time. A 2001 study conducted by the United Kingdom National Physical Laboratory found iris technology was capable of nearly 20 times more matches per minute than its closest competitor. Looking at speed in conjunction with accuracy, no other technology can deliver high accuracy authentication in anything close to the real-time performance of iris recognition [1].

Fingerprint searches, for example, are challenged by database size, adding time to searches or necessitating filtering as a search acceleration technique. Even so, fingerprint technology often returns multiple "possible matches," forcing introduction of human decision factors and increasing the potential for error in an authentication decision [1].

Scalable

Iris recognition is ideal for large-scale ID applications or enterprise physical security and applications characterized by large databases. As iris data templates require only 512 bytes of storage per iris [6], very large databases can be managed and speedily searched without degradation of performance accuracy [1].

Noninvasive

No bright lights or lasers are used in the imaging and iris authentication process. The user can stand as far as 10" away from the unit, and even wear glasses or contact lenses without compromising system accuracy. Unlike some other popular biometrics, iris authentication involves no physical contact. Not only does this mean "no touch" authentication, it also means the technology is ideally suited for use in environments where rubber gloves or other protective gear is used [1].

Iris recognition applications are generally opt-in—there is none of the surveillance stigma sometimes affiliated with facial recognition, which scans crowds looking for individuals. Nor is there any tie-in to the large fingerprint databases maintained by law enforcement agencies, which often gives a negative stigma to fingerprint-based systems [1].

Current and Future Use

The versatility of iris technology lends itself to virtually any application where identity authentication is required to enhance security. This includes service, elimination of fraud, and maximizing convenience [1].

Today

While the most common use of iris recognition to date is physical access control in private enterprise and government, the versatility of the technology will lead to its growing use in large sectors of the economy, such as transportation, health-care, and national identification programs. Although security is clearly a prime concern, iris recognition is also being adopted for productivity-enhancing applications like time and attendance [1].

Areas of Opportunity

Iris recognition technology is currently used in many locations and for many reasons. In 1996, Lancaster County Prison in Pennsylvania became the first correctional facility to use iris scanning. The facility sometimes needs to release prisoners on short notice and can't wait for fingerprint tests [2].

In the largest national deployment of iris recognition to date, the United Arab Emirates (UAE) Ministry of Interior requires iris recognition tests on all passengers entering the UAE from all 17 air, land, and sea ports. Their Iriscodes are then checked against those of deportees. Since its inception, the program has caught 10,586 deportees returning to that country [2].

Most uses of iris recognition technology are for business purposes of access to offices, laboratories, computers, and bank accounts. The Venerable Bede School in the United Kingdom uses this technology with its 1,200 students to check out library books and for cafeteria payments. Recently, iris recognition technology has been used in Pakistan to limit Afghan refugees to one cash grant each by the United Nations [2].

Tomorrow

Enterprise and government both acknowledge the convergence of physical and information security environments, but there are new security challenges on the horizon, such as just-in-time inventory control, sophisticated supply chain management, and even a phenomenon called "coopetition," in which companies that compete in some areas cooperate in others. Managing this convergence of physical and information security requirements now drives security system architecture design and implementation, and is an increasingly key factor in biometric technology selection. Managing convergence will only become a more complex task because, as IT and communications become increasingly wireless [7], the need for robust identity management will become more acute [2].

Finally, iris recognition technology is a natural "fit" in the physical, infosec, and wireless arenas. In the very near future, iris recognition technology will be deployed in ways that eliminate fraud, provide nonrepudiation of sales, authenticate funds transfers, provide signature verification, credit card authorization, and authorized access to healthcare records, intellectual property, and much more [2].

Summary/Conclusion

Iris-based personal identification (PI) or recognition uses the unique visible characteristics of the human iris (the tinted annular portion of the eye bounded by the black pupil and the white sclera) as its biometric. Most commercially available iris PI systems are based on research and patents held by Dr. John Daugman of the University of Cambridge, Cambridge, U.K. An iris PI system requires no intimate contact between the user and the image capture device. Typically, a conventional CCD camera is used to capture an image of an eye. Algorithms then isolate and transform the iris portion of the images into templates that offer exceptional matching performance for both FAR and FRR. Iris-based PI is one of the few biometric systems with proven "user identification" mode capability for large (national, international, and even planetary) template databases [3].

The human iris is composed of elastic connective tissue called the trabecular meshwork. The trabecular meshwork is completely developed by the eighth month of gestation. It consists of a host of visible features, namely, rings, furrows, and freckles, as well as several other features that require a medical degree and/or dictionary for explanation and comprehension. The color of the iris often changes during the first year of life; however, clinical evidence indicates that the trabecular pattern is stable throughout one's life span. The iris is immune to the environment except for the pupil's response to light. A remarkable fact about the iris (and one of the reasons that the iris image makes an excellent biometric) is that each possesses a highly detailed and unique visible texture. These textures are unique even when considering genetics. That is to say that, not only do identical twins have unique irises, the two irises of any individual are each unique and have uncorrelated textures [3].

Iris pattern–based PI is one of the few methodologies that has been proven to work well in user "identification mode." It exhibits a high degree of accuracy and therefore can be used in applications where high security is paramount. Although user interaction is required for an adequate image capture, the technology requires no physical contact and is basically nonintrusive. Once educated and acclimated, users have regularly accepted the technology for PI applications. If a PI system requires user "identification mode" over large template databases, this technology may be one of only two options (the other is retina scanning) for the PI system developer [3].

Finally, iris recognition technology is versatile. Systems can be relatively inexpensive at the cost of $\sim$\$8,000 each, depending on the application. Successful previous applications for high-security areas such as prisons, U.S. congressional offices, the U.S. Department of Treasury, and U.S. Vice President Cheney's offices provide powerful information on the reliability and accuracy of this biometric. Iris recognition is set to grow substantially from a $69 million industry in 2006 to a $811 million industry in 2011. Although there could be several ease-of-use issues when administering this technology into a school system, it is a viable option in education. This biometric technology could be an option for educational uses such as library check out; cafeteria payments; access to buildings for faculty, staff, students, and parents; parent verification for student pick-ups; access to computers; and attendance, to name a few [2].

References

1. "Iris Recognition," LG Electronics U.S.A., Inc., Iris Technology Division, 7 Clarke Drive, Cranbury, NJ 08512, 2003. Copyright © 2003 LG Electronics. All Rights Reserved.

2. Alex English, Christina Means, Kris Gordon and Kevin Goetz. "Biometrics: A Technology Assessment," Ball State University, 2000 W. University Ave., Muncie, IN 47306, 2006. Copyright © 2006.

3. "Iris Recognition," Electrical and Computer Engineering Department, University of Alabama in Huntsville, Huntsville, AL 35899, 2006.

4. John R. Vacca, *Optical Networking Best Practices Handbook*, John Wiley & Sons (2006).

5. John R. Vacca, *Identity Theft*, Prentice Hall (2002).

6. John R. Vacca, *The Essentials Guide to Storage Area Networks*, Prentice Hall, Professional Technical Reference, Pearson Education (2001).

7. John R. Vacca, *The Guide to Wireless Network Security*, Springer (2006).

How Retina Pattern Recognition Works

Retina pattern recognition technology captures and analyzes the patterns of blood vessels on the nerve on the back of the eyeball that processes light entering through the pupil. Retinal patterns are highly distinctive; even the eyes of identical twins are distinct. Although each pattern normally remains stable over a lifetime, they can be affected by disease such as glaucoma, diabetes, high blood pressure, and autoimmune deficiency syndrome. Because the retina is small, internal, and difficult to measure, capturing its image is more difficult than with most other biometrics. An individual must position the eye very close to the lens of the retina-scan device, gaze directly into the lens, and remain perfectly still while focusing on a revolving light while a camera scans the retina through the pupil. Any movement can interfere with the process. Enrollment can easily take more than a minute.

The retina (see Figure 5-1) can be described as a layer of complex blood vessels and nerve cells [1]. The complexity of this layer of the eye makes this form of biometrics one of the most reliable forms of verification. The key to retinal pattern recognition involves low-intensity lights (see Figure 5-2) shined directly at the test subject's retina, thus making the precise location of the subject one disadvantage of retinal pattern recognition [1]. Another disadvantage that has led to retinal pattern recognition not being implemented into a wide variety of applications is the expense that installing and maintaining this system entails [1], as well as the perceived public opinion on safety.

Retinal pattern recognition is not widely deployed for commercial applications like some other biometric technologies discussed in previous and forthcoming chapters. This is because of the costs involved, and the user invasiveness. But, despite this, retinal pattern recognition is considered by some to be the "ultimate" biometric of all because of its reliability and stability [2].

The Anatomy and the Uniqueness of the Retina

When talking about the eye, especially in terms of biometrics, there is often confusion between the iris and the retina of the eye. Although the iris and the

Figure 5-1 *The retina is a layer of complex blood vessels and nerve cells. (Source: Reproduced with permission from Ball State University.)*

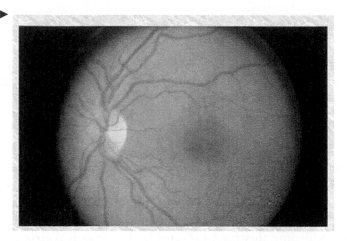

Figure 5-2 *Retinal pattern recognition involves low-intensity lights shined directly at the test subject's retina. (Source: Reproduced with permission from Ball State University.)*

retina can be grouped together into one broad category called "eye biometrics," the function of the two are completely different. The iris is the colored region between the pupil and the white region of the eye. The primary purpose of the iris is to dilate and constrict the size of the pupil. It is analogous to the aperture of a camera [2].

The retina is to the eye as film is to a camera. The retina is essentially sensory tissue that consists of multiple layers. The retina consists of literally millions of photoreceptors whose function is to gather the light rays that are sent to them and transform that light into electrical impulses that travel through the optic nerve into the brain, which then converts these impulses into images. The two distinct types of photoreceptors that exist within the retina are called rods and cones. The cones (there are about 6 million cones) help you to see different colors, and the rods (there are about 125 million rods) help with night and peripheral vision. It is the blood vessel pattern in the retina that forms the foundation for the science and technology of retinal pattern recognition [2].

Two famous studies confirmed the uniqueness of the blood vessel pattern of the retina. In 1935, a paper was published by Dr. Carleton Simon and Dr. Isodore Goldstein, in which they laid out their discovery that every retina possesses a unique blood vessel pattern. They later published a paper suggesting the use of photographs of these blood vessel patterns of the retina as a means to identify people. The second study was conducted in the 1950s by Dr. Paul Tower. He discovered that even among identical twins, the blood vessel patterns of the retina are unique [2].

The next part of the chapter examines the technology behind retinal pattern recognition, and the sources of problems encountered in retinal pattern recognition. Biometric performance standards are also covered [2].

The Technology Behind Retinal Pattern Recognition

The first major vendor for the research/development and production of retinal scanning devices was a company called EyeDentify, Inc., created in 1976. The first types of devices used to obtain images of the retina were called fundus cameras. These were instruments created for opthalmologists, adapted to obtain images of the retina. However, there were some problems in using this type of device. The equipment was considered to be very expensive and difficult to operate, and the light used to illuminate the retina was considered to be far too bright and uncomfortable for the user [2].

As a result, further research and development was conducted, which subsequently yielded the first true prototype of a retinal scanning device in 1981. This time, infrared light was used to illuminate the blood vessel pattern of the retina. Infrared light has been primarily used in retinal pattern recognition because the blood vessel pattern in the retina can "absorb" infrared light at a much quicker rate than the rest of the tissue in the eye. The infrared light and the illuminated blood vessel pattern is then reflected back to the retinal scanning device for processing. Various algorithms were developed for the extraction of the unique features. Further research and development gave birth to the first true retinal scanning device to be put on the market, the EyeDentification System 7.5 [2].

This retinal scanning device utilized a complex system of scanning optics, mirrors, and targeting systems in order to capture the blood vessel pattern of the retina. However, later research and development created devices with much simpler designs. These newer devices consisted of integrated retinal scanning optics, which sharply reduced the costs of production in comparison to the production costs of the EyeDentification System 7.5 [2].

The last known retinal scanning device to be manufactured by EyeDentify was the ICAM 2001. This device could store up to 3,000 templates, with a storage capacity of up to 3,300 history transactions [3]. However, this product was eventually taken off the market because of user concerns and its high price tag. To this date, this author is aware of only one vendor that is in the process of creating a retinal scanning device, which is Retinal Technologies, LLC. It is believed that the company is working on a prototype retinal scanning device that will be much easier to implement into commercial applications, and be much more user friendly.

There are three major components of a retinal scanning device:

1.   Imaging/Signal Acquisition/Signal Processing: This involves a camera capturing the retinal scan, and converting that scan into a digital format.

2.   Matching: A computer system for verification and identification of the user (as is the case with the other biometric technologies reviewed in previous chapters).

3.   Representation: The unique features of the retina are represented as templates [2].

The process of enrollment and verification/identification in a retinal scanning system is the same as the process for the other biometric technologies:

1.   Acquisition and processing of images

2.   Unique feature extraction

3.   Template creation [2]

The image acquisition and processing phase is the hardest phase to complete successfully. This is because the user has to be very cooperative in this phase. The user must first place his eye near a lens located in the retinal scanning device at extremely close range. It is very important that the user remain perfectly still at this point, in order to ensure that a robust image will be captured. Also, the user must remove any eyeglasses that he might be wearing, because any light reflection from the lens of the eyeglasses could cause interference with the signal of the retinal scanning device. Once the user is situated comfortably, he then will notice a green light embedded against a white background through the lens of the scanning device. Once the retinal scanning device is activated, this green light moves in a complete circle (360 degrees) and captures images of the blood vessel pattern of the retina through the pupil. At this phase, normally three to five images are captured. This phase can take over one minute to complete, depending upon how cooperative the user is, which is considered to be a very long time in comparison to the image acquisition and processing times of the other biometric technologies [2].

A strong advantage of retinal pattern recognition is that genetic factors do not dictate what the blood vessel pattern of the retina will be. This allows the retina to have very rich, unique features. As a result, it is possible that up to 400 unique data points can be obtained from the retina as opposed to other biometrics, such as fingerprint scanning, where only 30–40 data points (the minutiae) are available [2].

In the template creation phase, the unique features gathered from the blood vessel pattern of the retina forms the basis of the enrollment template. This template is only 96 bytes, and as a result, is considered to be one of the smallest biometric templates [2].

Sources of Problems (Errors) and Biometric Performance Standards

There are sources of problems that could affect the retinal scanning device from obtaining an accurate scan (as is the case with any other biometric technology), thus impacting its ability to successfully verify or identify users. Among the problems are:

1. The lack of cooperation on part of the user. As discussed earlier, the user must remain very still during the entire process, especially in the image acquisition phase. Any movement can seriously affect the alignment of the lens in the retinal scanning device.

2. The proper eye distance is not maintained while the user attempts to look into the lens of the retinal scanning device. In order for a high-quality scan to be captured, the user must focus his or her eye at an extremely close range to the lens. In this regard, iris scanning technology is much more user friendly, as a good-quality scan can be captured as far as three feet away from the lens of the scanning device to the iris of the user.

3. A dirty lens on the retinal scanning device. This will obviously interfere with the scanning process.

4.   Other types of light interference from the external environment.

5.   The pupil size of the user. A naturally small pupil that is constricted to a still smaller size because of a bright lighting environment can reduce the amount of light that reaches the retina via the pupil and vice versa. This can cause the system to have a higher rate of false rejection [2].

All types of biometric technology are rated against a set of performance standards. There are two performance standards that are most applicable to retinal pattern recognition: the false-reject rate and the ability-to-verify rate. These are described next [2].

False-Reject Rate (Also Known as Type 1 Errors)

The false-reject rate standard describes the probability of a legitimate user being denied authorization by the retinal scanning system. Of all the performance standards, retinal pattern recognition is most affected by this standard. This is because there are a number of factors that can impact the quality of a retinal scan (as described previously), and as a result, deny a legitimate user authorization [2].

Ability-to-Verify Rate

The ability-to-verify rate standard describes the probability of the overall user group that can be verified by the retinal scanning system on a daily basis. For retinal pattern recognition, this percentage has been as low as 85%. This can be attributed mostly to user concerns about using a retinal scanning device and having their eye scanned at a very close range [2].

The next part of the chapter examines the strengths and weaknesses of retinal pattern recognition. It also examines its applications in the commercial sector [2].

The Strengths and Weaknesses of Retinal Pattern Recognition

Retinal pattern recognition possesses its own set of strengths and weaknesses, just like all other types of biometric technology. The strengths can be described as follows:

1.   The blood vessel pattern of the retina hardly ever changes over the lifetime of an individual, unless he or she is afflicted by some disease of the eye, such as glaucoma or cataracts.

2.   The actual template is only 96 bytes, which is very small. This could result in quicker verification and identification processing times, as opposed to larger templates, which could slow down the processing times.

3.   Rich, unique features can be extracted from the blood vessel pattern of the retina—up to 400 data points.

4.   The retina is located inside the eye; thus, it is not exposed to the threats posed by the external environment, as other biometrics are, such as fingerprints, hand geometry, and so on [2].

In terms of weaknesses of retinal pattern recognition, most of them are inherent in user-based applications:

1.   The public perception of a health threat: There tends to be a public belief that a retinal scanning device can cause damage to your eye.

2.   The user unease of having their eye scanned at a very close distance.

3.   The motivational level of the user: Of all of the biometric technologies, there must be a high level of user motivation and patience to successfully use the retinal scanning device.

4.   Retinal scanning technology still cannot take into consideration eyeglasses; the user must remove them during the scanning process.

5.   At the current time, retinal scanning devices are very expensive to procure and implement [2].

The Applications of Retinal Pattern Recognition

Retinal pattern recognition systems are expensive to install and maintain because of user invasiveness. They have not been widely deployed as other biometric technologies have been (particularly fingerprint recognition, hand geometry recognition, facial recognition, and to a certain extent, iris recognition) [2].

The primary applications for retinal pattern recognition have been for physical access entry for high-security facilities. This includes military installations, nuclear facilities, and laboratories. One of the best-documented applications of the use of retinal pattern recognition was conducted by the state of Illinois in an effort to reduce welfare fraud. The primary purpose was to identify welfare recipients so that benefits could not be claimed more than once. Fingerprint recognition was also used in conjunction in this project. The use of retinal pattern recognition started in mid-1996, focusing upon two cities in the southern part of Illinois: Granite City and East Alton. The fingerprint recognition phase started after the retinal scanning program. A comparison was made between the two biometric technologies, and it was concluded that retinal scanning is not client- or staff-friendly and requires considerable time to secure biometric records. Based on these factors, retinal scanning technology is not yet ready for statewide adaptation to the Illinois welfare department and the use of retinal pattern recognition was terminated in that state [2].

Summary/Conclusion

Retinal pattern recognition may truly be the "ultimate" biometric of all, because of the rich and unique features of the blood vessel patterns of the retina. But, because of its high cost and user issues, it has not made its mark in the commercial sector. As technology advances, however, the day will come when retinal pattern recognition will make its mark, and user acceptance and public adoption will be widespread [2].

References

1. Barrett Key, Kelly Neal and Scott Frazier, "The Use of Biometrics in Education Technology Assessment," Ball State University, 2000 W. University Ave., Muncie, IN 47306, 2006. Copyright © 2006.

2. Ravi Das, "An Application of Biometric Technology: Retinal Recognition Series #1, 2 and 3," Technology Executives Club, Ltd., 1590 S Milwaukee, Suite 223, Libertyville, IL 60048, 2006. Copyright © 2006 Technology Executives Club, Ltd. All Rights Reserved.

3. John R. Vacca, *The Essentials Guide to Storage Area Networks*, Prentice Hall, Professional Technical Reference, Pearson Education (2001).

Part 3: How Biometric Facial Recognition Technology Works

6

How Video Face Recognition Works

Facial scan biometrics is an automated way of identifying a person by their distinct individual facial features. Facial scans have recently become a growing concern in this nation as they are being used to find and determine anyone who is known as a possible threat. However, this is not their only function; they are used to identify and verify people for many different applications. Facial scans are done via many different techniques, and involve advanced software to analyze and break down specific details and features of each face. Even though the idea of a face scan and the software used to complete them may be complicated, they can be achieved with a very simple store-bought camera. Although there are several different types of facial scans, they all use the same basic steps and similar procedures in a similar way. The steps are:

1. **Capture:** A raw biometric is captured by a sensing device.

2. **Process:** The distinguishing characteristics are extracted from the raw biometric sample and converted into a processed biometric identifier record.

3. **Enroll:** The processed sample (a mathematical representation of the biometric) is stored or registered, for comparison later during authentication.

4. **Verification:** Matching the sample against a record [1].

The three types of facial recognition techniques that are used are eigenface, eigenfeature, and thermal imaging. The three main parts of the face that usually don't change are some primary targets: the upper sections of the eye sockets, the area surrounding the cheekbones, and the sides of the mouth. Eigenface systems capture the image and change it to light and dark areas. Both the initial facial image and the facial image in question are also captured in a two-dimensional form. Then, the two images are compared according to the points of the two eigenface images (see Figure 6-1) [1]. Eigenfeature image systems work in a similar way, except it picks out certain features and calculates the distances

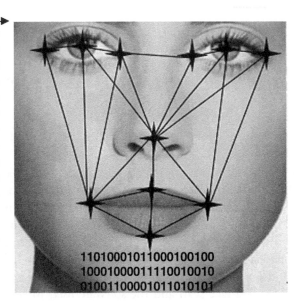

Figure 6-1 *Two images are compared according to the points of the two eigenface images. (Source: Reproduced with permission from Ball State University.)*

between them. The points are the facial features such as eyes, nose, mouth, bone curves, and other distinct features. However, many faces do change over the course of a person's lifetime, so the images in storage [4] need to reflect that. Some image systems do account for this change, but do not always do so correctly. This is where thermal imaging takes over. Thermal imaging takes a thermal image of the face that focuses on the blood vessels, because it is believed that even if the face changes, the blood vessel pattern does not. To accomplish this, an infrared camera must be used instead of any type of traditional camera. The 3D imaging on both the face in question and the stored face can help to bring out frauds in facial scans [1].

Some problems with face recognition are keeping information secure to avoid fraud, the necessity of cooperation to obtain scan for database, the difficulty of capturing a good image of the face at the correct angle to compare to the database, and ethical issues such as privacy. On the other hand, no contact is needed, meaning a suspect can be identified without their knowledge. Also, ordinary light is sufficient to complete the process and can be used for security purposes and identifying threats [1].

Now, let's look at some examples of how facial recognition works. Keep in mind that the use of video facial recognition in the United States is very limited compared to the widespread use of the technology in the United Kingdom, especially in London.

How Facial Recognition Technology Works

A ticket to Super Bowl XXXV in Tampa Bay, Florida, didn't just get you a seat at the biggest professional football game of the year. Those who attended the January 2000 event were also part of the largest police lineup ever conducted, although they may not have been aware of it at the time. The Tampa Police Department (http://www.tampagov.net/dept_Police/index.asp) was testing out a new technology, called FaceIt, that allows snapshots of faces from the crowd to be compared to a database of criminal mugshots (see Figure 6-2) [2].

The $100,000 system was loaned to the Tampa Police Department for one year. During that one-year period, no arrests were made using the technology. However, the 36 cameras positioned in different areas of downtown Tampa have allowed police to keep a more watchful eye on general activities, resulting in hundreds of arrests for various types of crimes. This increased surveillance of city residents and tourists has riled privacy rights groups [2].

People have an amazing ability to recognize and remember thousands of faces. In this chapter, you'll learn how computers are turning your face into computer code so it can be compared to thousands, if not millions, of other faces. The chapter also looks at how facial recognition software is being used in elections and criminal investigations and to secure your personal computer [2].

Figure 6-2
Facial recognition software can be used to find criminals in a crowd, turning a mass of people into a big lineup. (Source: Reproduced with permission from HowStuffWorks, Inc.)

The Face

Your face is an important part of who you are and how people identify you. Imagine how hard it would be to recognize an individual if all faces looked the same. Except in the case of identical twins, the face is arguably a person's most obvious unique physical characteristic. While humans have had the innate ability to recognize and distinguish different faces for millions of years, computers are just now catching up [2].

For example, Visionics, a company based in New Jersey, is one of many developers of facial recognition technology [2]. The twist to its particular software, FaceIt, is that it can pick someone's face out of a crowd, extract that face from the rest of the scene, and compare it to a database of stored images. In order for this software to work, it has to know what a basic face looks like. Facial recognition software is based on the ability to first recognize faces, which is a technological feat in itself, and then measure the various features of each face [2].

Disclaimer: The author and publisher do not endorse any of the products or vendors mentioned in this chapter or throughout the book. They are only mentioned here for illustration purposes and were selected at random for those purposes.

If you look in the mirror, you can see that your face has certain distinguishable landmarks. These are the peaks and valleys that make up the different facial features. Visionics terms these landmarks nodal points. There are about 80 nodal points on a human face. Here are a few of the nodal points that are measured by the software:

- Distance between eyes
- Width of nose
- Depth of eye sockets
- Cheekbones
- Jaw line
- Chin [2]

These nodal points are measured to create a numerical code, a string of numbers, that represents the face in a database. This code is called a faceprint. Only 14–22 nodal points are needed for the FaceIt software to complete the recognition process. Next, let's look at how the system goes about detecting, capturing, and storing faces [2].

The Software

Facial recognition methods vary, but they generally involve a series of steps that serve to capture, analyze, and compare your face to a database of stored images. Here is the basic process that is used by the FaceIt system to capture and compare images (see Figure 6-3) [2]:

- Detection
- Alignment
- Normalization
- Representation
- Matching

Detection

When the system is attached to a video surveillance system, the recognition software searches the field of view of a video camera for faces. If there is a face in the view, it is detected within a fraction of a second. A multiscale

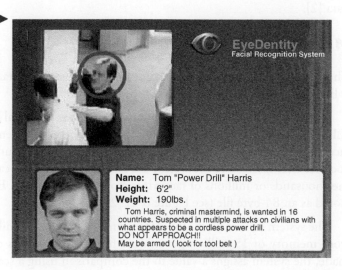

Figure 6-3 *To identify someone, facial recognition software compares newly captured images to databases of stored images. (Source: Reproduced with permission from HowStuffWorks, Inc.)*

EyeDentity
Facial Recognition System

Name: Tom "Power Drill" Harris
Height: 6'2"
Weight: 190lbs.
Tom Harris, criminal mastermind, is wanted in 16 countries. Suspected in multiple attacks on civilians with what appears to be a cordless power drill.
DO NOT APPROACH!!
May be armed (look for tool belt)

algorithm is used to search for faces in low resolution. The system switches to a high-resolution search only after a head-like shape is detected [2].

Note: An algorithm is a program that provides a set of instructions to accomplish a specific task.

Alignment

Once a face is detected, the system determines the head's position, size, and pose. A face needs to be turned at least 35 degrees toward the camera for the system to register it [2].

Normalization

The image of the head is scaled and rotated so that it can be registered and mapped into an appropriate size and pose. Normalization is performed regardless of the head's location and distance from the camera. Light does not impact the normalization process [2].

Representation

The system translates the facial data into a unique code. This coding process allows for easier comparison of the newly acquired facial data to stored facial data [2].

Matching

The newly acquired facial data is compared to the stored data. And, ideally, it is linked to at least one stored facial representation [2].

The heart of the FaceIt facial recognition system is the Local Feature Analysis (LFA) algorithm. This is the mathematical technique the system uses to encode faces. The system maps the face and creates a faceprint, a unique numerical code for that face. Once the system has stored a faceprint, it can compare it to the thousands or millions of faceprints stored in a database. Each faceprint is stored as an 84-byte file (see Figure 6-4) [2].

The system can match multiple faceprints at a rate of 60 million per minute from memory or 15 million per minute from hard disk. As comparisons are made, the system assigns a value to the comparison using a scale of one to 10.

Figure 6-4
Using facial recognition software, police can zoom in with cameras and take a snapshot of a face. (Source: Reproduced with permission from HowStuffWorks, Inc.)

If a score is above a predetermined threshold, a match is declared. The operator then views the two photos that have been declared a match to be certain that the computer is accurate [2].

Summary/Conclusion

The primary users of facial recognition software like FaceIt have been law enforcement agencies, which use the system to capture random faces in crowds. These faces are compared to a database of criminal mug shots. In addition to law enforcement and security surveillance, facial recognition software has several other uses, including:

- Eliminating voter fraud
- Check-cashing identity verification
- Computer security [2]

One of the most innovative uses of facial recognition is being employed by the Mexican government, which is using the technology to weed out duplicate voter registrations. To sway an election, people will register several times under different names so they can vote more than once. Conventional methods have not been very successful at catching these people [2].

Using the facial recognition technology, officials can search through facial images in the voter database for duplicates at the time of registration. New images are compared to the records already on file to catch those who attempt to register under aliases. The technology was used in the country's 2000 and 2006 presidential elections, and is expected to be used again in local elections [2].

Potential applications even include ATM and check-cashing security. The software is able to quickly verify a customer's face. After the user consents, the ATM or check-cashing kiosk captures a digital photo of the customer. The FaceIt software then generates a faceprint of the photograph to protect customers against identity theft [3] and fraudulent transactions (see Figure 6-5) [2]. By using facial recognition software, there's no need for a picture ID, bank card, or PIN to verify a customer's identity.

Figure 6-5
Many people who don't use banks use check-cashing machines. Facial recognition could eliminate possible criminal activity. (Source: Reproduced with permission from HowStuffWorks, Inc.)

Figure 6-6
Facial recognition software can be used to lock your computer. (Source: Reproduced with permission from HowStuffWorks, Inc.)

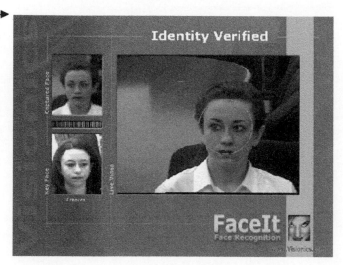

This biometric technology could also be used to secure your computer files. By mounting a webcam to your computer and installing the facial recognition software, your face can become the password you use to get into your computer (see Figure 6-6) [2]. IBM has incorporated the technology into a screensaver for its Thinkpad laptops [2].

While facial recognition can be used to protect your private information, it can just as easily be used to invade your privacy by taking your picture when you are entirely unaware of the camera. As with many developing technologies, the incredible potential of facial recognition comes with drawbacks [2].

References

1. Barrett Key, Kelly Neal and Scott Frazier, "The Use of Biometrics in Education Technology Assessment," Ball State University, 2000 W. University Ave., Muncie, IN 47306, 2006. Copyright © 2006.

2. Kevin Bonsor, "How Facial Recognition Systems Work," HowStuffWorks.com, c/o Convex Group, Inc., One Capital City Plaza, 3350 Peachtree Road, Suite 1500, Atlanta, GA 30326, 2006. Copyright © 1998–2006 HowStuffWorks, Inc.

3. John R. Vacca, *Identity Theft*, Prentice Hall (2002).

4. John R. Vacca, *The Essentials Guide to Storage Area Networks*, Prentice Hall, Professional Technical Reference, Pearson Education (2001).

7

How Facial Thermal Imaging in the Infrared Spectrum Works

Face recognition is gaining acceptance as a superior biometric method in access control and surveillance. It is touchless, highly automated, and the most natural method since it coincides with the mode of recognition that humans employ in their everyday affairs. Most of the research efforts in this area have focused on visible spectrum imaging. Despite progress, certain problems still remain. Some of them are due to the very nature of the legacy approaches.

Images in the visible band are formed primarily due to reflection [1]. Therefore, they depend on the existence of an external light source, which sometimes may be absent (e.g., nighttime). Imagery formed primarily due to reflection is also difficult to process because of the strong dependence on incident angle and light variation [1].

Recently, there has been an increased interest in face recognition in the thermal infrared spectrum. In this spectral region, images are formed primarily due to emission. Therefore, they do not depend on the existence and intensity of an external light source. They are also less dependent on the incident angle of radiation. Several efforts have been made to compare the performance of face recognition methodologies using visible and thermal infrared images. This research has highlighted several advantages of performing face recognition in the thermal infrared, along with some weaknesses [1].

In terms of algorithmic approaches, both appearance- and geometry-based methods have been applied in visible and thermal infrared cases. Appearance-based methods like principal component, independent component, and linear discriminant analysis treat the image simply as a matrix of numbers and impose the decision boundary without extracting any geometric features. Even though such approaches are computationally efficient, they do not perform well in challenging conditions such as variable poses and facial expressions. Geometry-based techniques and template-matching approaches extract certain features from the face and then impose probability models (or decision boundaries) on these features. Geometric approaches are usually more robust than appearance-based approaches, but at an additional computational cost [1].

One interesting approach decomposes the image into spectral rather than geometric features. This method prunes the hypothesis space by modeling the extracted spectral features through Bessel parametric forms. The algorithm is elegant and computationally efficient. However, the Bessel model is applied to the entire image, while only part of it contains facial information. The approach also does not yield a unique solution, but rather a set of highly likely solutions [1].

This chapter proposes a method that enhances and complements Srivastava's approach. First, the facial part of the image is segmented using adaptive fuzzy connectedness segmentation. Then, the Srivastava's algorithm is applied on the facial segment only, not the entire image. This application yields a pruned hypothesis space, but not a unique solution. Another method could take a step further and apply Bayesian classification on the pruned hypothesis space to find the exact match. It should be emphasized here that traditional adaptive fuzzy connectedness is not fully automated, as it requires a manual selection of a seed pixel [1].

Now, let's look at the methodology in some detail.

Methodology

Facial images are normally acquired under natural conditions, and it is common for such images to contain background. If the entire image is used to obtain features, it may affect the performance of the face recognition system [1].

Hence, the image needs to be segmented to remove the background and then decompose the facial segment into its spectral components using a bank of K Gabor filters. Bessel probabilistic models are then imposed on these spectral components to obtain $2K$ Bessel parameters, which are used to form the feature vector (see Figure 7-1(a)) [1].

The Bessel parameters of the database images can be computed off-line and stored for future use, as depicted in Figure 7-1(a) [1]. This reduces considerably the online computation cost of the algorithm. If a face image is given for testing, the algorithm needs to compute the Bessel parameters of the test image only, prune the hypothesis space, and find the exact match, as depicted in Figure 7-1(b) [1].

Segmentation

Adaptive fuzzy connectedness segmentation has been successfully applied to segment MRI medical images. A similar approach should be applied to segment

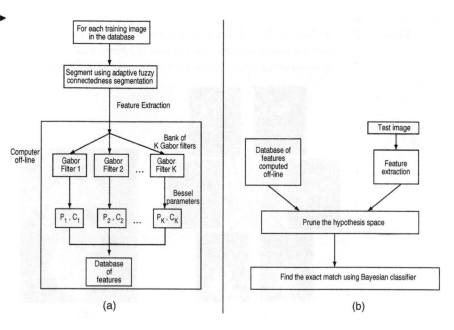

Figure 7-1
*Block diagram of
a face recognition
method.
(a) Training
phase. (b) Testing
phase. (Source:
Adapted with
permission from
the University of
Houston.)*

infrared facial images by providing facial skin pixels as seeds. Fuzzy affinity is assigned to other pixels with respect to these seed pixels [1].

Pednekar's algorithm assumes that the object to be segmented is relatively homogenized and requires the selection of a single seed. Facial segments in infrared images, however, are typically multimodal distributions. They feature hot and cold regions. Examples of hot regions include the area around the eyes and forehead. Examples of cold regions include the nose and ears. Here, Pednekar's algorithm has been expanded to select multiple seeds on the basis of sharp gradient changes on facial skin. Then, the algorithm checks for connectedness of other pixels to the respective seeds using the affinity functions. The resulting segmented parts are merged to obtain a complete segmented facial image, as depicted in Figure 7-2 [1].

Feature Extraction

Features should be extracted from segmented images, which can be used for pruning the hypothesis space. As shown in Figure 7-1 [1], this process involves dividing the segmented image into its spectral components using Gabor filters and then modeling these components using Bessel functionals. The Bessel functionals are completely characterized by the Bessel parameters, which form

Figure 7-2   *(a) Infrared facial image and intermediate multiseed segmentation results. The selected seeds are represented with cross marks. (b) Final result of adaptive fuzzy connectedness segmentation. (Source: Reproduced with permission from the University of Houston.)*

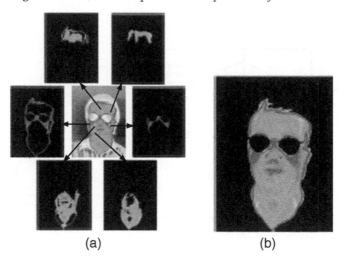

(a)   (b)

the feature vectors. The L^2-norm is applied to these feature vectors to shortlist the best matches [1].

Figure 7-3 depicts a segmented infrared facial image and one of its Gabor filtered components [1]. There is an advantage to representing the segmented infrared images via Bessel forms of their spectral components: The IR images can be compared by directly comparing their corresponding Bessel forms [1].

Results and Discussion

A face recognition method in the infrared spectrum should be tested against an Equinox facial database. This is the most extensive infrared facial database that is publicly available at the moment. The Equinox database has a good mix of subject images with accessories (glasses) as well as expressions of happiness, anger, and surprise, which account for pose variation. Figure 7-4 shows some examples from this database [1].

Finally, you should compare the identification performance with and without segmentation as well as with the eigenfaces method (an earlier appearance-based approach). You should also vary the number of test images to

Figure 7-3 *The (a) is defined as being a segmented infrared facial image. And the (b) is defined as corresponding to the Gabor filtered image at scale ó = 2 and orientation Ø = 30, with Bessel parameters of p = 0:5617 and c = 6694:202. (Source: Reproduced with permission from the University of Houston.)*

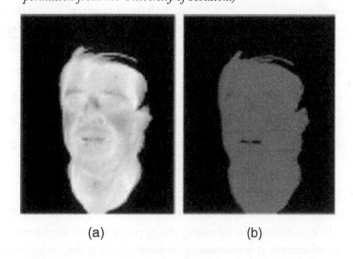

(a)   (b)

Figure 7-4
Sample images from the Equinox database. (Source: Reproduced with permission from the University of Houston.)

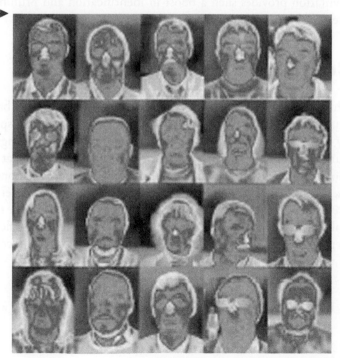

Table 7-1 *Performance of the Eigenfaces, Bessel Forms without Segmentation, and Bessel Forms with Segmentation Approaches at Varying Test/Training Ratios*

Test/Training Ratio	Eigenfaces	Face Recognition Method in the Infrared Spectrum on Nonsegmented Images	Face Recognition Method in the Infrared Spectrum on Segmented Images
2:1	81.21%	83.4%	89.6%
4:1	80.72%	83.01%	86.8%
6:1	79.32%	82.66%	85.6%
8:1	78.68%	81.85%	85.25%
10:1	76.72%	80.60%	84.72%

check the performance in different conditions. Table 7-1 provides performance results of the three approaches on the Equinox database at different training/test ratio conditions. In all cases, the approach with segmentation outperformed the other two. It is interesting to point out that the background in the Equinox database is typically uniform and simple. Therefore, it is remarkable that segmentation provides such a boost in identification and pruning performance, even in the case of mug shot type images [1].

Summary/Conclusion

This chapter presented a two-stage face recognition method based on infrared imaging and statistical modeling. In the first stage, the search space was reduced by finding highly likely candidates before arriving at a singular conclusion during the second stage. Previous work [1] has shown that Bessel forms a model accurately by the marginal densities of filtered components and can be used to find likely matches but not a unique solution [1]. An enhancement was presented to this approach by applying Bessel modeling on the facial region only, rather than the entire image; and by pipelining a classification algorithm to produce a unique solution. The detailed steps of the face recognition method in the infrared spectrum are as follows:

1. The faces are separated from the background using adaptive fuzzy connectedness segmentation.

2. Gabor filtering is used as a spectral analysis tool.

3.   The derivative filtered images are modeled using two-parameter Bessel forms.

4.   High-probability subjects are shortlisted by applying the L^2-norm on the Bessel models.

5.   The resulting set of highly likely matches is fed to a Bayesian classifier to find the exact match.

6.   The segmentation of the facial regions results in better hypothesis pruning and classification performance [1].

The choice of infrared makes the system less dependent on external light sources and more robust with respect to incident angle and light variation. The background is removed using adaptive fuzzy connectedness segmentation enhanced by automatic selection of multiple seeds. Features are computed by decomposing the segmented images into their spectral components and modeling them through Bessel forms. The parameters of the Bessel forms constitute the feature vectors, which are used for hypothesis pruning. A Bayesian classifier determines a unique solution out of the pruned subset. The face recognition method in the infrared spectrum compares favorably to older approaches such as eigenfaces. Experimental results also show that the face recognition method in the infrared spectrum performs better with segmentation rather than without. Finally, the face recognition method in the infrared spectrum involves a two-stage classification scheme that produces a unique solution and not a short list of candidates. Current work continues on establishing a thermal facial database and testing the face recognition method in the infrared spectrum further. In contrast to the Equinox database, a database will be temperature-calibrated and will include dynamic environmental conditions (changing environmental temperature and air flow) [1].

Reference

1. Pradeep Buddharaju, Ioannis Pavlidis and Ioannis Kakadiaris, "Face Recognition in the Thermal Infrared Spectrum," Computer Science Department, University of Houston, 4800 Calhoun Rd., Houston, Texas 77204-3010, 2005. Copyright © 2005.

Part 4: How Biometric Fingerscanning Analysis Technology Works

8

How Finger Image Capture Works

Among all the biometric techniques, fingerprint-based identification is the oldest method that has been successfully used in numerous applications. Everyone is known to have unique, immutable fingerprints. A fingerprint is made of a series of ridges and furrows on the surface of the finger (see Figure 8-1). The uniqueness of a fingerprint can be determined by the pattern of ridges and furrows as well as the minutiae points. Minutiae points are local ridge characteristics that occur at either a ridge bifurcation or a ridge ending [1].

What Is Finger Image Capture?

Finger image capture is the acquisition and recognition of a person's fingerprint characteristics for identification purposes. This allows the recognition of a person through quantifiable physiological characteristics that verify the identity of an individual [1].

There are basically two different types of fingerscanning technology that make this possible. One is an optical method [10], which starts with a visual image of a finger. The other uses a semiconductor-generated electric field to image a finger [1].

There are a range of ways to identify fingerprints. They include traditional police methods of matching minutiae, straight pattern matching, moiré fringe patterns, and ultrasonics [1].

Practical Applications for Finger Image Capture

There are a greater variety of fingerprint devices available than for any other biometric. Fingerprint recognition is the front-runner for mass-market biometric ID systems [1].

Finger image capture has a high accuracy rate when users are sufficiently educated. Fingerprint authentication is a good choice for in-house systems

Figure 8-1
Everyone is known to have unique, immutable fingerprints. (Source: Reproduced with permission from BioEnable Technologies Pvt. Ltd.)

where enough training can be provided to users and where the device is operated in a controlled environment. The small size of the fingerprint scanner, ease of integration (it can be easily adapted to keyboards), and most significantly, the relatively low costs make it an affordable, simple choice for workplace access security [1].

Plans to integrate fingerprint scanning technology into laptops using biometric technology include a single chip using more than 16,000 location elements to map a fingerprint of the living cells that lay below the top layers of dead skin. Therefore, the reading is still detectable if the finger has calluses, is damaged, worn, soiled, moist, dry, or otherwise hard to read—a common obstacle. This subsurface capability eliminates any attainment or detection failures [1].

Accuracy and Integrity

As with any security system, users will wonder whether a fingerprint recognition system can be beaten. In most cases, false negatives (a failure to recognize a legitimate user) are more likely than false positives. Overcoming a fingerprint system by presenting it with a false or fake fingerprint is likely to be a difficult deed. However, such scenarios will be tried, and the sensors on the market use a variety of means to circumvent them. For instance, someone may attempt to use latent print residue on the sensor just after a legitimate user accesses the system. At the other end of the scale, there is the gruesome possibility of presenting a finger to the system that is no longer connected to its owner. Therefore, sensors attempt to determine whether a finger is live, and not made of latex (or worse). Detectors for temperature, blood-oxygen level, pulse, blood flow, humidity, or skin conductivity can be integrated [1].

Unfortunately, no technology is perfect—false positives and spoiled readings do occur from time to time. But fingerprint scanners are worth looking into. It is estimated that 40% of help desk calls are password-related. Whether incorporated into the keyboard or mouse, or used as a stand-alone device, scanners are more affordable than ever, allowing encryption [6] of files keyed to a fingerprint, and, perhaps most importantly, helping minimize stress over that stolen laptop [1].

Fingerprint Matching

Fingerprint matching techniques can be placed into two categories: minutiae-based and correlation-based. Minutiae-based techniques first find minutiae points (see Figure 8-2) and then map their relative placement on the finger. However, there are some difficulties when using this approach. It is difficult to extract the minutiae points accurately (see Figure 8-3) when the fingerprint is of low quality. Also, this method does not take into account the global pattern of ridges and furrows. The correlation-based method (see Figure 8-4) is able to overcome some of the difficulties of the minutiae-based approach.

Figure 8-2
Minutiae-based techniques first find minutiae points and then map their relative placement on the finger. (Source: Reproduced with permission from BioEnable Technologies Pvt. Ltd.)

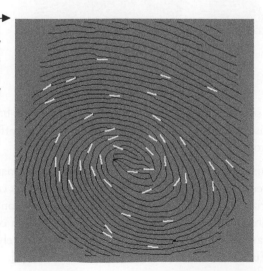

Figure 8-3 *It is difficult to extract the minutiae points accurately when the fingerprint is of low quality. (Source: Reproduced with permission from BioEnable Technologies Pvt. Ltd.)*

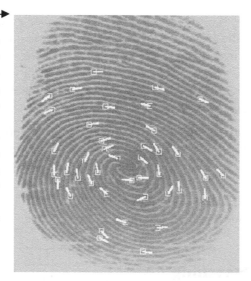

Figure 8-4 *The correlation-based method is able to overcome some of the difficulties of the minutiae-based approach. (Source: Reproduced with permission from BioEnable Technologies Pvt. Ltd.)*

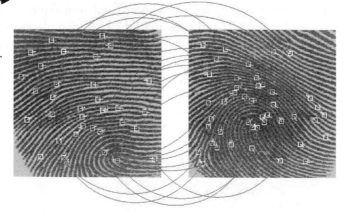

However, it has some of its own shortcomings. Correlation-based techniques require the precise location of a registration point and are affected by image translation and rotation [1].

Fingerprint matching based on minutiae has problems in matching different sized (unregistered) minutiae patterns. Local ridge structures can not be completely characterized by minutiae. An alternate representation of fingerprints may capture more local information and yield a fixed length code for the fingerprint. The matching will then hopefully become a relatively simple task of calculating the Euclidean distance between the two codes [1].

In addition, you should develop algorithms that are more robust to noise in fingerprint images and deliver increased accuracy in real time. A commercial fingerprint-based authentication system requires a very low false-reject rate for a given false-accept rate. This is very difficult to achieve with any one technique. Investigations are also under way to pool evidence from various matching techniques to increase the overall accuracy of the system. In a real application, the sensor, the acquisition system, and the variation in performance of the system over time is very critical. You should field test the system on a limited number of users to evaluate the system performance over a period of time [1].

Fingerprint Classification

Large volumes of fingerprints are collected and stored every day in a wide range of applications, including forensics [2], access control, and driver's license registration (see Figure 8-5). An automatic recognition of people based on fingerprints requires that the input fingerprint be matched with a large number of fingerprints in a database. (The FBI database contains approximately 140 million fingerprints!) To reduce the search time and computational complexity, it is desirable to classify these fingerprints in an accurate and consistent manner so that the input fingerprint is required to be matched only with a subset of the fingerprints in the database [1].

Fingerprint classification is a technique of assigning a fingerprint into one of the several pre-specified types already established that can provide an indexing mechanism. Fingerprint classification can be viewed as a coarse-level matching of the fingerprints. An input fingerprint is first matched at a coarse level to one of the pre-specified types and then, at a finer level, it is compared to the subset of the database containing that type of fingerprints only [1].

You should also develop an algorithm to classify fingerprints into five classes, namely, whorl, right loop, left loop, arch, and tented arch. The algorithm separates the number of ridges present in four directions (0 degrees, 45 degrees,

Figure 8-5 *Large volumes of fingerprints are collected and stored every day in a wide range of applications. (Source: Reproduced with permission from BioEnable Technologies Pvt. Ltd.)*

90 degrees, and 135 degrees) by filtering the central part of a fingerprint with a bank of Gabor filters. This information is quantized to generate a FingerCode that is used for classification. This classification is based on a two-stage classifier that uses a K-nearest neighbor classifier in the first stage and a set of neural networks in the second stage [1].

The classifier is tested on 4,000 images in the NIST-4 database. For the five-class problem, classification accuracy of 90% is achieved. For the four-class problem (arch and tented arch combined into one class), you should be able to achieve a classification accuracy of 94.8%. By incorporating a reject option, the classification accuracy can be increased to 96% for the five-class classification and to 97.8% for the four-class classification when 30.8% of the images are rejected [1].

Fingerprint Image Enhancement

A critical step in automatic fingerprint matching is to automatically and reliably extract minutiae from the input fingerprint images. However, the performance of a minutiae extraction algorithm relies heavily on the quality of the input fingerprint images. In order to ensure that the performance of an automatic fingerprint identification/verification system will be robust with respect to the quality of the fingerprint images, it is essential to incorporate a fingerprint enhancement algorithm in the minutiae extraction module (see Figure 8-6) [1].

You should also develop a fast fingerprint enhancement algorithm, which can adaptively improve the clarity of ridge and furrow structures of input fingerprint

Figure 8-6   *It is essential to incorporate a fingerprint enhancement algorithm in the minutiae extraction module. (Source: Reproduced with permission from BioEnable Technologies Pvt. Ltd.)*

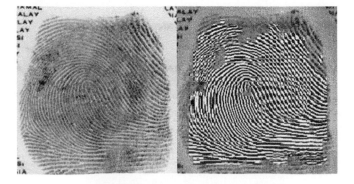

images based on the estimated local ridge orientation and frequency. In addition, you should evaluate the performance of the image enhancement algorithm using the goodness index of the extracted minutiae and the accuracy of an online fingerprint verification system. Experimental results show that incorporating the enhancement algorithms improves both the goodness index and the verification accuracy [1].

It has been more than 20 years since the introduction of commercial fingerprint authentication systems, yet they are just now gaining broad acceptance. You should not be surprised. Many technologies have required several years before the right combination of factors allowed them to become ubiquitous. If you look back to laptop computers, cell phones, fax machines, pagers, laser printers, and countless other everyday devices, you will realize most have had long gestation periods. Biometrics is now at the acceptance crossroads. What will propel it into common usage [1]?

Convenience First

There is the reason end-users should use fingerprint authentication in the IT world—security—and there is the reason they will use it—convenience. The simple fact is that passwords don't work very well. They are at "no cost" to establish, but very expensive to maintain. Just ask the help desk manager in a major corporation. Many help desk calls are related to passwords, either lost, forgotten, or otherwise useless. Count all the passwords you use every day and have to change once a month. Password administration is a nightmare for MIS managers and users. Fingerprint authentication eliminates the problem and the headaches [1].

Other authentication mechanisms such as tokens, smart cards, and so on require you to carry something. This is better than remembering a password, but easier to lose. Think about losing your credit card or driver's license. Losing your corporate network access card could be far worse: Information is valuable and harder to track than money [1].

Fingerprints can act as a simple, trusted, and convenient user interface to a well-thought-out security architecture. The two components need each other to provide truly effective security. A user authenticated via fingerprints can take advantage of a solid security system with minimal education [1].

Simple Truths

Users don't trust what they don't understand. Most IT security concepts are incomprehensible to the common user. Explaining public and private keys [5], key recovery systems, and digital certificates is beyond the skills of even experienced MIS professionals. Most users have no concept of encryption algorithms and their implementations, nor do they want to understand. Users want simple, trusted security [1].

Simple means put your finger down. It does not take a security professional to realize that 10 passwords on sticky notes attached to your monitor are poor security. Most breaches of security require doing the obvious, and are often done by insiders [1].

Trusted means having stood the test of time. Fingerprints have been used for identification for over 100 years. They are the standard, without question. In addition to signatures, fingerprints are the only form of identification that have legal standing. A key issue of trust is privacy [4]. The best way to maintain that is to store a template of unique fingerprint characteristics instead of the entire print. This is sufficient for one-to-one or one-to-many matching and eliminates the need for a database of searchable fingerprints [1].

Emerging Standards

IT professionals insist upon standards, multiple sources of supply, and endorsement by industry leaders. It's beginning to happen, but to think that a small biometrics company can set an industry standard is ludicrous. Yet many have tried [1].

Any CIO or MIS manager would not bet his or her job or company on a proprietary solution from a small biometrics company. These people want choice and standards to provide multiple sources of supply and fair competition among vendors. The one exception to this rule is when there has been a major catastrophe, such as a significant loss of money. However, it is tough to build a sustainable business chasing disasters [1].

Standards need to be set by the IT industry leaders such as Intel, Microsoft, Phoenix Technologies, and the top 10 computer companies. In the last year, many of these large organizations have banded together to begin the process of standardization. This is the first sign of an industry maturing [1].

Cost

In the early days of desktop computers, when a system cost more than $10,000, only a few people had systems. Now, when they cost less than $800, everybody has one. This same "order of magnitude" cost breakthrough has recently occurred with fingerprint technology. What cost $1,000 two years ago is now available for less than $100. Cost alone is not the answer, but it is a necessary component of broad market acceptance of this technology [1].

Complete Solutions

Lots of companies talk about "complete solutions," but what does this mean? It does not mean a custom, proprietary combination of fingerprint sensor, matching software, and application software—point products and closed solutions are not acceptable. It does mean an open architecture where the sensor, matching algorithm, and applications are interchangeable and leverageable. For example, Veridicom's OpenTouch architecture embraces this tenet and lets the user choose [1].

Measurable Usefulness

Being able to accurately gauge the usefulness of a fingerprint authentication solution is very important. This technology saves money in password administration, user up-time, and user support. More importantly, fingerprint authentication allows you to do more with a computer. Currently, remote secure network access is possible. Electronic commerce [3] makes sense when the authentication is trusted. It is has been found that 75% of all Internet [9] users are uncomfortable transmitting their credit card information over the public network. Imagine if this was never an issue. Fingerprint authentication is an enabling technology for trusted e-commerce [1].

All the signs are in the market for the acceptance of fingerprint authentication as a simple, trusted, convenient method of personal authentication. Industry leaders are validating the technology through standards initiatives. Cost and performance breakthroughs have transformed fingerprint biometrics from an interesting technology to an easy-to-implement authentication solution. Industry trends such as electronic commerce and remote computing exacerbate the need for better authentication. Most importantly, users understand and accept the concept. Passwords and tokens are universally disliked. You can't get much simpler than a fingerprint [1].

Biometric Versus Nonbiometric Fingerprinting

The aura of criminality that accompanies the term "fingerprint" has not significantly impeded the acceptance of fingerprint technology because the two authentication methods are very different. Fingerprinting, as the name suggests, is the acquisition and storage [7] of the image of the fingerprint. Fingerprinting was for decades the common ink-and-roll procedure, used when booking suspects or conducting criminal investigations. More advanced optical or noncontact fingerprinting systems (known as livescan), which normally utilize prints from several fingers, are currently the standard for forensic usage.

They require 250 kB per finger for a high-quality image. Fingerprint technology also acquires the fingerprint, but it doesn't store the full image. It stores particular data about the fingerprint in a much smaller template, requiring 250–1,000 bytes. After the data is extracted, the fingerprint is not stored. Significantly, the full fingerprint cannot be reconstructed from the fingerprint template [1].

Fingerprints are used in forensic applications: large-scale, one-to-many searches on databases of up to millions of fingerprints. These searches can be done within only a few hours, a tribute to the computational power of automated fingerprint identification systems (AFIS). AFIS—commonly referred to as "AFIS systems," a redundancy—is the term applied to large-scale, one-to-many searches. Although fingerprint technology can be used in AFIS on 100,000 person databases, it is much more frequently used for one-to-one verification within one-to-three seconds [1].

Many people think of forensic fingerprinting as an ink-and-paper process. While this may still be true in some locations, most jurisdictions utilize optical scanners known as livescan systems. There are some fundamental differences between these forensic fingerprinting systems (used in AFIS systems) and the biometric fingerprint systems used to log on to a PC [1].

When the differences between the two technologies are explained, nearly all users are comfortable with fingerprint technology. The key is the template—what is stored is not a full fingerprint, but a small amount of data derived from the fingerprint's unique patterns [1].

Response Time

AFIS may take hours to match a candidate, while fingerprint systems respond with seconds or fractions of seconds [1].

Cost

An AFIS capture device can range from several hundred to tens of thousands of dollars, depending on whether it is designed to capture one or multiple fingerprints. A PC peripheral fingerprint device generally costs less than $200 [1].

Accuracy

AFIS might return the top five candidates in a biometric comparison with the intent of locating or questioning the top suspects. Fingerprint systems are designed to return a single yes/no answer based on a single comparison [1].

Scale

AFIS are designed to be scalable to thousands and millions of users, conducting constant 1:N searches. Fingerprint systems are almost invariably 1:1, and do not require significant processing power [1].

Capture

AFIS are designed to use the entire fingerprint, rolled from nail to nail, and often capture all 10 fingerprints. Fingerprint systems use only the center of the fingerprint, capturing only a small fraction of the overall fingerprint data [1].

Storage

AFIS generally store fingerprint images for expert comparison once a possible match has been located. Fingerprint systems, by and large, do not store images, as they are not used for comparison [1].

Infrastructure

AFIS normally require a back-end infrastructure for storage, matching, and duplicate resolution. These systems can cost hundreds of thousands of dollars. Fingerprint systems rely on a PC or a peripheral device for processing and storage [1].

Fingerprint Market Size

Already the leading non-AFIS technology in the biometric market, fingerprint is poised to remain the leading non-AFIS technology through 2012, according to industry analysts. Because of the range of environments in which fingerprint can be deployed, its years of development, and the strong companies involved in the technology's manufacture and development, fingerprint revenues are projected to grow from $588.6m in 2006 to $6,775.3m in 2012. Fingerprint revenues are expected to comprise approximately 35% of the entire biometric market [1].

Fingerprint Growth Drivers and Enablers

A number of basic factors should combine to help drive fingerprint revenues. If and when biometrics become a commonly used solution for e-commerce and remote transactions, segments are expected to grow rapidly through 2012. The fingerprint will be a primary benefactor. Further, the fingerprint is a very strong desktop solution, and it is anticipated that the desktop will become a driver for biometric revenue derived from product sales and transactional authentication. Most middleware solutions leverage a variety of fingerprint solutions for desktop authentication [1].

The fingerprint is a proven technology capable of high levels of accuracy. The fingerprint has long been recognized as a highly distinctive identifier, and classification, analysis, and study of fingerprints has existed for decades. The combination of an innately distinctive feature with a long history of use as identification sets the fingerprint apart in the biometric industry. There are physiological characteristics more distinctive than the fingerprint (the iris and retina, for example), but technology capable of leveraging these characteristics has only been developed over the past few years, not decades [1].

Strong fingerprint solutions are capable of processing thousands of users without allowing a false match, and can verify nearly 100% of users with one or two placements of a finger. Because of this, many fingerprint technologies can be deployed in applications where either security or convenience is the primary driver [1].

Reduced size and power requirements, along with fingerprint's resistance to environmental changes such as background lighting and temperature, allow the technology to be deployed in a range of logical and physical access environments. Fingerprint acquisition devices have grown quite small—sensors slightly thicker than a coin, and smaller than 1.5 cm × 1.5 cm, are capable of acquiring and processing images [1].

Fingerprint Growth Inhibitors

Though radical changes in the composition of the marketplace would need to occur to undermine the fingerprint's anticipated growth, the technology does face potential growth inhibitors [1].

As opposed to technologies such as facial recognition and voice recognition, which can leverage existing acquisition devices, fingerprint's growth is contingent on the widespread incorporation of sensors in keyboards, peripherals, access control devices, and handheld devices. The ability to acquire fingerprints must be present wherever and whenever users want to authenticate. Currently, acquisition devices are present in but a tiny fraction of authentication environments [1].

A percentage of users, varying by the specific technology and user population, are unable to enroll in many fingerprint systems. Furthermore, certain ethnic and demographic groups have lower quality fingerprints and are more difficult to enroll. Testing has shown that elderly populations, manual laborers, and some Asian populations are more likely to be unable to enroll in some fingerprint systems. In an enterprise deployment for physical or logical security, this means that some number of users need to be processed by another method, be it another biometric, a password, or a token. In a customer-facing application, this may mean that a customer willing to enroll in a biometric system is simply unable to. In a large-scale 1:N application, the result may be that a user is able to enroll multiple times, as data from his or her fingerprints cannot be reliably acquired. If the system is designed to be more forgiving, and to enroll marginal fingerprints, then the common result is increased error rates [1].

Applications

Fingerprint technology is used by hundreds of thousands of people daily to access networks and PCs, enter restricted areas, and authorize transactions. The technology is used broadly in a range of vertical markets and within a range of horizontal applications, primarily PC/network access, physical security/time and attendance, and civil ID. Most deployments are 1:1, though there are a number of "one-to-few" deployments in which individuals are matched against modest databases, typically of 10–100 users. Large-scale 1:N applications, in which a user is identified from a large fingerprint database, are classified as AFIS [1].

Fingerprint Feature Extraction

The human fingerprint is comprised of various types of ridge patterns, traditionally classified according to the decades-old Henry system: left loop, right loop, arch, whorl, and tented arch. Loops make up nearly two-thirds of all fingerprints, whorls are nearly one-third, and perhaps 5%–10% are arches. These classifications are relevant in many large-scale forensic applications, but are rarely used in biometric authentication. The fingerprint shown in Figure 8-1 is a right loop [1].

Minutiae (see Figure 8-7), the discontinuities that interrupt the otherwise smooth flow of ridges, are the basis for most fingerprint authentication [1]. Codified in the late 1800s as Galton features, minutiae are at their most

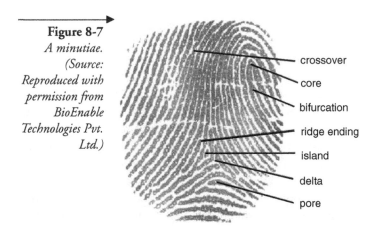

crossover

core

bifurcation

ridge ending

island

delta

pore

rudimentary ridge endings, the points at which a ridge stops, and bifurcations, the point at which one ridge divides into two. Many types of minutiae exist, including dots (very small ridges); islands (ridges slightly longer than dots and occupying a middle space between two temporarily divergent ridges); ponds or lakes (empty spaces between two temporarily divergent ridges); spurs (a notch protruding from a ridge); bridges (small ridges joining two longer adjacent ridges); and crossovers (two ridges that cross each other) [1].

Other features are essential to fingerprint authentication. The core is the inner point, normally in the middle of the print, around which swirls, loops, or arches center. It is frequently characterized by a ridge ending and several acutely curved ridges. Deltas are the points, normally at the lower left and right of the fingerprint, around which a triangular series of ridges center [1].

The ridges are also marked by pores, which appear at steady intervals. Some initial attempts have been made to use the location and distribution of the pores as a means of authentication, but very high resolution is required to capture pores consistently [1].

Once a high-quality image is captured, there are several steps required to convert its distinctive features into a compact template. This process, known as feature extraction, is at the core of fingerprint technology. Each of the 100 primary fingerprint vendors has a proprietary feature extraction mechanism; the vendors guard these unique algorithms very closely. What follows is a series of steps used, in some fashion, by many vendors. The basic principles apply even to those vendors that use alternative mechanisms [1].

The image must first be converted to a usable format. If the image is grayscale, areas lighter than a particular threshold are discarded, and those darker are made black. The ridges are then thinned from five to eight pixels in width down to one pixel, for precise location of endings and bifurcations [1].

Minutiae localization begins with this processed image. At this point, even a very precise image will have distortions and false minutiae that need to be filtered out. For example, an algorithm may search the image and eliminate one of two adjacent minutiae, as minutiae are very rarely adjacent. Anomalies caused by scars, sweat, or dirt appear as false minutiae, and algorithms locate any points or patterns that don't make sense, such as a spur on an island (probably false) or a ridge crossing perpendicular to two or three others (probably a scar or dirt). A large percentage of potential minutiae are discarded in this process [1].

The point at which a ridge ends and the point where a bifurcation begins are the most rudimentary minutiae, and are used in most applications. There is variance in how exactly to situate a minutia point: whether to place it directly on the end of the ridge, one pixel away from the ending, or one pixel within the ridge ending (the same concern applies to bifurcation). Once the point has been situated, its location is commonly indicated by the distance from the core, with the core serving as the 0,0 on an X,Y-axis. Some vendors use the far left and bottom boundaries of the image as the axes, correcting for misplacement by locating and adjusting from the core. In addition to the placement of the minutia, the angle of the minutia is normally used. When a ridge ends, its direction at the point of termination establishes the angle. (More complicated rules can apply to curved endings.) This angle is taken from a horizontal line extending rightward from the core, and can be up to 359° [1].

In addition to using the location and angle of minutiae, some vendors classify minutia by type and quality. The advantage of this is that searches can be quicker, as a particularly notable minutia may be distinctive enough to lead to a match. A vendor can also rank high-versus low-quality minutia and discard the latter. Those vendors who shy away from this methodology do so because of the wide variation from print to print, even on successive submissions. Measuring quality may only introduce an unnecessary level of complication [1].

Approximately 80% of biometric vendors utilize minutiae in some fashion. Those who do not utilize minutia use pattern matching, which extrapolates data from a particular series of ridges. This series of ridges used in enrollment is the basis of comparison, and verification requires that a segment of the same area be found and compared. The use of multiple ridges reduces dependence on minutiae points, which tend to be affected by wear and tear. The templates created in pattern matching are generally, but not always, two to three times larger than in minutia, usually 900–1,200 bytes [1].

Fingerprint Form Factors

Form factor is a term used to describe the manner in which a biometric sensor is imbedded into an acquisition device. Biometric sensors, in particular fingerprint sensors, can be imbedded on top of a device, on its side, recessed, or protruding. Some biometric devices require users to sweep their fingers across them, while others require that users place their fingers on the sensors and hold them still until they are authenticated [1].

Though the placement of the biometric sensor is important from an ergonomic standpoint, several other considerations are equally important form factors. One consideration is the type of device that the user interacts with. Several broad categories of device types are discussed next [1].

Desktop Peripherals

Desktop peripherals include biometrically enabled mice and other handheld devices that users interact with when they operate a desktop computer. Because the standard size of desktop peripheral is typically small, the biometric sensor must be small enough to fit on the device. However, the sensor's ability to acquire images effectively diminishes as it is made small enough to fit on the peripheral devices [1].

Embedded Desktop Solutions

Embedded desktop solutions include biometrically enabled keyboards and other primary components of computers that users interact with when they operate a desktop computer. Because the embedded desktop devices are larger than desktop peripherals, sensor size is not as significant a consideration—the sensors can be large enough to acquire images without compromising the ability of the device to operate effectively. Since desktop devices such as keyboards are typically cheap, the addition of an embedded sensor should not significantly increase its cost [1].

Embedded Physical Access Solutions

Embedded physical access solutions include biometrically enabled keypads and other devices that users interact with to gain access to restricted areas (opening doors). Because physical access solutions are often used to protect items of value, and because making the device small typically isn't a concern, the sensors can be large enough to meet this security requirement. Several other factors, including the location of the device (indoors or outdoors), the type of client (military, government, or commercial), and the purpose of the device (apartment access, protecting nuclear materials) will also be important in determining how the embedded physical access solution is deployed [1].

Embedded Wireless Handheld Solutions

Embedded wireless handheld solutions include biometrically enabled cell phones and other mobile personal communication devices that require owner authentication to use. Like desktop peripherals, embedded wireless handheld solutions [8] are small; consequently, the biometric sensor must be small enough to fit on the device. Similarly, the sensor's ability to acquire images effectively diminishes as it is made small enough to fit on the wireless device [1].

Ultimately, the type of application being deployed and the environment in which it is being rolled out will drive the form factor. In fact, these form factors will have implications on what type of sensor technology is used in the fingerprint device. Today there are three primary sensor technologies: optical, ultrasound, and silicon. Each sensor technology has its advantages and disadvantages. For example, optical sensors are durable and temperature-resistant, qualities that lend themselves well to usage in embedded military equipment solutions. However, because optical sensors must be large enough to achieve quality images, they are not well suited to embedded desktop solutions and embedded wireless handheld solutions. On the other hand, silicon sensors are able to produce quality images with less surface area and they are better suited to use in these compact devices [1].

Types of Scanners: Optical, Silicon, and Ultrasound

Acquiring high-quality images of distinctive fingerprint ridges and minutiae is a complicated task. The fingerprint is a small area from which to take measurements, and the wear of daily life affects which ridge patterns show most prominently. Increasingly sophisticated mechanisms have been developed to capture the fingerprint image with sufficient detail and resolution. The technologies in use today are optical, silicon, and ultrasound [1].

Optical technology is the oldest and most widely used. The finger is placed on a coated platen, usually built of hard plastic but proprietary to each company. In most devices, a charged coupled device (CCD) converts the image of the fingerprint, with dark ridges and light valleys, into a digital signal. The brightness is either adjusted automatically (preferable) or manually (more difficult), leading to a usable image [1].

Optical devices have several strengths: They are the most proven over time; they can withstand, to some degree, temperature fluctuations; they are fairly inexpensive; and they can provide resolutions up to 500 dpi. Drawbacks to the technology include size (the platen must be of sufficient size to achieve a quality image) and latent prints. Latent prints are leftover prints from previous users. This can cause image degradation, as severe latent prints can cause two sets of prints to be superimposed. Also, the coating and CCD arrays can wear with age, reducing accuracy [1].

Optical is the most implemented technology by a significant margin. Identicator and its parent company Identix are two of the most prominent fingerprint companies that utilize optical technology, much of which is developed jointly with Motorola. The majority of companies use optical technology, but an increasing number of vendors utilize silicon technology [1].

Silicon technology has gained considerable acceptance since its introduction in the late 1990s. Most silicon, or chip, technology is based on DC capacitance. The silicon sensor acts as one plate of a capacitor, and the finger as the other. The capacitance between platen and the finger is converted into an 8-bit grayscale digital image. With the exception of AuthenTec, whose technology employs AC capacitance and reads to the live layer of skin, all silicon fingerprint vendors use a variation of this type of capacitance [1].

Silicon generally produces better image quality, with less surface area, than optical technology. Since the chip is comprised of discrete rows and columns (200–300 lines in each direction on a 1 cm × 1.5 cm wafer), it can return exceptionally detailed data. The reduced size of the chip means that costs should drop significantly, now that much of the R&D necessary to develop the technology is bearing fruit. Silicon chips are small enough to be integrated into many devices that cannot accommodate optical technology [1].

Silicon's durability, especially in suboptimal conditions, has yet to be proven. Although manufacturers use coating devices to treat the silicon and claim that the surface is 100 times more durable than optical, this has to be proven. Also, with the reduction in sensor size, it is even more important to ensure that enrollment and verification are done carefully—a poor enrollment may not capture the center of the fingerprint, and subsequent verifications are subject to the same type of placement. Many major companies have recently moved into the silicon field. Infineon (the semiconductor division of Siemens) and Sony have developed chips to compete with Veridicom (a spin-off of Lucent), the leader in silicon technology [1].

Ultrasound technology, though considered perhaps the most accurate of the fingerprint technologies, is not yet widely used. It transmits acoustic waves and measures the distance based on the impedance of the finger, the platen, and air. Ultrasound is capable of penetrating dirt and residue on the platen and the finger, countering a main drawback to optical technology [1].

Until ultrasound technology gains more widespread usage, it will be difficult to assess its long-term performance. However, preliminary usage of products from Ultra-Scan Corporation indicates that this is a technology with significant promise. It combines a strength of optical technology, large platen size and ease of use, with a strength of silicon technology, the ability to overcome suboptimal reading conditions [1].

Summary/Conclusion

This chapter thoroughly discussed finger image capture technology, which is also called fingerprint scanning. This is the process of electronically obtaining and storing human fingerprints. The digital image obtained by such scanning is called a finger image. In some texts, the terms fingerprinting and fingerprint are used, but technically, these terms refer to traditional ink-and-paper processes and images [1].

Finger image capture technology is a biometric process, because it involves the automated capture, analysis, and comparison of a specific characteristic of the human body. There are several different ways in which an instrument can bring out the details in the pattern of raised areas (called ridges) and branches (called bifurcations) in a human finger image. The most common scanners are optical, silicon, and ultrasound.

Finally, biometric finger image capture offers improvements over ink-and-paper imaging. A complete set of fingerscans for a person (10 images, including those of the thumbs) can be easily copied, distributed, and transmitted over computer networks. In addition, computers can quickly analyze a fingerscan and compare it with thousands of other fingerscans, as well as with fingerprints obtained by traditional means and then digitally photographed and stored. This greatly speeds up the process of searching finger image records in criminal investigations [1].

References

1. "Fingerprint Recognition," BioEnable Technologies Pvt. Ltd., E-204, 3rd floor, Railway Station Complex, CBD Belapur, Navi Mumbai, MH, India, PIN: 400614, 2005. Copyright © 2005.

2. John R. Vacca, *Computer Forensics: Computer Crime Scene Investigation, 2nd ed.* Charles River Media (2005).

3. John R. Vacca, *Electronic Commerce, 4th ed.* Charles River Media (2003).

4. John R. Vacca, *Net Privacy: A Guide to Developing and Implementing an Ironclad eBusiness Privacy Plan*, McGraw-Hill (2001).

5. John R. Vacca, *Public Key Infrastructure: Building Trusted Applications and Web Services*, CRC Press (2005).

6. John R. Vacca, *Satellite Encryption*, Academic Press (1999).

7. John R. Vacca, *The Essentials Guide to Storage Area Networks*, Prentice Hall, Professional Technical Reference, Pearson Education (2001).

8. John R. Vacca, *Wireless Data Dymistified*, McGraw-Hill (2003).

9. John R. Vacca, *Practical Internet Security*, Springer (2006).

10. John R. Vacca, *Optical Networking Best Practices Handbook*, John Wiley & Sons (2006).

9

How Fingerscanning Verification and Recognition Works

As discussed in Chapter 8, use of the fingerprint by law enforcement for identification purposes is commonplace and widely accepted. However, the technology has diversified, migrating away from law enforcement and toward civil and commercial markets. In the context of commercial applications, the preferred term is "fingerscanning," which is the process of finger image capture [1].

There are a number of different types of fingerscanning systems on the market. Some analyze the distinct marks on the finger called minutiae points. Others examine the pores on the finger, which are uniquely positioned. Finger image density or the distance between ridges may be analyzed. The way in which the image is captured also differs among vendors. None involve the inking of the fingerprint that traditional law enforcement procedures often entail [1].

Fingerscanning can be used for both verification and recognition purposes. At present, the one-to-many identification IAFIS or AFIS applications are confined to law enforcement, government programs, and the military. However, there is mounting pressure to expand identification applications. In the areas of financial transactions, network security, and controlling the movement of individuals, fingerscanning is considered to be a highly mature biometric technology with a range of proven installations [1].

Verification and Recognition

Biometrics, the science of applying unique physical or behavioral characteristics to verify an individual's identity, is the basis for a variety of rapidly expanding applications for both data security and access control. Numerous biometric approaches currently exist, including voice recognition, retina scanning, facial recognition and others, but fingerscanning verification and recognition is increasingly being acknowledged as the most practical technology for its low cost, convenience, and reliable security [2].

The Basis of Fingerscanning Verification

Although fingerprints have been used as a means of identification since the middle of the 19th century, modern fingerscanning verification technology has little in common with the ink-and-roll procedure that most people associate with fingerprinting. In order to appreciate the distinction and understand modern fingerscanning verification technology, one needs to understand the basis of a fingerprint [2].

A fingerprint is composed of ridges, the elevated lines of flesh that make up the various patterns of the print, separated by valleys. Ridges form a variety of patterns that include loops, whorls, and arches as illustrated in Figure 9-1 [1]. Minutiae are discontinuities in ridges, and they can take the form of ridge endings, bifurcations (forks), and crossovers (intersections), among others [2].

Fingerscanning verification is based on a subset of features selected from the overall fingerprint. Data from the overall fingerprint is reduced (using an algorithm application that is usually unique to each vendor) to extract a dataset based on spatial relationships. For example, the data might be processed to select a certain type of minutiae or a particular series of ridges. The result is a data file that only contains the subset of data points. The full fingerprint is not stored, and cannot be reproduced from the data file. This is in contrast with ink-and-roll fingerprinting (or its modern optical equivalent [4]), which is based on the entire fingerprint [2].

Modern forensic fingerprinting [3], with files on the order of 250 kB per finger, is used in large-scale, one-to-many searches with huge databases, and can require hours for verification. Fingerscanning verification, using files of less than 1,000 bytes, is used for one-to-one verification and give results in a few seconds [2].

Figure 9-1   *Ridges form a variety of patterns that include loops, whorls, and arches. (Source: Reproduced with permission from Idwave Technologies.)*

How Fingerscanning Verification Works

In use, fingerscanning verification is very simple. First, a user enrolls in the system by providing a fingerprint sample. The sensor captures the fingerprint image. The sensor image is interpreted and the representative features extracted to a data file by algorithms either on a host computer or a local processor (in applications such as cellular handsets). This data file then serves as the user's individual identification template. During the verification process, the sequence is repeated, generating an extracted feature data file. A pattern-matching algorithm application compares the extracted feature data file to the identification template for that user, and the match is either verified or denied. State-of-the-art processor, algorithm, and sensor systems can perform these steps in a second or two [2].

Modern Fingerscanning Verification Technology

Fingerscanning verification can be based on optical, capacitance, or ultrasound sensors. Optical technology is the oldest and most widely used and is a demonstrated and proven technology, but has some important limitations. Optical sensors are bulky and costly, and can be subject to error due to contamination and environmental effects. Capacitance sensors, which employ silicon technology, were introduced in the late 1990s. These offer some important advantages compared to optical sensors and are being increasingly applied. Ultrasound, utilizing acoustic waves, is still in its infancy and has not yet been widely used for verification [2].

Silicon-Based Sensor Technology

Silicon-based sensors have a two-dimensional array of cells, as shown in Figure 9-2 [2]. The size and spacing of the cell is designed such that each

Figure 9-2   *Silicon-based sensors have a two-dimensional array of cells. (Source: Reproduced with permission from Fidelica Microsystems, Inc.)*

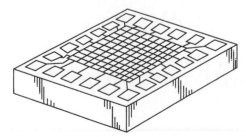

cell is a small fraction of the ridge spacing. Cell size and spacing are generally 50 microns, yielding a resolution of up to 500 dpi, the FBI's image standard. When a finger is placed on the sensor, it activates the transistors that underlay each individual cell capturing the image. Each cell individually records a measurement from the point on the finger directly above the cell, as shown in Figure 9-3 [2].

Though different vendors use different physical properties to make the measurement, the data is recorded as the distance, or spacing, between the sensor surface and that part of the finger directly above it. However, distance measurement has some inherent weaknesses [2].

The set of data from all cells in the sensor is integrated to form a raw, gray-scale fingerprint image, as shown in Figure 9-4 [2]. Fingerprint imaging using a continuum of distance measurements results in an eight-bit gray scale image, with each bit corresponding to a specific cell in the two-dimensional array of sensors. The extreme black and white sections of the image correspond to low and high points on the fingerprint. Only the high points on the fingerprint are of interest, since they correspond to the ridges on the fingerprint that are used to uniquely identify individuals. Therefore, the eight-bit gray-scale image must be converted into a binary, or bitonal, image using an additional procedure in the feature extraction algorithm. This process is a common source of error, since there could be many false high points or low points due to dirt, grease,

Figure 9-3
Each cell individually records a measurement from the point on the finger directly above the cell. (Source: Reproduced with permission from Fidelica Microsystems, Inc.)

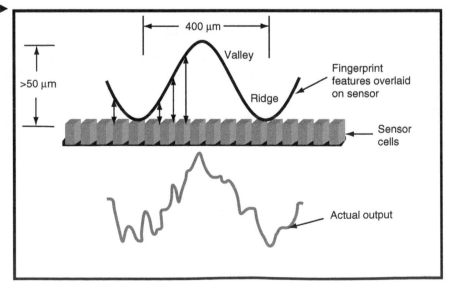

Figure 9-4   *The set of data from all cells in the sensor is integrated to form a raw, gray-scale fingerprint image. (Source: Reproduced with permission from Fidelica Microsystems, Inc.)*

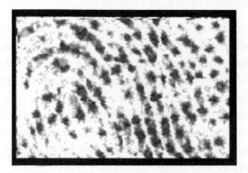

or other factors, each of which could result in a false minutia extraction and, hence, introduce additional error in the matching process [2].

The feature extraction algorithm is then used to extract the specific features from the fingerprint that make up the individual's unique data file. This data file serves as the user's individual identification template, which is stored on the appropriate device. During verification, the imaging and feature extraction process is repeated, and the resulting data file compared with the user's identification template by pattern-matching software to verify or deny the match [2].

As noted in the preceding, while there is more information in a gray-scale image, much of it is extraneous and must be filtered out. Both image types contain the same essential information content necessary for identification, specifically, the minutiae and ridges [2].

Indeed, for the first half-century that fingerprints were used by law enforcement, a bitonal ink image or photograph was the standard. The switch to gray-scale images is largely an artifact of the output of the optical sensors, the prevailing technology available when the transition was made to a digital database. The FBI fingerprint standard reflects this history: For purposes of commonality, a gray-scale fingerprint sensor is required when new templates are added to the database; however, the fingerprint sensor used to compare an individual with the database is not specified, and binary images are commonly used for the comparison [2].

Pressure-Sensing Science

Present sensor technology is unique among commercially available fingerscanning verification systems. The technology uses a thin-film-based sensor array that measures pressure to differentiate ridges from valleys on a fingerprint. This is in contrast to distance measurement, which is the basis of all other commercially available sensors, whether optical or capacitance (silicon-based) [2].

The sensor is architecturally and physically similar to the silicon-based sensors in terms of cell size and spacing, and therefore offers similar resolution. However, when a finger is placed over the sensor, only the ridges come in contact with the individual pressure-sensing cells in the two-dimensional array. No other part of the finger contacts the sensor. As a result, only those cells that experience the pressure from the ridges undergo a property change. To record the image, the array is scanned using proprietary electronic circuits. With an appropriate threshold setting, a distinction can be made between those cells that experience pressure and those that do not [2].

The sensor employs a resistive network at each cell location. Each cell incorporates a structure similar to those employed in the micro-electro-mechanical system industry. Upon the application of a fingerprint, the structures under the ridges of the fingerprint experience a deflection, and a change in resistance results. This change in resistance is an indication of the presence of a ridge above the cell being addressed. In principle, although the resistance value is an analog value, the difference between the resistance in the pressed and unpressed states is large enough that, with an appropriate threshold setting, one can easily distinguish between the presence or absence of a ridge with high resolution and accuracy [2].

Pressure measurement offers some inherently powerful performance advantages over the measurement of spacing. The first is improved accuracy of ridge and valley detection. Because the sensor detects pressure rather than distance, it readily differentiates between ridges and valleys. A valley exerts no pressure at all on the cell underneath it (as shown in Figure 9-5) [2], whereas all the cells underneath a ridge would record a pressure. With the appropriate threshold setting, this results in a "digital" response: The cell records either a ridge or a valley. In contrast, the spacing measurement technique used by other methods generates a continuum of measurements or a gray scale, which must be corrected for noise reduction, gray-scale adjustment, gain, and sensitivity adjustment [2].

As a result of using pressure rather than spacing to image the fingerprint, the sensor is considerably less sensitive to interference from dirt and grease on the finger or the sensor, wet or dry fingers, and other effects. In the presence of moisture, sweat, grease, or other oils, which are usually present as thin layers on the surface of the skin, there is usually no effect on a pressure-based sensor, whereas with a distance-based measurement, these thin layers cause significant distortion in the resulting image output. An example of a fingerprint image

Figure 9-5
A valley exerts no pressure at all on the cell underneath it. (Source: Reproduced with permission from Fidelica Microsystems, Inc.)

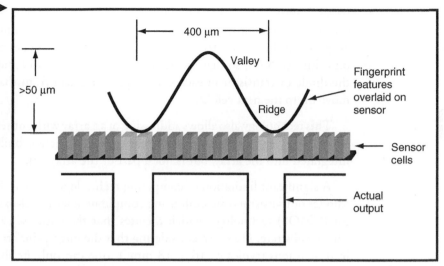

Figure 9-6 *An example of a fingerprint image under wet and dry conditions for a pressure sensor versus a competing sensor. (Source: Reproduced with permission from Fidelica Microsystems, Inc.)*

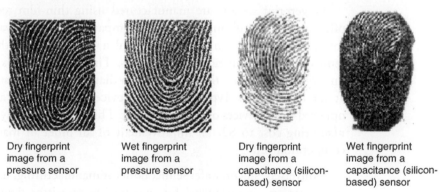

Dry fingerprint image from a pressure sensor

Wet fingerprint image from a pressure sensor

Dry fingerprint image from a capacitance (silicon-based) sensor

Wet fingerprint image from a capacitance (silicon-based) sensor

under wet and dry conditions for a pressure sensor versus a competing sensor is shown in Figure 9-6 [2].

Array-Addressing Scheme

In addition to the pressure-sensing technology, there is complementary technology for addressing a large two-dimensional array of cells using entirely passive means. Today, all competing silicon-based technologies must scan each cell of the array individually and then compile the data from all of the cells. To perform the scan, active switching devices such as diodes or transistors are built into each cell so that the addressing electronics can sequentially turn on the diode or transistor in each cell, acquire the data from that cell, and then move on to the next cell [2].

This technology also allows addressing of an array with entirely passive means within the array. As a result, all the electronic circuitry is built and physically integrated into the array, but is not a part of the array [2].

A significant limitation of competing technologies is that the active devices (diodes or transistors) are built using complementary metal-oxide semiconductor (CMOS) technology, which dictates that the entire sensor be built on a silicon substrate. However, considering that the fingerprint array is rather large in dimensions (typically, 16×18 mm^2), one can only fit about 20 to 25 of these devices on a typical 6- to 8-inch silicon wafer. Since the cost of processing a silicon wafer is typically in the range of $500–$600, the cost of a device to the sensor manufacturer is unlikely to fall below $20. Clearly, this limits the ability to deploy this device in mass-market applications such as computing, PDAs, cellphones, and smart-card readers [2].

Fingerprint sensors are manufactured using thin-film technology, which offers several significant advantages compared to silicon wafer semiconductor methods. Thin-film technology is substantially cheaper than conventional semiconductor manufacturing methods. Thin-film methods produce devices on large panels rather than 6- or 8-inch diameter single crystal silicon wafers, at far lower costs. Typically, 1,000 devices can fit onto a single panel as opposed to 25 devices on a silicon wafer. This lowers the fully assembled manufacturing cost to $2, an improvement of an order of magnitude over the competition [2].

Conversely, silicon wafer semiconductor manufacturing, using single crystal silicon wafers, is optimized for maximizing transistor density by minimizing the size of individual transistors. The economies of scale that have made silicon wafer processing the technology of choice for many semiconductor applications do not apply to fingerprint sensors. Modern memory chips, for example, contain on the order of a billion transistors on a 2×2 cm^2 chip. Fingerprint sensors require only 100,000 cells on a similar-sized device. Silicon wafer semiconductor facilities are designed to manufacture high-density integrated circuits, and costs are about the same per wafer to produce a low circuit density fingerprint sensor as it costs to produce a memory or processor chip [2].

Furthermore, thin-film methods are not restricted to single crystal silicon substrates. Sensors can be produced on glass, ceramic, plastic, and other substrates. Alternative substrates reduce cost, and allow greater flexibility for integration with all types of devices [2].

In addition to the enormous cost benefits that are allowed by the elimination of active switching devices in the array, the fingerprint sensor technology also allows greatly enhanced electrostatic discharge (ESD) reliability. ESD damage is common on CMOS-based circuits until they are packaged and sealed before incorporation into different products. However, in a fingerprint sensor, the CMOS device is directly exposed to the user for multiple uses in widely varying environments, which significantly increases the risk of ESD damage. Conversely, the elimination of CMOS-based active circuitry on the fingerprint sensor device drastically reduces the susceptibility to ESD damage [2].

The fingerprint sensor technology is also self-calibrating. A reference measurement is always made prior to or immediately following the actual fingerprint capture. This not only allows the sensor to automatically correct for effects such as environmental temperature or humidity, but also reduces the need for post-image gain or sensitivity adjustments based on ambient temperature. Other sensors depend on the presence of a finger to perform a measurement, and therefore it is impractical to determine a bad sensor from an invalid finger [2].

Verification Algorithms

As described earlier, a fingerprint sensor, by itself, has limited utility for a customer. Only the combination of a sensor with a verification algorithm adds value. Algorithms can be loosely divided into two broad classes: correlation-based and minutiae-based. Correlation-based verification algorithms use longer-length-scale information in the fingerprint image (the fingerprint ridges). These algorithms are preferred for convenience applications. Minutiae-based verification algorithms use shorter-length-scale information (for example, branching and terminations in fingerprint ridges). Minutiae-based verification algorithms are preferred for high-security applications [2].

Figure 9-7 shows a raw fingerprint image captured with a fingerprint sensor, and the same image after processing by using an algorithm with the minutiae locations highlighted [2]. However, depending on the application, some customers may prefer a correlation-based algorithm.

Figure 9-7   *A raw fingerprint image captured with a fingerprint sensor, and after processing by using an algorithm with the minutiae locations highlighted. (Source: Reproduced with permission from Fidelica Microsystems, Inc.)*

Summary/Conclusion

A fingerprint sensor solves the size, cost, and reliability problems that have limited the widespread application of fingerscanning verification. These are the most important criteria to any verification system:

- Size
- Cost
- Reliability and sensitivity
- Comparisons [2]

Size

Sensor chips are small (about the size of a postage stamp). They can be integrated into practically any device—cellphone, keyboard, mouse, door lock, and nearly any security application imaginable [2].

Cost

A sensor chip is thin-film-based, rather than silicon-based, and can be manufactured on plastic, glass, and many other substrates. Thin-film manufacturing is substantially less expensive than other methods. Sensor chips can be produced and distributed for less than $5 in quantity [2].

Reliability and Sensitivity

Thin-film technology, combined with unique control circuitry, yields a more durable and reliable sensor. Unique pressure-sensor technology is 10 times more sensitive than other methods, resulting in more reliable identification that is less affected by dirt and moisture. A fingerprint sensor also eliminates semiconductor cell addressing circuitry, which improves reliability and reduces the potential of ESD damage [2].

Comparisons

Finally, extensive comparisons should be conducted of the performance, such as the false-acceptance rate and false-rejection rate, of bitonality and gray-scale fingerprint images using several commercial pattern matching and verification algorithms, as well as a minutiae-based algorithm. The bitonal fingerprint image from a pressure-based sensor resulted in performance that equaled or exceeded that offered by other sensors that produced a gray-scale image [2].

References

1. "What Is Biometrics?," IDWAVE Technologies, 440 McMurchy Avenue South, Suite #1401, Brampton, Ontario L6Y 2N5, Canada, 2004. Copyright © 2004 by Idwave Technologies. All Rights Reserved.

2. "Fingerscanning Verification," Fidelica Microsystems, Inc., 1585 McCandless Drive [423 Dixon Landing Rd.], Milpitas, CA 95035, 2006.

3. John R. Vacca, *Computer Forensics: Computer Crime Scene Investigation, 2nd ed.* Charles River Media (2005).

4. John R. Vacca, *Optical Networking Best Practices Handbook*, John Wiley & Sons (2006).

Part 5: How Biometric Geometry Analysis Technology Works

How Hand Geometry Image Technology Works

The hand geometry biometric approach uses the geometric form of the hand for confirming an individual's identity (see Figure 10-1) [1]. Because human hands are not unique, specific features must be combined to assure dynamic verification [1].

Some hand-scan devices measure just two fingers, while others measure the entire hand. These features include characteristics such as finger curves, thickness, and length; the height and width of the back of the hand; the distances between joints; and overall bone structure [1].

It should be noted that although the bone structure and joints of a hand are relatively constant traits, influences such as swelling or injury can disguise the basic structure of the hand. This could result in false matching and non-false matching; however, the amount of acceptable distinctive matches can be adjusted for the level of security needed [1].

To register in a hand-scan system, a hand is placed on a reader's covered flat surface. This placement is positioned by five guides or pins that correctly situate the hand for the cameras. A succession of cameras captures 3D pictures of the sides and back of the hand. The attainment of the hand scan is a fast and simple process. The hand-scan device can process the 3D images in five seconds or less and the hand verification usually takes less than one second. The image capturing and verification software and hardware can easily be integrated within stand-alone units. Hand-scan applications that include a large number of access points and users can be centrally administered, eliminating the need for individuals to register on each device [1].

Applications for Hand Scanning

Many airports use hand-scan devices to permit frequent international travelers to bypass waiting lines for various immigration and customs systems. Employers use hand-scan devices for entry/exit, recording staff movement,

Figure 10-1
The geometric form of the hand confirms an individual's identity. (Source: Reproduced with permission from International Biometric Group, LLC.)

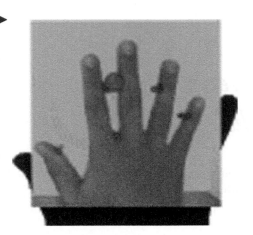

and time/attendance procedures. This can go a long way to eradicating the age-old problem of buddy-clocking and other deceptive activities [1].

Combining Biometric Methods

Hand scanning can be easily combined with other biometrics such as fingerprint identification. A system where fingerprints are used for infrequent identification and hand scanning is used for frequent verification would create a two-tiered structure. The hand-scan component, used frequently, allows identity verification or one-to-one verification that ensures the user is who they claim they are. The fingerprint identification component, used infrequently, confirms who the user is and accurately identifies the user in a one-to-many identification that is compared with numerous records [1].

Some anthropologists suspect that human intelligence has evolved due in large part to the shape of the hand. While the hand hasn't changed much in a long time, it's now being put to a new use: to verify its owner's identity [1].

How It Works

Hand geometry scanners take over 90 measurements of the length, width, thickness, and surface area of the hand and four fingers—all in just one second [1].

The technology uses a 32,000-pixel charged coupled device (CCD) digital camera to record the hand's three-dimensional shape from silhouetted images projected within the scanner. The scanner disregards surface details, such as fingerprints, lines, scars, and dirt, as well as fingernails, which may grow or be cut from day to day [1].

When a person uses the scanner, the scanner compares the shape of the user's hand to a template recorded during an enrollment session. If the template and the hand match, the scanner produces an output—it may unlock a door, transmit data to a computer, verify identification, or log the person's arrival or departure time [1].

During enrollment, which takes approximately 30 seconds, the user places their right hand in the reader three times. The unit's internal processor and software convert the hand image to a nine-byte mathematical template, which is the average of the three readings [1].

> **Note:** The user's template may reside in internal memory (capable of holding over 27,000 users), or on other media such as a hard disk or smart-card chip.

The method for capturing the biometric sample is fairly straightforward. To enroll, the user places his or her hand palm-down on the reader's surface. The user then aligns his or her hand with the five pegs designed to indicate the proper location of the thumb, forefinger, and middle finger. Three placements are required to enroll on the unit; the enrollment template is a representation of the most relevant data from the three placements [1].

The units use a 32,000-pixel CCD digital camera, inferring the length, width, thickness, and surface area of the hand and fingers from silhouetted images projected within the scanner. Over 90 measurements are taken, and the hand and fingers' characteristics are represented as a nine-byte template. Other technology is similar, but draws on the shape and characteristics of the index and middle finger. The data is saved as a 20-byte template [1].

Hand geometry is a relatively accurate technology, but does not draw on as rich a data set as finger, face, or iris biometrics. A decent measure of the distinctiveness of a biometric technology is its ability to perform one-to-many searches; that is, the ability to identify a user without the user first claiming an identity. Hand geometry does not perform one-to-many identification, as similarities between hands are not uncommon. Where hand geometry does have an advantage is in its failure to enroll (FTE) rates, which measure the likelihood that a user is incapable of enrolling in the system. Fingerprint, by comparison, is prone to FTEs due to poor quality fingerprints; facial recognition requires consistent lighting to properly enroll a user. Since nearly all users will have the dexterity to use hand geometry technology, fewer employees and visitors will need to be processed outside the biometric. Hand geometry is occasionally misunderstood as "palm reading," as the placement of the hand palm-down on the reader can be confusing to those unfamiliar with the technology [1].

Hand Geometry Strengths and Weaknesses

The following are hand geometry strengths:

- Ease of use
- Resistance to fraud
- Template size
- User perceptions

Ease of Use

The submission of the biometric is straightforward, and with proper training it can be done with few misplacements. The only misplacements may be elderly clientele or those with arthritic hands, who may be unable to easily spread their fingers and place their hand on the unit's surface. The unit also works fairly well with dirty hands [1].

Resistance to Fraud

Short of casting a model of an enrolled person's hand and fingers, it would be difficult and time-consuming to submit a fake sample. Since much of the value of hand scan is as a deterrent in time and attendance scenarios, it would rarely be worth the effort to attempt a fake submission [1].

Template Size

A template size of nine bytes is extremely small, orders of magnitude smaller than most other biometric technologies. By contrast, finger-scan biometrics require 250–1,000 bytes and voice-scan biometrics commonly require 1,500–3,000 bytes. This facilitates storage [3] of a large number of templates in a stand-alone device, which is how many hand-scan devices are designed to work. It also facilitates card-based storage, as even magstripe cards have ample room for nine-byte samples [1].

User Perceptions

As opposed to facial-scan or eye-based technologies, which can encounter some resistance, the use of hand geometry is not problematic for the vast majority of users. It bears very little of the stigma of other authentication methods [1].

The following are hand geometry weaknesses:

- Static design
- Cost
- Injuries to hands
- Accuracy

Static Design

As opposed to other biometrics, which can take advantage of technological breakthroughs like silicon development or camera quality, hand scanning has remained largely unchanged for years. Its size precludes it from being used in most logical access scenarios, where compact design may be a prerequisite [1].

Cost

Hand-scan readers cost approximately $2,100 to $2,700, placing them toward the high end of the physical security spectrum. Finger-scan readers, whatever strengths and weaknesses they may have, can be much less expensive, in the $1,500 to $1,900 range [1].

Injuries to Hands

As with all biometrics, physiological changes can cause users to be rejected falsely. Injuries to hands are fairly common, and would make use of such systems impossible [1].

Accuracy

Although generally more reliable than behavioral biometrics such as voice or signature, hand geometry, in its current incarnation, cannot perform one-to-many searches, but instead is limited to one-to-one verification. This limits its use in many different applications [1].

Enhanced Biometric Technology

Enhanced biometric technology for hand scanners occurs by maintaining a low false-reject rate (the probability that the device will reject an authorized user), while maintaining a high deterrent to unauthorized access. These units process large numbers of people with minimal delays [1].

The crossover of false-reject and false-accept rates for hand geometry readers is 0.1%. These optimal error rates were documented in independent testing at Sandia National Laboratories. Subsequent field results from thousands of users and hundreds of thousands of transactions confirmed the Sandia findings [1].

Highest User Acceptance

Among biometric technologies, Sandia reported that hand geometry had the highest user acceptance of all devices tested. With a high level of security, ease of use, and nonthreatening technology, hand geometry has become the most widely accepted biometric technology in use today [1].

Applications

Hand geometry scanners verify identity at the front entrances of over half the nuclear power plants in the United States. At the 1996 Olympic Games, hand geometry scanner units were integrated with the Olympic Village security system to process millions of transactions, with minimum delay [1].

Note: The U.S. Immigration and Naturalization Service (INS) uses hand geometry scanners to allow over 130,000 frequent travelers to bypass immigration lines (through the INSPASS program).

The drastic reductions in cost of microprocessors in recent years has brought affordable hand geometry technology to the commercial market. Biometrics are no longer found only in nuclear power plants. Daycare centers, athletic clubs, obstetrics wards, and police departments now use hand geometry scanners [1].

Tomorrow will find ever-expanding applications for this thoroughly time-tested technology. These could be applications for financial transactions, ticketless travel, and new business and residential applications where high security is a major concern [1].

As opposed to more exotic biometric technologies, whose implementations may be quite few and far between, hand scanning is used reliably at thousands of places of employment, universities, apartment buildings, and airports—any place requiring reasonably accurate, nonintrusive authentication. The nature of hand geometry technology is such that most projects are fairly small scale and involve only a handful of readers, but there are some projects that incorporate dozens of readers.

Perhaps the most frequently used and most successful hand-scan project is the Immigration and Naturalization Service Passenger Accelerated Service System (INSPASS) project. This project allows frequent travelers to circumvent long immigration lines at international airports in Los Angeles, Miami, Newark, N.J., New York City, Washington, San Francisco, Toronto, and Vancouver. Qualified passengers, after enrolling in the service, receive a magstripe card encoded with their hand-scan information. Instead of being processed by passport control personnel, INSPASS travelers swipe their card, place their hand, and proceed with their I-94 to the customs gate. Nearly 120,000 people have enrolled in the service, and approximately 90,000 verifications take place every month. Travelers from 100 different countries are qualified to register for INSPASS; pending budgetary constraints, the near-term objective is to roll out the INSPASS project to over 90 U.S. airports [1].

Implemented in 2000 is another high-profile hand-scan project in Israel known informally as "Basel." Designed to control access to a road connecting the Gaza Strip and the West Bank, Basel incorporates both hand-scan and facial scan; an overriding objective in the design of this biometric system is to provide maximum security while allowing for authentication under challenging environmental conditions. Hundreds of biometrically enabled turnstiles feature proximity-based smart cards, hand-scan readers, and cameras to perform facial scan matching. Although neither face nor hand is ideal for this application, the combination of the two allows for maximum efficiency in processing over 100,000 Palestinian workers each day. Another implementation of hand geometry is Disney's verification of season-pass-holders via two-finger geometry. This is both a convenience measure and a deterrent, as season-pass-holders are able to circumvent long lines, but cannot give their season passes to friends [1].

Hand Geometry Market Size

Although the technology for biometrics is mature, hand geometry is projected to be a slow-growing biometric technology through 2014. Because the range of applications of hand geometry is typically limited to access control and time and attendance, it will draw a progressively smaller percentage of biometric revenue. Overall, according to industry analysts, hand geometry revenues are projected to grow from $72.2m in 2005 to $142.9m in 2010. Hand geometry revenues are expected to comprise approximately 7.0% of the entire biometric market [1].

Summary/Conclusion

Handprint recognition scans the outline or the shape of a shadow, not the handprint. It can be used for many types of access, but accuracy does tend to be a problem. Since it is a fast and semireliable method of verifying identity, many companies use it, but many people have similar hand shapes and sizes so this system is not considered 100% secure. The National Center for State Courts indicates that the use of hand geometry is not enough to identify an individual, but by combining various individual features like fingerprints, this method can be very effective [2].

Finally, hand geometry is good for places where quick verification is a positive. Scanning a hand takes anywhere from five to six seconds. As stated before, although it is not a completely secure method of identification it is very fast, and combined with other areas can be very effective [2].

References

1. "Hand Geometry," International Biometric Group, LLC, One Battery Park Plaza, New York, NY 10004, 2006. Copyright © 2006 International Biometric Group.

2. Barrett Key, Kelly Neal and Scott Frazier, "The Use of Biometrics in Education Technology Assessment," Ball State University, 2000 W. University Ave., Muncie, IN 47306, 2006. Copyright © 2006.

3. John R. Vacca, *The Essentials Guide to Storage Area Networks*, Prentice Hall, Professional Technical Reference, Pearson Education (2001).

How Finger Geometry Technology Works

A few biometric vendors use finger geometry or finger shape to determine identity. Analysis generally focuses on the geometry of one or two fingers. Unique finger characteristics, such as finger width, length, thickness, and knuckle size are measured.

One technique has the user insert the index and middle finger into a reader, and a camera then takes a three-dimensional image. A second technique requires the user to insert a finger into a tunnel so that sensors can take three-dimensional measurements.

Finger geometry systems are very accurate, simple to use, and are impervious to fraudulent deception. However, public acceptance is somewhat lower than for finger scanning, apparently because users must insert their fingers into a reader.

In any event, finger geometry machines scan the dimensions of one or more digits. These machines typically cost about $2,200. In 1995, The Walt Disney Company contracted with the Swiss firm BioMet Partners to test a 3D, two-finger geometry device (the Digi-2) at entrance turnstiles for holders of season passes to Walt Disney World in Orlando, Florida. The following discussion covers the test results of that technology, how it works, and its present use today.

One or Two Fingers?

When visitors step up to the gates of the four Disney World theme parks, the Magic Kingdom, Epcot, Animal Kingdom, or the MGM Studios, they will encounter something unexpected and largely foreign to them. Disney has embarked on a program to use an established biometric technology (finger geometry) to secure its valuable passes. Ostensibly, this new security is for the benefit of the pass owner. However, it is also being implemented to secure Disney's pricing structure and marketing strategy. It has not come without controversy—and at least a bit of confusion [1].

So What Is Finger Geometry?

Hand geometry has been aptly described as "the 'granddaddy' of all biometric technology devices." It is essentially based on the fact that virtually every individual's hand is shaped differently than another individual's hand, and over the course of time, the shape of the person's hand does not significantly change. Operationally, finger- or hand-scanning systems capture the physical, geometric characteristics of an individual's hand—with most systems having the capacity to do so in less than a second. From these measurements, a profile or "template" is constructed that will be used to compare against subsequent readings by the user. Finger geometry and hand geometry are considered somewhat interchangeable terms. However, hand geometry evaluates the person's entire hand form as a biometric identifier, while finger geometry looks only at a subset of the five fingers to form the identifier (see sidebar, "Advantages of Using 3D Geometry of Two Fingers") [2]. In either case, such geometry does not entail the taking of a person's fingerprints. In a recent study, the National Academies of Science found that while a person's finger geometry is indeed far less distinctive than his or her fingerprints, hand or finger biometrics is indeed suitable as an identifier for a wide variety of circumstances, where one in a thousand uniqueness is sufficient [1].

Advantages of Using 3D Geometry of Two Fingers

The most proven and accurate biometric technology for fast, positive verification of a person's identity is the use of highly accurate optical scans [7] of the three-dimensional geometry of two fingers of either hand. So, why use a three-dimensional geometry of two fingers for highest performance? Let's take a look at the major advantages:

- User friendly
- Low cost
- Flexible
- Totally proven

User Friendly

When being enrolled, users quickly understand that finger geometry is unique and private. The 3D picture of two fingers cannot be used with any (criminal or other) recorded information (fingerprints, facial photos, etc.). There is absolutely zero invasion of privacy [3].

Low Cost

Because of speed and accuracy, the total cost of the installation will be far less than with any other biometric verification technology. Throughput of up to 10 users per minute means fewer entry installations (turnstiles, gates, entry doors), less infrastructure, less computer power, and less data storage [6] required (only 20 bytes per user).

Flexible

Two-finger readers can be used on a stand-alone basis with the user's 20-byte finger photo data stored on any type of entry card or in the reader without use of cards. A networked system of any number of entry/exit readers will be far less costly than systems using any other type of biometric identification or verification. Finger geometry readers can be installed in ATMs, border "green card" checkpoints, police cars, and mobile payment vehicles for distribution of public benefit funds, as well as large offices, factories, stadiums, or theme parks for entry control.

Totally Proven

There is no guesswork about the performance and user acceptance of two-finger photo technology. Developed nearly 15 years ago, there are many thousands of two-finger biometric terminals (see sidebar, "Two-Finger Biometric Terminals: . . .") installed worldwide with more than 100 million users enrolled. These installations range from offices, laboratories, banks, and government facilities to the largest theme parks and stadiums for user-friendly control of season ticket holders, as well as major airports for security control of passenger identity.

Performance Measurements

The following are the performance measurements for 3D geometry of two fingers:

- Speed and throughput
- Accuracy
- Counterfeiting

Speed and Throughput

At the rate of 10 users per minute, no other biometric verification technology can come close.

Accuracy

The two-finger scanners can be adjusted to a wide range of accept/reject threshold levels, depending upon the degree of security required. Less than 0.3% false rejects and 0.7% false accept retries are required at the standard factory threshold setting of 150. To obtain near-zero false acceptances a threshold setting of 240 is used, and still only 1 in 25 users would be required to make a second try to pass.

Counterfeiting

Unlike fingerprints, which can be copied from a drinking glass or even a piece of paper and replicated in gelatin or latex, there is virtually no way the three-dimensional configuration of a person's two fingers (length, thickness, shape, and knucklebone characteristics) can be copied and reproduced. User voice patterns, signatures, and facial photos are much easier to copy, hence to counterfeit. The two-finger geometry is by far the most "counterfeit resistant" biometric ID/verification technology available today [2].

The Two-Finger Biometric Terminals: The Ultimate Synergy of Smart Cards, Biometrics, and Advanced Encryption

The two-finger biometric terminal is the efficient combination of smart cards (contact or contactless), biometric verification, and the higher security provided by public key cryptography [4] to achieve highly accurate authentication of the cardholder's identity. The marriage of the technologies provides for maximum system protection against both fraud and invasion of personal privacy. The two-finger biometric terminal is also the most effective solution for a wide range of positive personal identity applications, enhancing presently used basic systems for access control, medical records, computerized timekeeping, passenger ticketing, welfare payment systems, driver's licenses, security, and border control.

The Ultimate Solution for Border Control if It's Enforced

The currently discussed initiatives for border control call for countries to implement a visitor/identity system requirement, which includes fingerprints and "facial or eye recognition." This has created a public outcry mainly because of the enormous cost (it does not warrant the term "investment") over alternative and better solutions and the time to implement under international accord, as well as the public outrage at having their biometric data integrated with existing police and other fingerprint databases of criminals and illegal residents.

The two-finger biometric terminal provides a perfect alternative to this badly flawed visitor identity system, with the advantages of:

- No invasion of personal privacy;
- High accuracy, overcoming the problem that up to 18% of people do not have electronically verifiable fingerprints;
- Avoiding billions in government expense on highly complex networks and databases;
- Faster clearance of passengers and at border crossings;

- Major reduction in cost of hardware and staff at airports, seaports, and border entries;

- Far better total security control.

How Does It Work?

First of all, the cardholder inserts their special smart card into the two-finger biometric terminal. This tells a FingerFoto camera module (see sidebar, "FingerFoto OEM Camera") to "wake up" and capture a virtual three-dimensional image of the cardholder's two fingers (of either hand) and pass the image data to the smart card.

Second, the smart card processes the image using its on-board engine with proprietary algorithms and compresses the data into a 20-byte template, which is immediately encrypted and stored. The cardholder's two-finger data is not kept in the terminal and is not transmitted to any database or anywhere else—the person's encrypted data is only on the smart card.

Third, there is nothing to "hack" or copy, no memory in the terminal, no communications lines to tap, no central database. If the card is lost or stolen, it is totally useless to another person.

Fourth, the cardholder can go to a two-finger biometric terminal anywhere in the world and insert their card. The two-finger biometric terminal captures the two-finger data and passes it onto the smart card. The processor on the card is programmed to carry out the "compare function" and confirm that the new template is the same as the one stored on the smart card. If it is, the "open" command is given to the terminal's door or gate controller. Meanwhile, on the smart card, the two templates are combined (template updating) and the new template is immediately randomly re-encrypted on the smart card.

Finally, there is no trace or track anywhere that can be compromised. The cardholder's template is never the same twice and is in a different encrypted format between each verification [2].

FingerFoto OEM Camera

The FingerFoto three-dimensional two-finger geometry camera has been carefully designed for easy integration into a wide spectrum of host systems (see Figure 11-1) [2]. The host equipment provides the electronic and mechanical housing for the camera. For OEMs, this provides a great reduction in the incremental cost of adding a well-proven biometric feature to their existing terminals and systems.

By selectively combining software versions and built-in options (commands) to achieve specific customer requirements, the performance of the FingerFoto can be optimized for each host product application. The FingerFoto can easily be integrated into a wide spectrum of host systems, including financial terminals (ATM, point of sale, interbank); confidential databases and security workstations;

Figure 11-1
The FingerFoto three-dimensional two-finger geometry camera. (Source: Reproduced with permission from BioMet Partners Inc.)

personnel screening systems, access control and time and attendance systems; public entry/ticketing systems; and automatic benefit payment kiosks.

Where required, custom-designed extra components can be supplied to complete the integration. The host frequently already has facilities such as card readers, displays, keypads, and 12 VDC power and communications capabilities that can be used by the FingerFoto cameras, thus greatly reducing the incremental cost of adding the "Best by Test" biometrics feature into the OEM's host product.

Finger Geometry Use

Finger geometry has been used successfully since its commercial introduction in 1975, when it was brought to Wall Street for security purposes by the investment firm of Shearson Hamill. Over the years, it has been utilized to provide secure access and verify one's identity in a wide variety of settings, including:

- Athletes at Olympic Villages
- Members of the Colombian National Legislature
- Employees of over 90% of all nuclear power plants in the United States
- Military officers

- Prisoners
- Parents at daycare centers
- Donors at sperm clinics [1].

Probably the widest use of finger or hand scanning is in the corporate realm. Such scanning is used to complement employee badges, passes, and ID cards to prevent payroll fraud, a seemingly intransigent problem that has been estimated to cost employers in the United States alone hundreds of billions of dollars each year. While other forms of biometrics may be growing more rapidly, there is still substantial growth potential for hand and finger scanning.

Giving Disney Your Fingers

Disney has moved over the past decade to use automatic identification in various forms. In 1996, the company moved away from a hard plastic laminated pass for all holders of multiday or annual passes, which contained both a barcode identifier and a photo of the passholder. In its place, Disney began issuing Mylar paper passes. These new passes had no photo identifier and, indeed, contained only minimal visual evidence of ownership, essentially only the guest's name and the expiration date of the pass. Beginning in June 2005, all Walt Disney World parks began using finger scanning at its park entrances to complement the security measures embedded in its Mylar passes. When a Disney guest presents his or her pass at the turnstile, he or she is asked to insert the pass into a reader, and after doing so, to make a "peace sign" with his or her index and middle fingers and insert those fingers into a scanning area. During the scan, a camera takes a picture of several points on each person's index and middle fingers and assigns a numerical value to the image. The scan (which is accomplished in less than a second) measures the length and width of the individual's two fingers and the spread distance between the digits. Once the scan is taken (and all adults are required to complete the scan), the pass is returned to the guest [1].

For a number of years now, Disney's marketing approach has been to shrewdly push the sale of multiday and annual passes to its theme parks that comprise the Disney World complex. (Disney passes are not interchangeable between its parks in Anaheim, California and Orlando, Florida.) The pricing structure at Disney World is transparently meant to encourage its visitors to buy passes for longer stays at its Orlando properties. In fact, the daily price of a Disney park visit drops significantly as longer-lasting park passes are purchased, dropping by half at the seven-day mark and by almost two-thirds at the 10-day mark. To put it quite simply, Disney makes about $200 more by selling five separate two-day tickets than by selling a single 10-day pass. So, to protect its revenue stream, Disney does not allow its annual or multiday passes to be shared or transferred. They don't want people to buy a 10-day pass, use it for two days, and then resell the pass to another buyer to use for the remaining days [1].

Longer stays mean that families visiting Disney World will have more opportunities to spend more money on food and beverages, souvenirs and trinkets, and other experiences, such as breakfast with Cinderella, while on Disney property. Perhaps even more importantly, the passes serve to "lock in" guests to focus their Orlando visits on Disney parks, rather than spending their time (and money) at the competitors' parks and other entertainment experiences available in this burgeoning family resort area [1].

A Mixed Reaction

From Disney's perspective, the ticket tag is a necessary security measure that does not violate its customers' privacy. In fact, the company does not maintain a permanent database of scans, as the information is purged from its systems after the individual's pass expires. Disney has not disclosed the vendor for its biometric system [1].

However, Disney's move to finger scanning has generated some degree of controversy since its implementation. Since Disney defines an "adult" park guest as being 10 or older, many minors are being subjected to finger scanning. Leading privacy groups have also attacked Disney's move. The American Civil Liberties Union recently called the addition of biometric technology "a step in the wrong direction." The Electronic Privacy Information Center (EPIC) recently issued a blistering attack on Disney for its use of finger scanning. It called the practice a "a gross violation of privacy rights," as there is little notice given to consumers as to why their biometric information is being collected, how it will be used, and the protection afforded to the data. EPIC criticized Disney's move based on the legal principle known as "the proportionality test," which can be encapsulated as whether the amount and type of information being collected equals the level of security being sought. To date, however, there have been no lawsuits filed against Disney over its use of finger scanning technology [1].

Finally, both at ticket sales' locations and at the actual park entry points, Disney has not seen fit to post information on exactly what is being done when the park patrons are asked to make the peace sign and insert their digits into the reading machine. Most patrons (and even some public interest groups and media covering the developments at Disney) have assumed that the company is fingerprinting park visitors and matching the passholder's print to the pass—and perhaps even other databases, such as criminal records, sex offender registries, and terror watch lists. This has led some industry observers to criticize Disney for having a corporate communications problem in not explaining the "whys" for the use of the technology to its patrons, while others have seen fit to call upon Disney to find creative ways to leverage the technology (and the data it collects) beyond gate security to provide better in-park customer experiences for its guests [1].

Summary/Conclusion

What is certain is that you will see more use of finger geometry in the future, as Disney is by no means alone in exploring this established technology in the theme park industry. Indeed, several of the company's principal competitors are looking to implement similar pass protection technology to their valuable tickets and passes in 2007, including Universal Orland, SeaWorld Adventure Parks, and Paramount Theme Parks [1]. Finally, such biometric scanning may be a necessary tool for the entire theme park industry. The introduction of this type of solution will be used more broadly in the industry in the future [1].

Finally, for now, the introduction of finger scanning seems to present Disney with an operational challenge to get visitors used to the new requirement. Overall, it's good. But it seems to make the queues longer. No one seems to put their fingers in all the way on the first try [1].

References

1. David Wyld, "Biometrics at the Disney Gates," AVISIAN Publications, March 2, 2006.

2. "Important Advantages of Biomet's Technology," Biomet Partners Inc., Feldmattstrasse 40, CH-3213 Kleinbösingen, Switzerland, 2006. Copyright © 2002 BioMet Partners Inc. All Rights Reserved.

3. John R. Vacca, *Net Privacy: A Guide to Developing and Implementing an Ironclad eBusiness Privacy Plan*, McGraw-Hill (2001).

4. John R. Vacca, *Public Key Infrastructure: Building Trusted Applications and Web Services*, CRC Press (2005).

5. John R. Vacca, *Satellite Encryption*, Academic Press (1999).

6. John R. Vacca, *The Essentials Guide to Storage Area Networks*, Prentice Hall, Professional Technical Reference, Pearson Education (2001).

7. John R. Vacca, *Optical Networking Best Practices Handbook*, John Wiley & Sons (2006).

Part 6: How Biometric Verification Technology Works

12

How Dynamic Signature Verification
Technology Works

Dynamic signature verification is an extremely active research area. Although systems exist on the market, there are few that can promise sufficiently high accuracy rates at a reasonable level of efficiency. Few have reached correct recognition rates above 96% and those that have, are extremely time-consuming. What will be explored in this chapter is what new dynamic signature verification technology is doing to solve this problem.

Requirements of a Dynamic Signature Verification System

Signature verification is the process used to recognize an individual's handwritten signature. When considering a dynamic signature verification system, there are many factors that go into making that system effective. First, data collection has to be sufficiently accurate. Next is the identification of the signature as the correct signature, to make sure that the signed name is the correct name. Following that, there must be some determination as to whether or not the given signature is accurate or a forgery. Currently data collection is largely a solved problem; data tablets exist at a sufficient resolution, both spatially and temporally, to acquire accurate data. Solutions for the other two problems have been attempted with moderate to reasonable success, but there is still room for improvement. Accuracy and/or processing time for the given systems must be improved in order to make any dynamic signature verification system a distributable and marketable technology.

In order to have a full understanding of signature verification and its possible uses, it is important to look at the different types of signature verification: capture and dynamic [1].

Capture

Capture signature systems are those systems that allow a person's signature to be "captured" electronically for the purpose of transaction validation and then forwarded to remote locations. Most people who use credit cards have experienced signature capture when signing for purchases on a small electronic tablet device. The signature is captured with the receipt and can be forwarded anywhere via e-mail. This type of device can only serve to validate a transaction and does not serve as a means to identify the signer. Another means of identification is needed to verify that the person signing is actually that person. Many banks use a type of capture signature verification that allows them to quickly compare signed checks against stored signature cards. This system uses signature comparisons and has helped banks cut down on forgery, but it has limited uses for means of identification [1].

As previously mentioned, signature verification is the process used to recognize an individual's handwritten signature. Dynamic signature verification is a biometric technology that is used to positively identify a person from their handwritten signature [2].

Dynamic

The second type of signature verification is called dynamic signature verification. There is an important distinction between simple signature comparison and dynamic signature verification. Both can be computerized, but a simple comparison only takes into account what the signature looks like. Dynamic signature verification takes into account how the signature was made. With dynamic signature verification it is not the shape or look of the signature that is meaningful; it is the changes in speed, pressure, and timing that occur during the act of signing. Only the original signer can re-create the changes in timing and X, Y, and Z (pressure) (see Figure 12-1 [2]). This unique means of analyzing a signature makes it virtually impossible for another person to duplicate the timing changes in X, Y, and Z. A pasted bitmap, a copy machine, or an expert forger may be able to duplicate what a signature looks like, but the natural motion of the original signer would be required to complete the signature [1].

In other words, dynamic signature verification utilizes the unique way in which a handwritten signature is made to identify or recognize an individual. This is done by analyzing the shape, speed, stroke, pen pressure, and timing information during the act of signing the signature (see Figure 12-2) [2].

There will always be slight variations in a person's handwritten signature, but the consistency created by natural motion and practice over time creates a recognizable pattern that makes the handwritten signature a natural for biometric identification [2].

Figure 12-1
The original signer can re-create the changes in timing and X, Y, and Z. (Source: Reproduced with permission from Cyber SIGN Incorporated.)

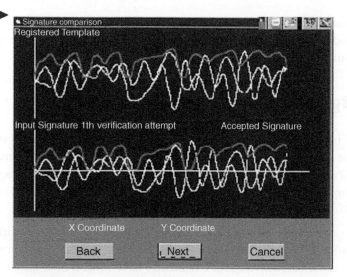

Figure 12-2
Dynamic signature verification. (Source: Reproduced with permission from Cyber SIGN Incorporated.)

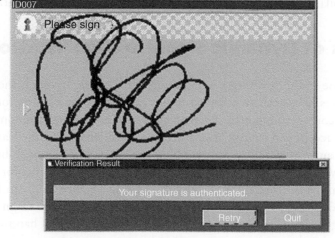

Signature verification is natural and intuitive. The technology is easy to explain and trust. The primary advantage that signature verification systems have over other types of biometric technologies is that signatures are already accepted as the common method of identity verification. This history of trust means that people are very willing to accept a signature-based verification system [2].

Dynamic signature verification technology uses the behavioral biometrics of a handwritten signature to confirm the identity of a computer user. Unlike the older technologies of passwords and keycards (which are often shared or easily forgotten, lost, and stolen), dynamic signature verification provides a simple method for increased computer security and trusted document authorization [2].

Dynamic Signature Verification Costs

The costs of implementing a dynamic signature verification system appear to be on the low end as compared with other biometric systems. Software, user license, and a graphic tablet or PDA are all that is required to operate a dynamic signature verification system. Cyber Sign, for example, charges $1,500 per 10-user agreement, which breaks down to $150 per user [1].

Although dynamic signature verification appears to be within a reasonable price range, it is still too expensive for most public schools across the country. Federal cutbacks and state mandates have most school budgets at their limits. Public opinion does not appear to be on the side of spending thousands of dollars to implement these new systems for the increased security they offer [1].

Advantages of Dynamic Signature Verification

One of the advantages of this type of biometric system is the fact that signatures have been an accepted means of identity verification for centuries. This encourages a biometric technology that is easy for organizations and consumers to accept and to trust. Signatures are considered to be the least intrusive form of authentication and enjoy a high degree of social acceptability. Thus, a move toward adopting dynamic signature verification biometrics for electronic transactions is expected to be smooth. This acceptance seems to be supported by the U.S. government, which in 2000 passed the U.S. eSign Bill and related legislation. This legislation extended the legal status of handwritten signatures to the electronic equivalent, thereby promoting the growth of e-transactions nationwide [1].

Another advantage of a dynamic signature verification system is in the replacement of passwords, PINs, or keycards. Identification that can be stolen, lost, or forgotten is eliminated and replaced with a simple signature. Logging on to a secure computer system from a remote location could be done with a graphics tablet and a signature. Other advantages are as follows:

- Low total error rate (about 1.5% per session);
- Forgery is detected even when the forger has managed to get a copy of the authentic signature;

- Possible detection of inconsistent user during enrollment stage;

- Fast and simple training;

- Cheap hardware (almost any tablet device is allowable);

- Little storage requirements [3];

- Fast response (about one second per signature on the "old" 486DX-33 computer);

- The results do not depend on the native language of the user;

- You can use any kind of information as your signature: name, second name, or even nice curves;

- Very high compression rate (100–150 bytes are needed to keep the shape of the signature);

- The system represents a natural way to prove authenticity [1].

Disadvantages of Dynamic Signature Verification

Dynamic signature verification is designed to verify subjects based on the traits of their unique signature. As a result, individuals who do not sign their names in a consistent manner may have difficulty enrolling and verifying in dynamic signature verification. Individuals with muscular illness and people who sometimes sign with only their initials might result in a higher false-rejection rate, which measures the likelihood that a system will incorrectly reject an authorized user [1].

Finally, signatures can be affected by behavioral factors; stress or distractions could cause a person to vary from their normal signature sequence, thereby generating a rejection. There has been presented information regarding the frequency-of-use factor. It has been stated that a person's signature could vary over a period of years. Therefore the person not using a dynamic signature verification system for an extended period of time may develop different characteristics in their signature, thereby causing the signature to be rejected [1].

Summary/Conclusion

In an effort to address the questions concerning the use of biometrics in a school, university, or an online class situation, the conclusions drawn are as follows: First of all, dynamic signature verification would be an excellent way of securing computers and computer labs. Having students sign in would mean that the student was actually there, and that it was not in fact another student who had the correct password. Student sign in could also be useful in other areas of everyday operations, such as signing in and out of a secure building, or verifying identity in the cafeteria or bookstore [1].

As far as an online class situation, there seems to be no easy solution, short of surveillance cameras. The point being made here is that anyone could sit down at the computer once the correct student has logged in. The bottom line is that there is nothing available at this time that can positively guarantee that the person operating a computer is who he or she says they are. The conclusions then must be that there are some situations in which this type of biometrics may be useful, while in other areas biometrics are not foolproof [1].

Finally, one last point to consider. Although there has been a major focus into the recent intrusions into your constitutional rights (the current Bush administration's illegal wiretapping and phone surveillance by NSA of U.S. citizens) due to the heightened security issues here in the United States, dynamic signature verification in its simplest form is older than the U.S. Constitution. For centuries man has been legally identified by his or her signature or mark. It has been an accepted means both here and around the world and, as stated earlier, dynamic signature verification is considered to be one of the least intrusive forms of authentication and enjoys a high degree of social acceptability. Being able to verify a person's identity is rapidly becoming an important part of modern society. In these times of ongoing terrorist threats, you may have to sacrifice some part of your constitutional rights for the protection of your country, your families, and your way of life. Anyway, that's what the present administration would like you to think!

References

1. Alex English, Christina Means, Kris Gordon and Kevin Goetz, "Biometrics: A Technology Assessment," Ball State University, 2000 W. University Ave., Muncie, IN 47306, 2006. Copyright © 2006.

2. "Biometrics," Cyber SIGN Incorporated, 180 Montgomery St., Suite 925, San Francisco, CA 94104, 2006.

3. John R. Vacca, *The Essentials Guide to Storage Area Networks*, Prentice Hall, Professional Technical Reference, Pearson Education (2001).

13

How Voice Recognition Technology Works

The human voice is unique because of the "physiological and behavioral aspects of speech production." The shape of the vocal tract in humans is what makes the voice unique. The vocal tract's location is depicted by the shaded area shown in Figure 13-1 [1].

Voice recognition is an automated system that analyzes and interoperates with human vocal patterns and characteristics to recognize words and identify or verify one's identity. This is a more recent biometric and has not been implemented as long as many others. Some points about the technology are as follows:

- Voice-related biometrics should not be confused with speech recognition computer software that recognizes words as they are spoken.

- Biometric systems involve the verification of the speaker's identity based on numerous characteristics, such as cadence, pitch, and tone.

- Voice recognition is considered a hybrid behavioral and physiological biometric, because the voice pattern is determined to a large degree by the physical shape of the throat and larynx, although it can be altered by the user.

- One-to-one verification is the preferred application.

- The technology is easy to use and does not require a great deal of user education.

- However, background noise greatly affects how well the system operates.

- Voice recognition works with a microphone or with a regular telephone handset.

- It is well suited to telephone-based applications where identity has to be verified remotely.

- Detailed menus are available for designing a voice recognition system.

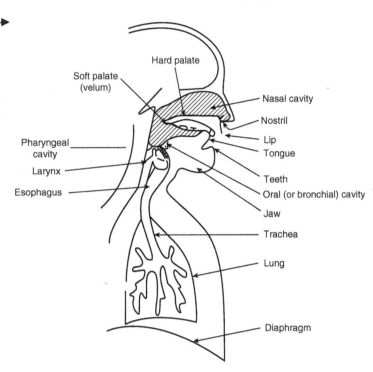

Figure 13-1
Vocal tract.
(Source:
Reproduced with
permission from
Ball State
University.)

In other words, voice recognition is classified as a behavioral biometric and analyzes many different parts of human speech. Some areas analyzed include language, speech pathologies, and the physical and emotional state of the speaker. Voice recognition technology is one of the cheaper biometrics because it can work with any microphone, such as a microphone that comes with or could be plugged into your computer or a telephone handset. However, the better the microphone, the higher quality and better accuracy of the system. Some applications for voice recognition are recording attendance, granting access to sensitive areas, over-the-phone verification such as banking, and forensic purposes. Currently, over-the-phone voice recognition is the most commonly used [2].

There are two different types of information that exist with voice recognition: low-level and high-level. High-level deals with characteristics that humans also use to recognize and distinguish one person from the next. Some examples of these are dialect, accents, talking style, and content. Low-level information is used mainly in voice recognition system to analyze speech. These are pitch periods, rhythm, tone, spectral magnitude, frequencies, and bandwidths [2].

Voice recognition systems are classified in many different ways. One distinction that is used to classify these is text dependency. There are two different types: text-dependent and text-independent. Text-dependent or fixed-text systems require the speaker to say a certain phrase or password. This text usually comes from a training system that uses words to better clarify the person's individual characteristics. To accomplish this, a person must first have their voice recorded repeating the chosen word or words. Text-independent or free-text systems have no required word or phrase that must be said. Any speech can be captured and analyzed. This has several advantages over the text-dependent system. First, there are no passwords to remember, so a person does not have to struggle with remembering one. Second, it helps to eliminate possible fraud because the phrase is not the same every time and cannot be recorded and perfected to grant fraudulent access. However, text-independent systems have many problems with mismatches and granting authorized people access. This is why they mainly remain in the experimental stage [2].

Another way to classify voice recognition is speaker-identification and speaker-verification systems. Speaker identification is a system that attempts to identify a speaker from a file of known speaker characteristics to determine if a known person is speaking. This can be done from a person speaking alone or in a group environment. To accomplish this, the speaker voice must have been previously recorded and analyzed. Speaker verification is a system that attempts to confirm if a person is who they say they are. This is done by a person stating a password or their identity into the microphone. False rejection and false approval are two types of error that can occur when using speaker verification [2].

There are two models used to make a voiceprint that can be stored in a database to verify at a later time: stochastic model and template model. In briefs, the stochastic model takes random words and samples the speech to determine parameters, and the template model takes samples from the same word spoken multiple times. Both of these models are much more complex than this, but for the purposes of this chapter it is not necessary to understand in detail how these models work. However, it is important to know that different voice recognition software solutions use one of these two models [1].

There are several things that can affect and cause problems in a voice recognition system. The environment can be an issue for concern. Background noises in the environment cause the system to result in failure. Error in the speaker's pronunciation of the phrase or password can cause problems. A person's mental and physical health variations can cause a problem in recognizing their voice patterns on a consistent basis. The speed and attitude can also have effects on

Figure 13-2
Flow chart of a voice recognition system. (Source: Adapted with permission from Ball State University.)

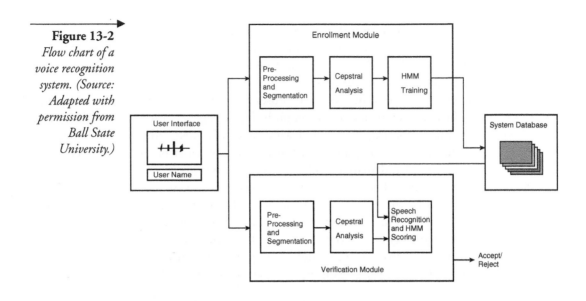

the process. All of these can cause problems if the speaker is not in the same environment, condition, and mood as when they recorded their voice [2].

Finally, a voice recognition system matches the pattern of a voice that was obtained using one of the preceding mentioned models and compares it to a database of voiceprints. Figure 13-2 shows this process [1].

Voice Recognition: Security and Costs

Like fingerprints, the human voice is specific to a single person. Therefore, no one voice can be "altered, forged or stolen." Aside from being very secure, voice recognition is a very cost-effective solution. The reason voice recognition is cost effective is because, unlike other biometric solutions, voice recognition does not require any special hardware to make it work. Verification can be done using standard sound cards and microphones. The software will cost $1,995 and up, depending on the number of biometric profiles to be stored. Cost effectiveness is very important when considering a large deployment [1].

Another feature that makes voice recognition a viable biometric solution is its integrity. With the proper voice recognition technology, the likeliness of using tape-recorded voices to gain access to secure information/areas is greatly reduced. The proper technology will test for liveness. The way this technology works is by listening for acoustic patterns that can be found in a voice that has been tape-recorded. This type of technology helps to prevent users who are not physically at a location from gaining access [1].

Areas of Opportunity: Education

The features previously mentioned make voice recognition a viable solution in the education field. Voice recognition requires virtually no new hardware because most PCs built today already have the necessary sound cards and microphones built in. This makes voice recognition a cost-effective solution, especially for educational institutions with limited budgets [1].

There is also very little time and effort required to record a person's voice to use for authentication. A user simply needs to speak into a microphone and they are done, although there might be some additional information that would need to be entered (Social Security number, name, address). Another feature is that you do not have to remember your pass code [1].

This technology could be used to log users on to computers and allow them access to rooms without the need for a physical key, which could be lost or stolen. Voice recognition could also be used in environments where a large number of students are taking a test to verify who they are before they are allowed to receive the test [1].

Finally, voice recognition could be used for distance-learning situations. Voice recognition tools have recently been developed that can be used over the World Wide Web to transmit a voiceprint back to a remote server for verification. By using this technology, students taking an online course could be required to use their voice to verify who they are before they are allowed to participate in the class [1].

Summary/Conclusion

Voice recognition technology is a viable solution to securely and inexpensively authenticate users both at a physical location and remotely. By using technology that is already available in most PCs on the market today, voice recognition limits costs for hardware. Even if new hardware is required, it is relatively inexpensive. Voice recognition is secure because every human has a distinct voice that is virtually impossible to duplicate, even using tape-recording devices [1].

References

1. Alex English, Christina Means, Kris Gordon and Kevin Goetz, "Biometrics: A Technology Assessment," Ball State University, 2000 W. University Ave., Muncie, IN 47306, 2006. Copyright © 2006.

2. Barrett Key, Kelly Neal and Scott Frazier, "The Use of Biometrics in Education Technology Assessment," Ball State University, 2000 W. University Ave., Muncie, IN 47306, 2006. Copyright © 2006.

How Keystroke Dynamics Technology Works

The idea behind keystroke dynamics has been around since World War II (see Figure 14-1) [1]. It was well documented during the war that telegraph operators on many U.S. ships could recognize the sending operator. Known as the "Fist of the Sender," the uniqueness in the keying rhythm (even for Morse code) could distinguish one operator from another [1].

Since then, many adaptations of this phenomenon have been studied. Currently, several patents exist in the field of keystroke dynamics: 4621344, 4805222, 4962530, 4998279, and 5056141 [1].

What Is Keystroke Dynamics?

As discussed, biometrics is the statistical analysis of biological observations and phenomena. Biometric measurements can be classified as physical or behavioral. Some points specific to keystroke dynamics follow:

- Typing biometrics are more commonly referred to as keystroke dynamics.

- Verification is based on the concept that how a person types is distinctive, particularly their rhythm.

- Keystroke dynamics are behavioral and evolve over time as users learn to type and develop their own unique typing pattern.

- The National Science Foundation and the National Bureau of Standards in the United States have conducted studies establishing that typing patterns are unique.

- The technique works best for users who can "touch type."

- The health and fatigue of users, however, can affect typing rhythm.

- This technology has experienced a recent resurgence with the development of software to control computer and Internet access [5].

Figure 14-1
History of keystroke dynamics. (Source: Adapted with permission from WSTA.)

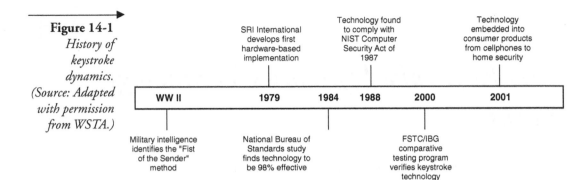

One system creates individual profiles according to how users enter their passwords, accounting for factors such as hand size, typing speed, and how long keys are held down.

Reportedly, the technology can be used with any keypad, "from computer keyboards to ATM machines to telephones."

Previously, differences in keyboards had been one of the problems that had limited the implementation of keystroke dynamics.

Detailed menus exist for designing a keystroke dynamics system [1].

Behavioral biometrics defines characteristic traits exhibited by a person that can determine identity. Measurements are considered dynamic, which results in a "confidence match." The quality of this measurement varies by behavioral as well as external factors of the subject being measured. Examples of behavioral biometrics are: handwriting, speech, language removal, gait, gesture, and typing patterns [1].

Keystroke dynamics, being a behavioral measurement, is a pattern exhibited by an individual using an input device in a consistent manner. Raw measurements already available via the standard keyboard can be manipulated to determine dwell time (the time one keeps a key pressed) and flight time (the time it takes a person to jump from one key to another). Variations of algorithms differentiate between absolute versus relative timing. The captured data is analyzed to determine aggregate factors such as cadence, content, spatial corrections, and consistency. These data are then fed through a signature processing routine, which deduces the primary (and supplementary) patterns for later verification. Signature processing is not unique to biometrics; many of these algorithms are present in actuarial sciences, from economic trending to quantum mechanics [1].

How Effective Is Keystroke Dynamics?

It is a widely held belief that physical biometrics is more effective than its behavioral counterpart. But how does keystroke dynamics fare in the realm of behavioral sciences? All biometrics are validated by several important criteria:

- False-acceptance rate (FAR)
- False-rejection rate (FRR)
- Crossover [1]

False-Acceptance Rate (FAR)

FAR determines how often an intruder can successfully bypass the biometric authentication. A lower rate is more secure; for example, an FAR of 0.01% states that the chance of fooling the system is 1:10,000 [1].

False-Rejection Rate (FRR)

FRR signifies how often a real user will not be verified successfully. A high rate translates into more user retries; hence, usability suffers [1].

Crossover

The relationship between FAR and FRR is converse, although not always linear in behavioral biometrics. Crossover is when the FAR and FRR are equal. The best technologies have the lowest crossover rate [1].

Figure 14-2 shows where keystroke dynamics falls with respect to physical biometrics (such as fingerprint) and other behavioral biometrics [1]. One critical point to note is that all behavioral biometrics, because they are confidence-based measurements, have the capability to be "tweaked" for specific applications. This allows the implementer to explicitly trade usability for security (or vice versa). Physical biometrics do not have this capability. As a consequence, the FAR and FRR for all behavioral biometrics are dynamic, and crossover can vary between implementations of the same biometric method [1].

Keystroke Dynamics in Corporate Use

The advantages to using keystroke dynamics versus other enhanced security mechanisms are twofold and rest with the end-user as well as the implementer [1].

Figure 14-2
Where keystroke dynamics falls with respect to physical biometrics and other behavioral biometrics. (Source: Adapted with permission from WSTA.)

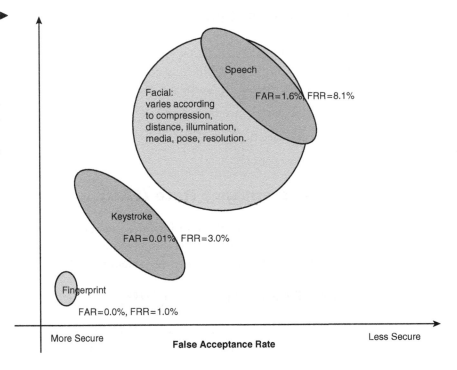

Usage/Acceptance

Biometric verification is a time-consuming operation in the computer world. Many applications for biometrics need to identify the user prior to obtaining the biometric sample, simply to limit the number of biometric templates they need to verify against. With keystroke dynamics, the identification can be in the captured sample, so verification is limited to a single template [1].

Every workstation has a keystroke dynamics input device (a.k.a., keyboard); thus, the technology can really be seen as software-only. With a software-only biometric solution, users are not limited to individual or specific workstations.As stated previously, behavioral measurements have more flexibility than conventional physical biometrics:

- Behavioral measurements accommodate different thresholds for acceptance—risk versus reward.

- Behavioral biometrics can adapt to changing behavior, for example, by merging each successful verification into the master template, thus constantly refining the accuracy of the user over time.

- Keystroke dynamics, by design, has a noninvasive user interface. It can be implemented to silently capture user typing during normal operation, thus making enrollment "invisible" [1].

Implementation/Deployment

The comparison between keystroke dynamics and other biometric solutions (software-only versus hardware-based) emerges as a strong advantage when translated into real savings from an implementation and deployment perspective. As a software-only solution:

- This technology requires no physical hardware to install, and no manpower is needed for client-side installations or upgrades;

- The technology can be embedded in any in-house software application to augment entitlement-based access;

- The implementation provides a seamless method to harden remote-access security;

- It can be wrapped into your corporation's existing single-sign-on solution as a secondary authentication mechanism;

- Simplified templates can be embedded into documents as a biometric signature (different from digital signature) and then verified from anywhere [1].

In addition, as a biometric solution:

- This technology does not require changes to existing network access policies; it more effectively enforces these policies;

- This technology provides better audit control and promotes proper use of application licensing;

- Logging of biometric access creates better forensic evidence [2] and can deter many internal threats to network security [1].

Markets for Keystroke Dynamics

Keystroke dynamics has already found its way into many areas in the past two years. For corporations, this technology has found uses in network security (single-sign-on, multipassword management, RADIUS, application access, and document control management) as well as asset identification (online training, document signing, software licensing, and PKI) [1, 3].

The future of this technology in the corporation will hardly be seen by the end users. It will be embedded into many aspects of network infrastructure, and invisibly so [1].

Finally, the consumer market has seen some integration of this technology in personal information security (individual document encryption [4], online purchase verification, and secure laptop access). The greatest growth for this technology will be seen in the consumer market. As refinements in this technology allow verification with less input and alternative forms of input (stylus, for example), it will find its way into PDAs, tablet PCs, RIM, ATMs, cellphones, and home security access pads [1].

Summary/Conclusion

In summary, the most useful inventions in history have one of two qualities. Some create an undeniable impression on their audience. It is easy to remember the beginning of the computer age and all its media attention. Others simply create no attention at all, because they extend what their audience has perceived to being there all along. How many of you can remember the exact year when video tape players also became recorders? Keystroke dynamics is one of those subtle technologies that will raise the bar on access security without users ever knowing it [1].

References

1. John C. Checco, "Keystroke Dynamics & Corporate Security," WSTA, 241 Maple Avenue, Red Bank, NJ 07701, 2006. Copyright © 2006, WSTA. All Rights Reserved.

2. John R. Vacca, *Computer Forensics: Computer Crime Scene Investigation, 2nd ed.*, Charles River Media (2005).

3. John R. Vacca, *Public Key Infrastructure: Building Trusted Applications and Web Services*, CRC Press (2005).

4. John R. Vacca, *Satellite Encryption*, Academic Press (1999).

5. John R. Vacca, *Practical Internet Security*, Springer (2006).

15

How Palm Print Pattern Recognition Technology Works

Palm print pattern recognition implements many of the same matching characteristics that have allowed fingerprint recognition to be one of the most well-known and best publicized biometrics. Both palm and finger biometrics are represented by the information presented in a friction ridge impression. This information combines ridge flow, ridge characteristics, and ridge structure of the raised portion of the epidermis. The data represented by these friction ridge impressions allows a determination that corresponding areas of friction ridge impressions either originated from the same source or could not have been made by the same source. Because fingerprints and palms have both uniqueness and permanence, they have been used for over a century as a trusted form of identification. However, palm recognition has been slower in becoming automated due to some restraints in computing capabilities and live-scan technologies. This chapter provides a brief overview of the historical progress of and future implications for palm print biometric recognition [1].

History

In many instances throughout history, examination of handprints was the only method of distinguishing one illiterate person from another, since they could not write their names. Accordingly, the hand impressions of those who could not record a name but could press an inked hand onto the back of a contract became an acceptable form of identification. In 1858, Sir William Herschel, working for the Civil Service of India, recorded a handprint on the back of a contract for each worker to distinguish employees from others who might claim to be employees when payday arrived. This was the first recorded systematic capture of hand and finger images that were uniformly taken for identification purposes [1].

The first known AFIS system built to support palm prints is believed to have been built by a Hungarian company. In late 1994, latent experts from the United States benchmarked the palm system and invited the Hungarian company to the 1995 International Association for Identification conference.

The palm and fingerprint identification technology embedded in the palm system was subsequently bought by a U.S. company in 1997 [1].

In 2004, Connecticut, Rhode Island, and California established statewide palm print databases that allowed law enforcement agencies in each state to submit unidentified latent palm prints to be searched against each other's database of known offenders. Australia currently houses the largest repository of palm prints in the world. The new Australian National Automated Fingerprint Identification System (NAFIS) includes 5.9 million palm prints. The new NAFIS complies with the ANSI/NIST international standard for fingerprint data exchange, making it easy for Australian police services to provide fingerprint records to overseas police forces such as Interpol or the FBI, when necessary [1].

Over the past several years, most commercial companies that provide fingerprint capabilities have added the capability for storing and searching palm print records. While several state and local agencies within the United States have implemented palm systems, a centralized national palm system has yet to be developed. Currently, the FBI Criminal Justice Information Services (CJIS) Division houses the largest collection of criminal history information in the world. This information primarily utilizes fingerprints as the biometric, allowing identification services to federal, state, and local users through the Integrated Automated Fingerprint Identification System (IAFIS). The federal government has allowed maturation time for the standards relating to palm data and live-scan capture equipment prior to adding this capability to the services offered by the CJIS Division. The FBI Laboratory Division has evaluated several different commercial palm AFIS systems to gain a better understanding of the capabilities of various vendors. Additionally, state and local law enforcement have deployed systems to compare latent palm prints against their own palm print databases. It is a goal to leverage those experiences and apply them toward the development of a National Palm Print Search System [1].

In April 2002, a government report on palm print technology and IAFIS palm print capabilities was submitted to the Identification Services (IS) Subcommittee, CJIS Advisory Policy Board (APB). The Joint Working Group then moved for strong endorsement of the planning, costing, and development of an integrated latent print capability for palms at the FBI CJIS Division. This effort should proceed along parallel lines as when IAFIS was developed, and should be integrated into the CJIS technical capabilities [1].

As a result of this endorsement and other changing business needs for law enforcement, the FBI announced the Next Generation IAFIS (NGI) initiative. A major component of the NGI initiative is the development of the requirements for and deployment of an integrated National Palm Print Service. Law enforcement agencies indicate that at least 30% of the prints lifted from crime scenes (knife hilts, gun grips, steering wheels, and windowpanes) are of palms, not fingers. For this reason, capturing and scanning latent palm prints is becoming an area of increasing interest among the law enforcement community. The National Palm Print Service is being developed on the basis of improving law enforcement's ability to exchange a more complete set of biometric information, making additional identifications, quickly aiding in solving crimes that formerly may have not been possible, and improving the overall accuracy of identification through the IAFIS criminal history records [1].

Approach: Concept

Palm identification, just like fingerprint identification, is based on the aggregate of information presented in a friction ridge impression. This information includes the flow of the friction ridges (Level 1 detail), the presence or absence of features along the individual friction ridge paths and their sequences (Level 2 detail), and the intricate detail of a single ridge (Level 3 detail). To understand this recognition concept, one must first understand the physiology of the ridges and valleys of a fingerprint or palm. When recorded, a fingerprint or palm print appears as a series of dark lines that represent the high, peaking portion of the friction-ridged skin, while the valley between these ridges appears as a white space and represents the low, shallow portion of the friction-ridged skin (see Figure 15-1) [1].

Figure 15-1
Fingerprint ridges (dark lines) versus fingerprint valleys (white lines).

Palm recognition technology exploits some of these features. Friction ridges do not always flow continuously throughout a pattern and often result in specific characteristics such as ending ridges or dividing ridges and dots. A palm recognition system is designed to interpret the flow of the overall ridges to assign a classification and then extract the minutiae detail—a subset of the total amount of information available, yet enough information to effectively search a large repository of palm prints. Minutiae are limited to the location, direction, and orientation of the ridge endings and bifurcations (splits) along a ridge path. Figure 15-2 presents a pictorial representation of the regions of the palm, two types of minutiae, and examples of other detailed characteristics used during the automatic classification and minutiae extraction processes [1].

Hardware

A variety of sensor types (capacitive, optical, ultrasound, and thermal) can be used for collecting the digital image of a palm surface; however, traditional live-scan methodologies have been slow to adapt to the larger capture areas required for digitizing palm prints. Challenges for sensors attempting to attain high-resolution palm images are still being dealt with today. One of the most common approaches, which employs the capacitive sensor, determines each pixel value based on the capacitance measured. This is made possible because an area of air (valley) has significantly less capacitance than an area of palm (ridge). Other palm sensors capture images by employing high-frequency ultrasound or optical devices that use prisms to detect the change in light reflectance related to the palm. Thermal scanners require a swipe of a palm across a surface to measure the difference in temperature over time to create a digital image. Capacitive, optical, and ultrasound sensors require only placement of a palm [1].

Figure 15-2
Palm print and close-up showing two types of minutiae and other characteristics.

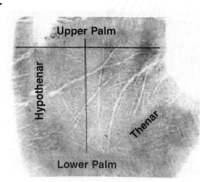
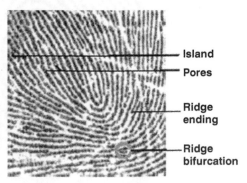

Software

Some palm recognition systems scan the entire palm, while others require the palm to be segmented into smaller areas to optimize performance. Reliability within either a fingerprint or palm print system can be maximized by searching smaller data sets. While fingerprint systems often partition repositories based upon finger number or pattern classification, palm systems partition their repositories based upon the location of a friction ridge area. Latent examiners are very skilled in recognizing the portion of the hand from which a piece of evidence or latent lift has been acquired. Searching only this region of a palm repository rather than the entire database maximizes the reliability of a latent palm search [1].

Like fingerprints, the three main categories of palm matching techniques are minutiae-based matching, correlation-based matching, and ridge-based matching. Minutiae-based matching, the most widely used technique, relies on the minutiae points previously described, specifically the location, direction, and orientation of each point. Correlation-based matching involves simply lining up the palm images and subtracting them to determine if the ridges in the two palm images correspond. Ridge-based matching uses ridge pattern landmark features such as sweat pores, spatial attributes, geometric characteristics of the ridges, and/or local texture analysis, all of which are alternates to minutiae characteristic extraction. This is a faster method of matching and overcomes some of the difficulties associated with extracting minutiae from poor-quality images [1].

The advantages and disadvantages of each approach vary based on the algorithm used and the sensor implemented. Minutiae-based matching typically attains higher recognition accuracy, although it performs poorly with low-quality images and does not take advantage of textural or visual features of the palm. Processing using minutiae-based techniques may be time-consuming because of the time associated with minutiae extraction. Correlation-based matching is often quicker to process but is less tolerant to elastic, rotational, and translational variances and noise within the image. Some ridge-based matching characteristics are unstable or require a high-resolution sensor to obtain quality images. The distinctiveness of the ridge-based characteristics is significantly lower than the minutiae characteristics [1].

United States Government Evaluations

Unlike several other biometrics, a large-scale government-sponsored evaluation has not been performed for palm recognition. The amount of data currently available for test purposes has hindered the ability for not only the federal government but also the vendors in efficiently testing and benchmarking commercial palm systems. The FBI laboratory is currently encoding its hard-copy palm records into three of the most popular commercial palm recognition systems. This activity, along with other parallel activities needed for establishing a National Palm Print Service, will address these limitations and potentially provide benchmark data for U.S. government evaluations of palm systems [1].

Standards Overview

Just as with fingerprints, standards development is an essential element in palm recognition because of the vast variety of algorithms and sensors available on the market. Interoperability is a crucial aspect of product implementation, meaning that images obtained by one device must be capable of being interpreted by a computer using another device. A major standards effort for palm prints currently under way is the revision to the ANSI NIST ITL-2000 Type-15 record. Many, if not all, commercial palm AFIS systems comply with the ANSI NIST ITL-2000 Type-15 record for storing palm print data. Several recommendations to enhance the record type are currently being "vetted" through workshops facilitated by the National Institute for Standards and Technology. Specifically, enhancements to allow the proper encoding and storage of major case prints, which essentially are any and all friction ridge data located on the hand, are being endorsed to support the National Palm Print Service initiative of NGI [1].

Summary/Conclusion

Even though total error rates are decreasing when comparing live-scan enrollment data with live-scan verification data, improvements in matches between live-scan and latent print data are still needed. Data indicates that fully integrated palm print and fingerprint multibiometric systems are widely used for identification and verification of criminal subjects as well as in security access applications. But there remain significant challenges in balancing accuracy with system cost. Image-matching accuracy may be improved by building and using larger databases and by employing more processing power, but then purchase and maintenance costs will most certainly rise as the systems become larger and more sophisticated. Future challenges require balancing the need for more processing power with more improvements in algorithm technology to produce systems that are affordable to all levels of law enforcement [1].

Reference

1. "Palm Print Recognition," Office of Science & Technology Policy (OSTP), NSTC Subcommittee on Biometrics, National Science & Technology Council (NSTC) Committees on Technology and Homeland & National Security, Washington, DC, March 27, 2006.

16

How Vein Pattern Analysis Recognition Technology Works

Vascular pattern recognition, also commonly referred to as vein pattern authentication, is a fairly new biometric in terms of installed systems. Using near-infrared light, reflected or transmitted images of blood vessels of a hand or finger are derived and used for personal recognition. Different vendors use different parts of the hand, palms, or fingers, but rely on a similar methodology. Researchers have determined that the vascular pattern of the human body is unique to a specific individual and does not change as people age. Claims for the technology include that it:

- Is difficult to forge: Vein patterns are difficult to re-create because they are inside the hand and, for some approaches, blood needs to flow to register an image.

- Is contact-free: Users do not touch the sensing surface, which addresses hygiene concerns and improves user acceptance.

- Has many and varied uses: It is deployed in ATMs, hospitals, and universities in Japan. Applications include ID verification, high-security physical access control, high-security network data access, and point-of-sale access control.

- Is capable of 1:1 and 1:N matching: Users' vascular patterns are matched against personalized ID cards/smart cards or against a database of many scanned vein patterns [1].

History

Potential for the use of this technology can be traced to optical trans-body imaging and potential optical CT scanning applications [5]. The use of vein patterns for biometric recognition is a technology that uses the subcutaneous blood vessel pattern in the back of the hands; it became the first commercially available vascular pattern recognition system in 2000. Additional research has further improved the technology. The introduction of this technology inspired additional research and commercialization into finger- and palm-based systems (see sidebar, "Palm Vein Recognition").

Palm Vein Recognition

Fujitsu Laboratories Ltd., based in Tokyo, Japan, rolled out what they are billing as "the world's first contactless biometric authentication system" that can verify a person's identity by recognizing the pattern of blood veins in the palm (see Figure 16-1) [2]. Like fingerprints, the pattern of blood veins in the palm is unique to every individual and, apart from size, this pattern will not vary over the course of a person's lifetime. The fact that this pattern lies under the skin makes it that much harder for others to read, so palm-vein pattern biometrics are an especially secure method of verification. In contrast to contact-based biometric technology, this new contactless technology alleviates concerns about hygiene associated with having many people touching the same sensor device, making it ideal for widespread use.

Figure 16-1
Contactless biometric authentication system. (Source: Reproduced with permission from Biometric Watch.)

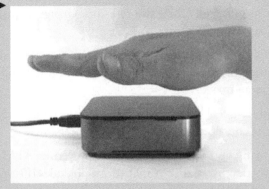

Vein recognition was first developed by Joseph Rice. In 1984 he had his identity stolen, leading to fraudulent use of his bank account. He decided to do something about it, which led to his first vein recognition prototype around 1985. He constructed the first unit in his kitchen using light-emitting diodes and photodiodes and a BBC-B computer. Rice's patents expired in 2004, and he does see plenty of future potential for vein recognition technology. The future of biometrics and particularly the vein biometric lies in personal biometric authenticators that people own and wear, and use to access their homes, transportation, and possessions and to digitally sign for their transactions and transmissions.

The Fujitsu Group is not new to biometrics. They have developed biometric authentication technologies based on fingerprints, voice, facial features, and vein patterns in the palm, and have also combined two or more of these capabilities in multibiometric authentication systems. Fujitsu maintains that for biometrics to gain wider acceptance, it needs to be considered less intrusive, and concerns about hygiene need to be addressed. Their focus in developing this new sensor thus was to address a market need.

Fujitsu's new technology is a combination of a device that can read the pattern of blood vein patterns in the palm without making physical contact and software that can authenticate an individual's identity based on these patterns. Infrared light is used to capture an image of the palm as the hand is held over the sensor device. The software then extracts the vein pattern and compares it against patterns already stored in the database. The palm floats in mid-air with contactless-type systems, and there are no restrictions as to height and side positioning of the palm to a reader.

The technology works by using infrared light to scan for hemoglobin found in our blood. The veins absorb the infrared rays, and on the resulting image, shows up as black, as shown in Figure 16-2 [2]. The rest of the hand structure appears as white.

Figure 16-2
The veins absorb the infrared rays and, on the resulting image, show up as black. (Source: Reproduced with permission from Biometric Watch.)

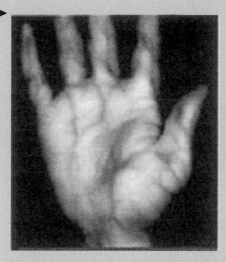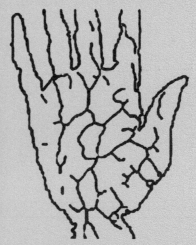

The scanner took two years for Fujitsu's engineers to develop (see Figure 16-3) [2]. Fujitsu newly developed a technology that, even if a sensor device is installed in several different locations, can detect palm position and verify palm vein patterns at high speeds and reliability, as well as a technology that can optimally control the lightning and the capturing of a variety of palm positions.

Tested in 2003 with the cooperation of 700 people aged 10 to 70 from different walks of life, a total of 1,400 palm profiles were collected. In terms of authentication precision, Fujitsu reports the system had a false rejection rate of 1% and a false acceptance rate of 0.5%, and had an equal error rate of only 0.8%. The sensor device used in this system can be embedded in a variety of equipment. Embedded in a wall, it could be used for access control to secure areas. Integrated into electrical equipment, such as personal digital assistants, it could be used to authorize user access to the device. In public spaces or medical facilities where hygiene is a particular concern, the contactless feature of this system may make it especially appropriate [2].

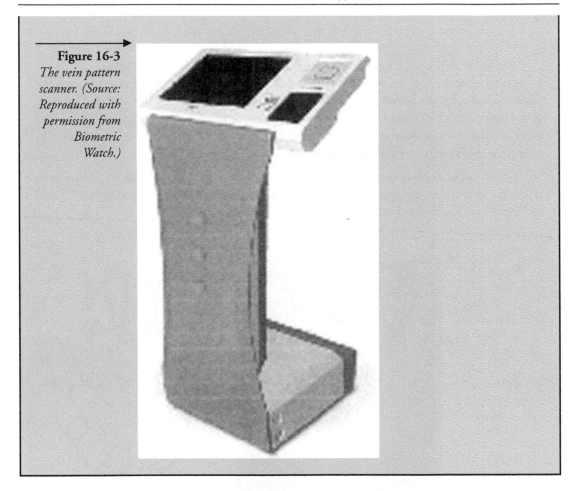

Figure 16-3
The vein pattern scanner. (Source: Reproduced with permission from Biometric Watch.)

Approach: Vascular Pattern in the Back of Hands

Near-infrared rays generated from a bank of light-emitting diodes (LEDs) penetrate the skin of the back of the hand. Due to the difference in absorbance of blood vessels and other tissues, the reflected near-infrared rays produce an image on the sensor. The image is digitized and further processed by image-processing techniques producing the extracted vascular pattern. From the extracted vascular pattern, various feature data such as vessel branching points, vessel thickness, and branching angles are extracted and stored as the template [1].

Vascular Pattern in Fingers

The basic principle of this technology is shown in Figures 16-4 and 16-5 [1]. Near-infrared rays generated from a bank of LEDs penetrate the finger or hand

Figure 16-4
Transmittance images of a hand. (Source: Office of Science and Technology Policy (OSTP).)

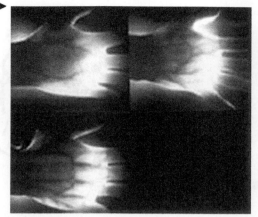

Figure 16-5
Principle of transmittance imaging. (Source: Office of Science and Technology Policy (OSTP).)

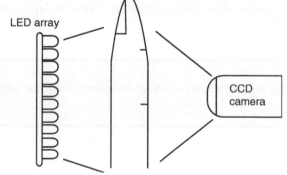

and are absorbed by the hemoglobin in the blood. The areas in which the rays are absorbed (veins) appear as dark areas, similar to a shadow in an image taken by a CCD camera. Image processing can then construct a vein pattern from the captured image. Next, this pattern is digitized and compressed so that it can be registered as a template [1].

Vein biometric systems record subcutaneous infrared (IR) absorption patterns to produce unique and private identification templates for users. The technology is a vascular "barcode" reader for people. Veins and other subcutaneous features present large, robust, stable, and largely hidden patterns. Subcutaneous features can be conveniently imaged within the wrist, palm, and dorsal surfaces of the hand [3].

Vein pattern IR gray scale images are binarized, compressed, and stored within a relational database of 2D vein images (see Figure 16-6) [3]. Subjects are verified against a reference template [3].

Figure 16-6
*Vein pattern IR
gray scale images.
(Source:
Reproduced with
permission from
Biometric
Watch.)*

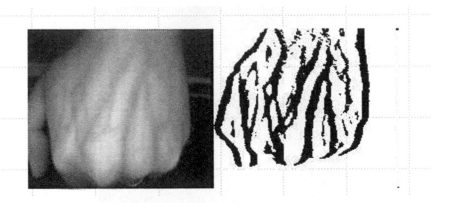

The technology can be applied to small personal biometric systems (biowatches and biokeys) and to generic biometric applications including intelligent door handles, door locks and so on. Banks in Japan are now using this technology [3].

Note: The U.S. government has not performed technology evaluations of vascular pattern recognition biometrics at this time.

Summary/Conclusion

Vein pattern recognition has gained sponsorship from companies that have developed reputations for developing products that compete successfully in global markets (see sidebar, "First Laptop with Built-In Vein Scanner"). There appears to be some testing and validation by third parties. Standards work will need to be accomplished before this technology can grow to broader acceptance [1].

First Laptop with Built-In Vein Scanner

In Japan, Hitachi has unveiled its Lora SE210 security laptop computer featuring a biometric security device that uses vein recognition. The security device is placed below the keyboard and uses infrared light to scan a finger's tissue for vein patterns.

The laptop is the first mobile computer using this technology and targets users handling sensitive or secure information. It is designed to be used as a networked client. It has no hard drive, but uses flash memory to store the Windows XP and 2003 embedded operating systems.

Biometric security devices, including vein recognition, fingerprint readers, and iris scanners, use unique body features for authentication. While vein recognition is a less common form of biometric security, the technology's backers claim that it is easy to perform a vein scan and that it is very hard to damage or change the pattern. The technology also offers an advantage over fingerprint scanners because identical twins have highly similar (although not identical) fingerprints but different vein patterns.

Vein scanning technology is already used in Japan on ATM machines and USB vein scanners are now available on the market. Hitachi also showed off vein recognition in a door handle that would enable the owner to open his or her car without a key.

The Lora SE210 features a 600 MHz Intel Celeron processor and 256 Mb of memory. It sells in Japan for ¥185,000 without tax ($1,550). A model without the security device is available for ¥160,000 [4].

References

1. "Vascular Pattern Recognition," Office of Science & Technology Policy (OSTP), NSTC Subcommittee on Biometrics, National Science & Technology Council (NSTC) Committees on Technology and Homeland & National Security, Washington, DC, March 27, 2006.

2. Dave Mintie, "Contactless Palm Vein Pattern Biometric Authentication System," Biometric Watch, PO Box 3707, Milford, CT 06460, April 30, 2005.

3. Joseph Rice, "Vein Recognition," Biometric Watch, PO Box 3707, Milford, CT 06460, March 12, 2005.

4. Tom Sanders, "Hitachi Points to Vein Recognition," VNU Business Publications, November 30, 2005. Copyright © 1995–2006. All Rights Reserved.

5. John R. Vacca, *Optical Networking Best Practices Handbook*, John Wiley & Sons (2006).

17

How Ear-Shape Analysis Technology Works

The ear is a viable new class of biometrics since it has desirable properties such as universality, uniqueness, and permanence. For example, the ear is rich in features; it is a stable structure that does not change with age; it doesn't change its shape with facial expressions, cosmetics, and hairstyles. Although it has certain advantages over other biometrics, the ear has received little attention compared to other popular biometrics such as face, fingerprint, and gait. Current research has used intensity images; therefore, the performance of the system is greatly affected by imaging problems such as lighting and shadows. Range sensors that are insensitive to imaging problems can directly provide 3D geometric information. Therefore, it is desirable to design a human ear recognition system from 3D side-face range images obtained at a distance. Human ear detection is the first task of a human ear recognition system, and its performance affects the overall quality of the system [1].

This chapter proposes a simple ear-shape model-based technique (see sidebar, "New Type of Ear-Shape Analysis") for locating human ears in side-face range images. The ear-shape model is represented by a set of discrete 3D vertices corresponding to ear helix and antihelix parts. Since the two curves formed by ear helix and antihelix parts are similar for different people, the small deformation of two curves between different persons is not taken into account, which greatly simplifies the model. Given side-face range images, step edges are extracted; then the edge segments are dilated, thinned, and grouped into different clusters, which are potential regions containing ears. For each cluster, the ear-shape model with the edges is registered. The region with the minimum mean registration error is declared as the detected ear region; the ear helix and antihelix parts are meanwhile identified [1].

New Type of Ear-Shape Analysis

A new type of ear-shape analysis could see ear biometrics surpass face recognition as a way of automatically identifying people. The technique could be used to identify people from CCTV footage or be incorporated into cellphones to identify the user.

Ears are remarkably consistent. Unlike faces, they do not change shape with different expressions or age, and remain fixed in the middle of the side of the head against a predictable background. Hair is a problem. But that might be solved by using infrared images.

In an initial small-scale study Involving 63 subjects (all taken from a database of face profiles), researchers from the University of Southampton found their method to be 99.2% accurate. This is a great starting point, but in theory the method could be greatly improved. There are more fixed features available in an ear than the researchers have been measuring.

Order of Magnitude

Much larger populations are needed to determine how reliably the technology could be implemented. But an initial analysis of the decidability index (a measure of how similar or dissimilar each of the ears were) indicates how unique an individual ear might be.

The researchers found that this index was an order of magnitude greater than for face analysis, but not as large as for iris biometrics. Ears have been used to identify people before now, but previous methods have used an approach similar to face recognition. This involves extracting key features, such as the position of the nose and eyes or, in the case of the ear, where the channels lie. These are then represented as a vector, describing where features appear in relation to each other. The new approach instead captures the shape of the ear as a whole and represents this in code, allowing the whole ear shape to be compared.

Ear Print

But despite the promising results, researchers may have a job convincing people. In 1998 an ear print left on a window led to the conviction of Mark Dallagher for murdering a 94-year-old woman.

This conviction, however, was overturned in January 2004 because the evidence relating to the ear print was found to be flawed. But researchers note that the original evidence was not, strictly speaking, biometric—it relied on a subjective opinion of an ear expert [2].

Related Work

Very little research has been done in dealing with object detection from range images. This part of the chapter gives a brief review of object detection techniques from range images [1].

There does exist a template matching a based detection method for extracting ears from side-face range images. The model template is represented by an average histogram of shape index of ears. However, this method cannot identify the ear region accurately [1].

There also exists a unique signature of the 3D object by the Fourier transform of the phase-encoded range image at each specific rotation. The signature defined in a unit sphere permitted the detection of 3D objects by correlation techniques [1].

There is another method for segmenting temporal sequences of range and intensity images. The fusion of range and intensity data for segmentation was solved by clustering 4D intensity/position features. Kalman filters were then used to stabilize tracking by predicting dynamic changes in cluster positions [1].

Furthermore, there is a method to detect lanes and classify street types from range images. First, you calculate the lane's width, curvature, and relative position to the car, compare those to a prior knowledge of construction rules of different street types, and finally achieve street type based on the mean value of the lane's width [1].

There also exists a fuzzy logic system for automatic target detection from LADAR images. Two fuzzy logic detection filters are used and one statistical filter to create pixel-based target confidence values, which are fused by the fuzzy integral to generate potential target windows. Features extracted from these windows were fed to a neural network post-processor to make a final decision [1].

Finally, there is a method to separate image features into ground and road obstacles by assuming the road is flat. Obstacles and road pixels are distinguished by using the separating plane. The plane model is updated by fitting a plane through all pixels marked as ground. Connected component analysis is used to partition detected obstacles into different objects [1].

Motivation

The anatomical structure of the ear is shown in Figure 17-1 [1]. The ear is made up of standard features. These include the outer rim (helix) and ridges (antihelix) parallel to the helix, the lobe, and the concha, which is the hollow part of the ear. From Figure 17-1 [1], you can clearly see that two curves formed by the ear helix and antihelix parts are easily identified. You can use these two curves to guide the procedure to locate the ear in side-face range images [1].

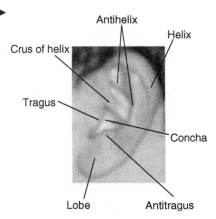

Contributions

The contributions of this chapter are a proposal for an ear-shape model for locating 3D ears in side-face range images. Next is the development of an effective approach to detect human ears from side-face range images [1].

Technical Approach: Ear-Shape Model Building

Considering the fact that the curves formed by the ear helix and antihelix parts are similar for different people, the ear-shape model from one person only is constructed in this chapter. The plan here is to work on building a generic ear model from multiple persons. You need to extract ear helix and antihelix parts by running a step edge detector with different thresholds, to choose the best extraction result, and to do the edge thinning. By running a connected component labeling, you can extract the edges that correspond to ear helix and antihelix parts. You also need to define the ear-shape model s as 3D coordinates $\{x, y, z\}$ of n vertices that lie on the ear helix and antihelix parts. The s is represented by a $3n \times 1$ vector $\{x_1, y_1, z_1, x_2, y_2, z_2, , x_n; y_n; z_n\}$. Figure 17-2 shows the 3D side-face range image with textured appearance, over which the ear-shape model s that is marked by gray vertices is overlaid [1].

Step Edge Detection and Thresholding

One example of the step edge magnitude image is shown in Figure 17-3(b) [1]. In Figure 17-3(b) [1], larger magnitudes are displayed as brighter pixels. You can clearly see that most of the step edge magnitudes are small values. To get

Figure 17-2
The textured 3D face and overlaid ear-shape model. (Source: Reproduced with permission from University of California at Riverside.)

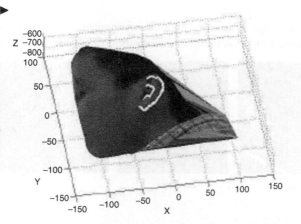

Figure 17-3 *(a) Original side-face range image. (b) The step edge magnitude image. (c) The step edge image. (Source: Reproduced with permission from University of California at Riverside.)*

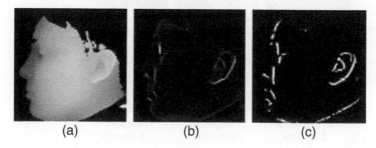

edges, the step edge magnitude image must be segmented using a threshold operator. The selection of threshold value is based on the cumulative histogram of the step edge magnitude image. You can easily determine the threshold by investigating the cumulative histogram. The thresholded binary image is shown in Figure 17-3(c) [1], while the original side-face range image is shown in Figure 17-3(a) [1].

Edge Thinning and Connected Component Labeling

Since some step edge segments are broken, you should dilate the binary image to fill the gaps. The dilated image is shown in Figure 17-4(a). After doing edge thinning, the resulting image is shown in Figure 17-4(b). The edge segments are labeled by running the connected component-labeling algorithm, and some small edge segments (less than 15 pixels) are removed. The left edge segments are shown in Figure 17-4(c) [1].

Figure 17-4   *(a) Dilated edge image. (b) Thinned edge image. (c) Left edge segments. (Source: Reproduced with permission from University of California at Riverside.)*

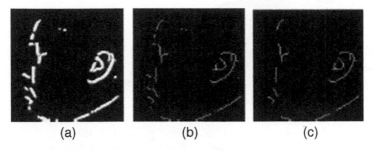

(a)   (b)   (c)

Figure 17-5
Examples of edge clustering results. (Source: Reproduced with permission from University of California at Riverside.)

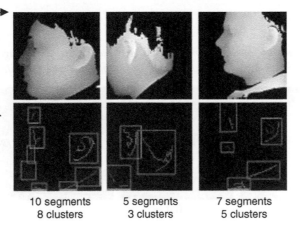

10 segments   5 segments   7 segments
8 clusters   3 clusters   5 clusters

Clustering Edge Segments

Since the ear region contains several edge segments, you need to group the edge segments that are close to each other into different clusters. Three examples of clustering results are shown in the second row of Figure 17-5 [1], where each cluster is bounded by a rectangular box. The first row of Figure 17-5 shows side-face range images [1].

Locating Ears by Use of the Ear-Shape Model

For each cluster obtained, as discussed previously, you should extract step edges around the ear helix and antihelix parts. The problem when locating ears is to minimize the mean square error between the ear-shape model vertices and their corresponding edge vertices in the bounded rectangular box [1].

The iterative closest point (ICP) algorithm is a well-known method for aligning 3D shapes. However, ICP requires that every point in one set should have a corresponding point on the other set. You can't guarantee that the edge vertices in the potential regions satisfy this requirement. Therefore, you need to use a modified ICP algorithm to register the ear-shape model with the edge vertices. The steps of a modified ICP algorithm to register a test shape Y to a model shape X are:

1.   Initialize the rotation matrix R_0 and translation vector T_0.

2.   Given each point in Y, find the closest point in X.

3.   Discard pairs of points that are too far apart.

4.   Find the rigid transformation (R, T) such that E is minimized.

5.   Apply the transformation (R, T) to Y.

6.   Report from step (2) until the difference $|E_k - E_{k-1}|$ in two successive steps falls below a threshold or the maximum number of iterations is reached [1].

By initializing the rotation matrix R_0 and the translation vector T_0 to the identity matrix and the difference of centroids of two vertex sets respectively, you should run the ICP iteratively. Finally, you should get the rotation matrix R and translation vector T, which brings the ear-shape model vertices and edge vertices into alignment. The cluster with the minimum mean square error is declared as the detected ear region. Meanwhile, the ear helix and antihelix parts are identified [1].

Experimental Results: Data Acquisition

You should use real range data acquired by Minolta Vivid 300. During the acquisition, you should have 52 subjects sit on a chair at about 0.55–0.75 m from the camera. The first shot should be taken when the subject's left-side face is approximately parallel to the image plane; two shots should be taken when the subject is asked to rotate his or her head to the left and right side within ±35° with respect to his or her torso. The same acquisition procedure should be repeated. Six images per subject should be recorded. Therefore, you then have 312 images in total. Each range image contains 200×200 grid points, and each grid point has a 3D coordinate (x; y; z). The ear-shape model is built from

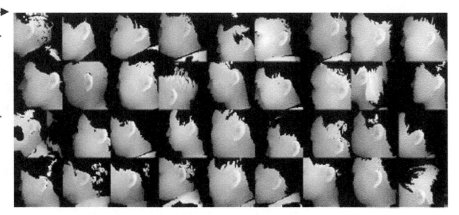

a side-face range image described earlier. Examples of side-face range images
are shown in Figure 17-6 [1].

Results

Finally, you should test the proposed detection method on the 312 side-face
range images. If the ear-shape model is aligned with the ear helix and antihelix
parts, you should classify it positive detection; otherwise, false detection. In
various experiments, the number of vertices of the ear-shape model has been
113; the average number of edge segments has been six, and the average number
of clusters has been four. The average time to detect an ear from a side-face
range image is 6.5 seconds with Matlab implementation on a 2.4G Celeron
CPU. Examples of positive detection results are shown in Figure 17-9 [1].
In Figure 17-9, the transformed ear-shape model marked by light points is
superimposed on the corresponding textured 3D face. From Figure 17-9, you
can observe that the ear is correctly detected and the ear helix and antihelix parts
are identified from side-face range images. The distribution of mean square error
for positive detection is shown in Figure 17-7 [1]. The mean of mean square
error is 1:79 mm. You should achieve a 92.6% detection rate. For the failed
cases, notice that there are some edge segments around the ear region caused
by hair, which brings more false edge segments or results in the cluster that
cannot include the ear helix and antihelix parts. Since the ICP algorithm cannot
converge due to the existence of outliers, false detection happens, as shown in
Figures 17-8 and 17-10 [1]. The original face-range images and corresponding
edge clusters are shown in Figure 17-8 [1]. In this figure, the first row shows face
images; the second row shows edge clustering results. The textured 3D faces
with the overlaid detected ear helix and antihelix are shown in Figure 17-10 [1].

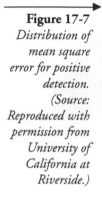

Figure 17-7
Distribution of mean square error for positive detection. (Source: Reproduced with permission from University of California at Riverside.)

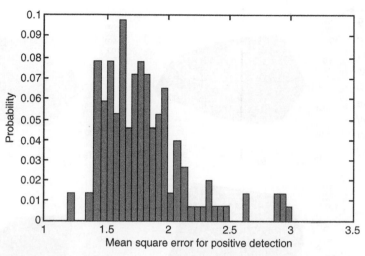

Figure 17-8
Examples of failed cases. (Source: Reproduced with permission from University of California at Riverside.)

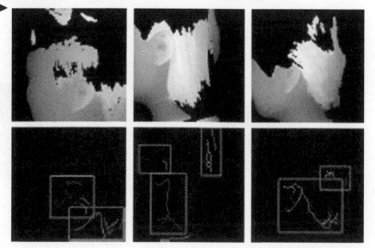

Summary/Conclusion

Ear detection is an important part of an ear recognition system. This chapter proposes a shape model–based technique for locating human ears in side-face range images. The ear-shape model is represented by a set of discrete 3D vertices corresponding to the ear helix and antihelix parts. Given side-face range images, step edges are extracted, considering the fact that there are strong step edges around the ear helix part. Then the edge segments are dilated, thinned, and grouped into different clusters, which are potential regions containing ears.

Figure 17-9
Examples of positive detection results. (Source: Reproduced with permission from University of California at Riverside.)

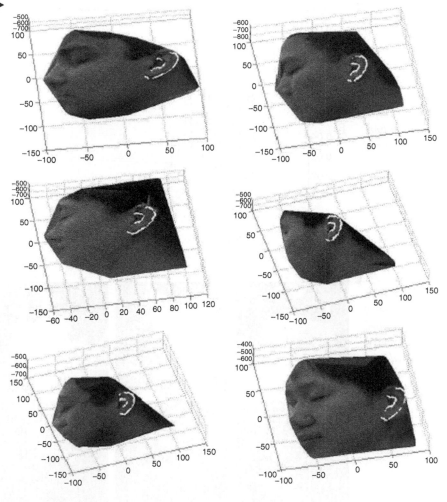

Figure 17-10 *Examples of false detection results. (Source: Reproduced with permission from University of California at Riverside.)*

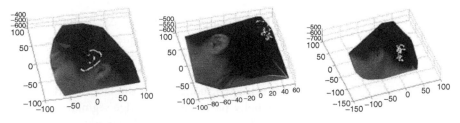

For each cluster, you should register the ear-shape model with the edges. The region with the minimum mean registration error is declared as the detected ear region. Meanwhile, the ear helix and antihelix parts are identified. Experiments are performed with a large number of real-face range images to demonstrate the effectiveness of the approach. The contributions of this chapter are:

1.   An ear-shape model for locating 3D ears in side-face range images;

2.   An effective approach to detect human ears from side-face range images;

3.   Experimental results on a large number of ear images [1].

References

1. Hui Chen and Bir Bhanu, "Shape Model–Based 3D Ear Detection from Side Face Range Images," Center for Research in Intelligent Systems, University of California, Riverside, CA 92521, 2006.

2. Duncan Graham-Rowe, "Ear Biometrics May Beat Face Recognition," New Scientist, Lacon House, 84 Theobald's Road, London WC1X 8NS, July 14, 2005. Copyright © 2006, Reed Business Information Ltd.

18

How Body Odor and/or Scent Analysis Technology Works

Lost in the July 2005 London bombings, along with innocent lives, was any illusion that today's surveillance technology can save one from evildoers. Britain has 6 million video cameras monitoring streets, parks, and government buildings, more than any other country. London alone has 700,000 cameras watching for signs of illicit activity. Studying camera footage helped link the bombings with four men—but only after the fact. The disaster drove home some painful reminders: Fanatics bent on suicide aren't fazed by cameras. And even if they are known terrorists, most video surveillance software won't pick them out anyway [1].

Tomorrow's surveillance technology may be considerably more effective. But, each uptick in protection will typically come at the cost of more intrusion into the privacy of ordinary people [3]. For now, the public seems to find that trade-off acceptable, so scientists around the world have intensified efforts to perfect the art of surveillance, hoping to catch villains before they strike [1].

Research laboratories envision tools that could identify and track just about every person, anywhere, and sound alarms when the systems encounter hazardous objects or chemical compounds. Many such ideas seem to leap from the pages of science fiction: An artificial nose in doorways and corridors sniffs out faint traces of explosives on someone's hair. Tiny sensors floating in reservoirs detect a deadly microbe and radio a warning. Smart cameras identify people at a distance by the way they walk or the shape of their ears. And a little chemical lab analyzes the sweat, body odor, and skin flakes in the human thermal plume—the halo of heat that surrounds each person [1].

All of these projects have been on a fast track since September 11, 2001. Meanwhile, consumer demand is speeding their development by lowering the cost of the underlying technologies. Camera phones, "nanny cams," and even satellite photos are commonplace. Biological sensors are flooding into households in the form of tests for HIV, pregnancy, and diabetes (some of which can relay data to a doctor), and soon there will be far more sensitive DNA-based tests (see Chapter 19). Next up are radio-frequency identification (RFID) tags. They're showing up in stores to help track inventory, and 500 people in the

United States have had them implanted under their skin to broadcast their ID and medical data in case of an emergency [1].

Together, these developments herald a high-tech surveillance society that not even George Orwell's book *1984* could have imagined—one in which virtually every advance brings benefits as well as intrusions. Rapid DNA-based probes, for example, could help protect you from bioweapons and diagnose diseases, but they might also reveal far too much about you to health insurers or prospective employers. The trade-offs are uncomfortable, in part, because corporations and governments will continue to wield the most advanced surveillance systems. But ordinary citizens will also gain capabilities to monitor their surroundings with consumer technologies, from webcams to Internet search and tracking tools, allowing the watched to observe other watchers [1].

One great worry is that those who stand out from the norm or express unpopular views—minorities, the poor, or just the ill-mannered—may get stomped in new and surprising ways. A recent incident in South Korea shows how this can play out. A subway commuter posted on the Internet some cellphone photos he took of a passenger who had refused to clean up after her dog, which had relieved itself during the ride. In no time, a vigilante mob on the Web identified her by her face and the purse she was carrying, and she became the object of national villification. You can move into a surveillance society one tiny camera at a time [1].

If terrorism becomes endemic in Europe and the United States, emerging surveillance tools may be abused in even more egregious ways. At the same time, the overhead burdens of a police state, from the dossier-building to the endless security checkpoints, could impose crippling costs on a free-market economy. Witness the U.S. clampdown on foreign student visas, which could end up crimping universities' ability to do advanced research. In other words, you could bankrupt yourself, much like the Soviet Union did [1].

Experts disagree about when the most visionary tools to thwart terrorist acts will arrive on the market—and whether they will deliver on their promise. Sensors that can detect bombs, radiation, and toxins exist today, and will be far more sophisticated a decade from now. But strewing them across every U.S. city would cost untold billions of dollars. High-tech electronic eavesdropping on communications networks can be effective, but only if terrorists use telecom systems. And even with improvements in cameras, biometric devices such as iris scans, bomb sniffers, and tracking software, it will be years before these technologies can pick a terrorist out of a crowd. In short, the march toward a surveillance society may be inevitable, but no simple cost-benefit equation can assure you that the sacrifices will be worth it. We'll be debating the point for decades to come [1].

So, while keeping the preceding in mind, a problem of human recognition through the odor authentication is presented in this chapter. This is the prospective technique, and it is still under development. There are no available commercial applications on the market yet. While human odor recognition is not available, odor recognition is widely used nowadays. Odor recognition is realized by electronic noses (ENoses). The main components of ENoses are considered in this chapter. Both types of applications, current and future, are analyzed [6].

Body Odor

In the quest to sort bad guys from good, scientists are poking ever more intimately at the core of each person's identity—right down to DNA. One day, people's distinctive body odor (see sidebar, "Body Odor Recognition"), breath, or saliva could serve as an identifier, based on the subtle composite of chemicals that make up a person's scent or spit. One's smell is a cocktail of hundreds of molecules. The question is whether it's a gin and tonic or a margarita. While some of these sensors perform well in the lab, the real world may be different: The technology is still in its infancy [1].

Body Odor Recognition

Each unique human smell is made up of chemicals knows as volatiles. These can be converted into a template by using sensors to capture body odor from nonintrusive parts of the body such as the back of the hand.

Body odor recognition is a contactless physical biometric that attempts to confirm a person's identity by analyzing the olfactory properties of the human body scent. According to the University of Cambridge (http://www.cam.ac.uk), the sensors that have been developed are capable of capturing body scent from nonintrusive body parts, such as the hand. Each chemical of the human scent is extracted by the biometric system and converted into a unique data string.

Science today is hard put to identify smells that a beagle could nail in an instant. There is a set of underlying odors in people independent of perfume and what they ate that day. But surveillance is just one objective. The more immediate goal is to use a biochemical understanding of human odor to diagnose diseases. Specific chemicals are associated with certain illnesses—carbon disulfide with some forms of mental illness, for instance, and nitric oxide with cancer [1].

Significance of Olfaction (Smell)

Olfaction has an extremely high importance for human beings. It is one of the five main senses: sight, smell, taste, hearing, and touch. Philosophers and scientists have been trying to comprehend the sense of smell for several thousand years. It is a difficult task, because people often have a problem with finding words even to describe their smell sensations. However, odorants influence your life deeply through mood. Reactions like discomfort, attraction, and other sensations are hard to extinguish since neurons of the nose are connected straight to a part of the brain called the olfactory bulb. The olfaction mechanism is still unknown [6].

The main problem associated with odor perception is that there is no physical continuum, like sound frequency in hearing or Newton's circle in color vision. From this point of view, stimuli based only on the intuitive experience must be chosen. Therefore, there is absolutely no guarantee that the chosen stimuli will span the whole olfactory perception space. It is possible to say that there are no tests to appraise the quality of smell during experiments [6].

The main purpose in human odor recognition is to try to build an electronic system that is as sensitive as possible. This kind of electronic system is assumed to be created on the human olfactory model. Thus, before you create this device, the human olfactory model must be thoroughly comprehended [6].

Human Olfactory Model

Anything that has an odor constantly evaporates tiny quantities of molecules that produce the smell of so-called odorants. A sensor that is capable of detecting these molecules is called a chemical sensor. In this way, the human nose is a chemical sensor and smell is a chemical sense [6].

The human's ability to smell is not impressive in comparison with animals. A human brain devotes only 4–8 cm^2 to the entire olfactory apparatus. At the same time, a dog uses 65 cm^2 and a shark utilizes 2–3 m^2 [6].

Despite our inferiority, a human has about 40 million olfactory nerves. This allows detecting slight traces of some chemical components. Some odorants can be detected even if the concentration in the air is only one part per trillion [6].

Odor information processing in a human model is a tremendously complicated task. Humanity knows much about the functional characteristics and structure of the brain and can comprehend at least some of its information-processing mechanisms. However, overall dynamical properties of the brain are still unknown. To catch the behavior of the olfactory system, it can be helpful to understand how other parts are involved [6].

Different methods have been used to understand olfaction. In the olfactory bulbs, each neuron participates in the generation of olfactory perception, and no one receptor type alone identifies a specific odor. Thus, the main operations of olfaction can be divided roughly in five parts: sniffing, reception, detection, recognition, and cleansing [6].

Sniffing

The olfaction begins with sniffing, which mixes the odorants into a uniform concentration and delivers these mixtures to the mucous layer in the upper part of nasal cavity. Next, the molecules are dissolved in this layer and transported to the cilia of the olfactory receptor neurons [6].

Reception

The reception process includes binding of these odorant molecules to the olfactory receptors. Odorant molecules are temporarily bound to proteins that transport molecules across the receptor membrane with the simultaneous stimulation of the receptors. During this stimulation, the chemical reaction produces an electrical stimulus [6].

Detection

These electrical signals from the receptor neurons are transported to the olfactory bulb. From the olfactory bulb, the receptor response information is forwarded to the olfactory cortex (detection) [6].

Recognition

The odor recognition part takes place in the olfactory cortex. Then, the information is transmitted to the cerebral cortex. Keep in mind that there are no individual receptors or parts of the brain capable of recognizing specific odors. The brain is the key component associated with the collection of olfactory signals with a specific odor [6].

Cleansing

Cleansing finishes the olfaction process. For this purpose, the breathing of fresh air that removes the odorant molecules from the olfactory receptors is required [6].

To grasp the mechanism of the olfactory perception, a model of your nose can be considered. A schematic view of the human nose is presented in Figure 18-1 [6].

As seen in Figure 18-1, inside each side of the nose is an air chamber, the nasal cavity [6]. Air, including odorants, is inhaled through the nostril and flows down. During the sniffing, air swirls up into the top of the cavity. Figure 18-1 shows a small patch of about 10 million specialized olfactory cells [6]. They have long microhairs, or cilia, sticking out from them. Odor particles in the air stick to the cilia and make the olfactory cells produce nerve signals, which travel to the olfactory bulb. This is a pre-processing center that partly sorts the signals before they go along the olfactory tract to the brain, where they are recognized as smells [6].

Figure 18-1
Human olfactory model. (Source: Reproduced with permission from the Lappeenranta University of Technology.)

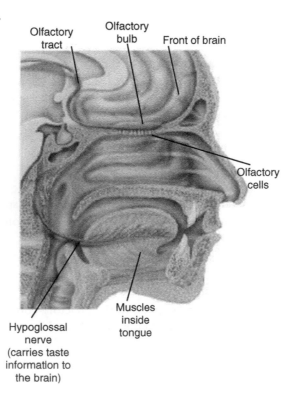

Electronic Olfactory Model

Keep in mind that the main task in odor recognition is to create a model as similar to the human model as possible. From this point of view, electronic/artificial noses (so-called ENoses) are being developed as a system for the automated detection and classification of odors, vapors, and gases [6].

ENose is represented as a combination of two components: the sensing system and the pattern-recognition system. The schematic representation of ENose can be found in Figure 18-2 [6].

The sensing system is represented as an array of chemical sensors in which each sensor measures a different property of the sensed chemical; or as a single sensing device; or as a hybrid of both. The major task of this component is to catch the odor. Each odorant presented to the sensing system produces a signature of the characteristic pattern of the odorant. The database of signatures is built up by presenting many different odorants to the sensing system. It is used further to create the odor recognition system [6].

The pattern-recognition system is utilized to recognize procedure. The goal of this process is to train and create the recognition system that will be capable of producing a unique classification or clustering of each odorant, so that an automated identification can be implemented. This process incorporates several approaches: statistical, artificial neural network (ANN), and neuromorphic [6].

The creation of a mathematical model of the dynamics in the olfactory bulb is an arduous problem. Modeled after the human nose, the ENose relies on the interactions of sniffed chemicals with an array of sensing films that create an identifiable pattern [6]. The two components of ENose are described in detail next [6].

Figure 18-2   *Schematic diagram of ENose. (Source: Adapted with permission from the Lappeenranta University of Technology.)*

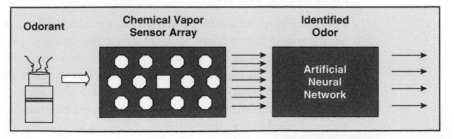

Sensing System

The sensing system allows the tracing of the odor from the environment. This system can be a single sensing device, like a gas chromatograph and spectrometer. In that case, it produces an array of measurements for each component. The second type of sensing system is an array of chemical sensors. This is more appropriate for complicated mixtures, because each sensor measures a different property of the sensed chemical. Hybrids of a single sensing device and array of chemical sensors is also possible [6].

Each odorant presented to the sensing system produces a characteristic pattern of the odorant. By presenting a mass of sundry odorants to this system, a database of patterns is built up. It is used then to construct the odor recognition system [6].

There are five available categories of sensors. A brief description of all these types is given hereinafter [6].

Categories of Sensors: Conductivity Sensors

There are two types of conductivity sensors: metal oxide and polymer. They exhibit a change in resistance when exposed to volatile organic compounds. Both of these classes are widely available commercially because of their low cost. These sensors respond to water vapor and humidity difference, but are not too sensitive for specific odorants. Currently, many research groups work under the enhancement of this type of sensor. Conducting polymer sensors are commonly used in electronic nose systems. Because conducting polymer sensors operate at ambient temperature, they do not need heaters and thus are easier to make. The electronic interface is straightforward, and they are suitable for portable instruments [6].

Piezoelectric Sensors

The piezoelectric family of sensors (quartz crystal microbalance, surface acoustic wave devices) can measure temperature, mass changes, pressure, force, and acceleration. During an operation, a gas sample is adsorbed at the surface of the polymer, thus increasing the mass of the disk-polymer device and thereby reducing the resonance frequency. The reduction is inversely proportional to the odorant mass adsorbed by the polymer. In the electronic nose, these sensors are configured as mass-change-sensing devices [6].

Metal-Oxide-Silicon Field-Effect-Transistor (MOSFET)

MOSFET odor sensing devices are based on the principle that volatile odor components in contact with a catalytic metal can produce a reaction in the metal. The reaction's products can diffuse through the gate of a MOSFET to change the electrical properties of the device. By operating the device at different temperatures and varying the type and thickness of the metal oxide, the sensitivity and selectivity can be optimized [6].

Optical-Fiber Sensors

Optical-fiber sensors utilize glass fibers with a thin, chemically active material coating on their sides or ends. A light source at a single frequency (or at a narrow band of frequencies) is used to interrogate the active material, which in turn responds with a change in color to the presence of the odorant to be detected and measured [6].

Arrays of these devices with different dye mixtures can be used as sensors for an ENose. The main application for such kind of ENoses is in medicine [6].

Spectrometry-Based Sensors

Spectrometry-based sensors use the principle that each molecule has a distinct infrared spectrum. Usually, devices based on these sensors are large and expensive [6].

Pattern-Recognition System

A pattern-recognition system is the second component of electronic nose used for odor recognition. Its goal is to train or to build the recognition system to produce a unique classification or clustering of each odorant through the automated identification [6].

Unlike human systems, electronic noses are trained to identify only a few different odors or volatile compounds. This is a very strong restriction in using these noses for human recognition. The state of the art does not make it possible to identify all components of the human body precisely [6].

Recognition process incorporates several approaches: statistical, ANN, and neuromorphic. Many of the statistical techniques are complementary to ANNs and are often combined with them to produce classifiers and clusters. This includes PCA, partial least squares, discriminant and cluster analysis. PCA breaks apart data into linear combinations of orthogonal vectors based on axes that maximize variance. To reduce the amount of data, only the axes with large variances are kept in the representation [6].

When an ANN is combined with the sensor array, the number of detectable chemicals is generally greater than the number of unique sensor types. A supervised approach involves training a pattern classifier to relate sensor values to specific odor labels. An unsupervised algorithm does not require predetermined odor classes for training. It essentially performs clustering of the data into similar groups based on the measured attributes or features [6].

Neuromorphic approaches center on building models of olfaction based on biology and implementing them in electronics. Unfortunately, there is a lack of realistic mathematical models of biological olfaction. Thus, the area of neuromorphic models of the olfactory system lags behind vision, auditory, and motor control models. Olfactory information is processed in both the olfactory bulb and in the olfactory cortex. The olfactory bulb performs the signal pre-processing of olfactory information, including recording, remapping, and signal compression. The olfactory cortex performs pattern classification and recognition of the sensed odors [6].

There are two competing models of olfactory coding. The selective receptor comes from recent experimental results in molecular biology. It can be thought of as an odor mapper. This approach is similar to a visual system, with the idea of receptive fields of olfactory receptors and mitral cells in the olfactory bulb. The second approach is a nonselective receptor, distributive-coding model that comes from data collected by electrophysiology and imaging of the olfactory bulbs [6].

Note: The neuromorphic approach has an advanced feature that incorporates temporal dynamics to handle identification of combinations of odors.

Olfactory Signal Processing and Pattern-Recognition System

The goal of an electronic nose is to identify an odorant sample and to estimate its concentration (in human recognition cases). It entails signal processing and a pattern-recognition system. However, those two steps may be subdivided into pre-processing, feature extraction, classification, and decision making. All these

Figure 18-3   *Signal processing and pattern recognition systems stages. (Source: Adapted with permission from the Lappeenranta University of Technology.)*

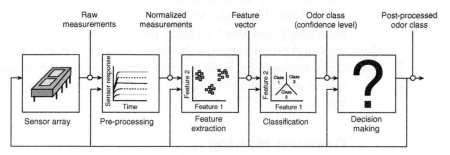

substeps are shown in Figure 18-3 [6]. But first, keep in mind that a database of expected odorants must be compiled, and the sample must be presented to the nose's sensor array [6].

You should consider all signal processing and pattern recognition stages (see Figure 18-3), particularly the following stages:

- Pre-processing
- Feature extraction
- Classification
- Decision making

Pre-Processing

Pre-processing compensates for sensor drift, compresses the response of the sensor array, and reduces sample-to-sample variations. Typical techniques include normalization of sensor response ranges for all the sensors in an array, and compression of sensor transients [6].

Feature Extraction

Feature extraction has two purposes: to reduce the dimensionality of the measurement space, and to extract information relevant for pattern recognition. Feature extraction is generally performed with linear transformations such as the classic PCA [6].

Classification

The commonly used method for performing the classification task is an artificial neural network (ANN). An artificial neural network is an information-processing system that has certain performance characteristics in common with biological neural networks. It allows the electronic nose to function in a way that is similar to the brain function when it interprets responses from olfactory sensors in the human nose. A typical ANN classifier consists of two or more layers [6].

Decision Making

The classifier produces an estimate of the class for an unknown sample along with an estimate of the confidence placed on the class assignment. A final decision-making stage may be used if any application-specific knowledge is available, such as confidence thresholds or risk associated with different classification errors. The decision-making module may modify the classifier assignment and even determine that the unknown sample does not belong to any of the odorants in the database [6].

Prototype Electronic Nose

Electronic nose research groups have developed a number of prototype electronic noses. Some of them are illustrated in Figure 18-4 [6].

Usually, during an operation, a chemical vapor is blown across the array; the sensor signals are digitized and fed into the computer. Then the ANN (implemented in software) identifies the chemical. This identification time is limited only by the response time of the chemical sensors, which is on the order of a few seconds [6].

Human Odor Recognition

Biometric tools are becoming more popular as a form of identification as technology needs become increasingly sophisticated and less expensive. Already, vendors are selling fingerprint recognition technology on computer keyboards or iris recognition for automated teller machine manufacturers [6].

Can you identify people by their odor? Sound like a snorter? It's not. Scientists already have linked a collection of immunity genes with unique

Figure 18-4 *The 4440B (Agilent Technologies), Prometheus (Alpha Mos), and A320 (Cyrano Sciences) electronic noses. (Source: Reproduced with permission from the Lappeenranta University of Technology.)*

human body odor. And with ENoses, now sensitive enough to test beer, perfume samples, and uncover pollution and disease, it may be only a matter of time before an ENose will be able to identify persons [6].

Problems

Now it's absolutely clear that people with differing genes produce different body odors, but scientists do not know how that happens. And even if researches knew exactly which compounds to look for, artificial noses are not yet sophisticated enough to do this job [6].

First of all, today's smell sensors are not sensitive to a wide variety of compounds. Although scientists have cameras that can see beyond the spectrum of the human eye and microphones that can detect a vibration a mile away, in terms of chemical sensing, researchers are still far away from what biology can do [6].

Computers are not as smart or flexible as dogs or humans or other biological creatures. If you get a brand-new scent that you've never smelled before, you can learn what it means and recognize it the next time you encounter it. Machines aren't very good at being able to adjust to new conditions. Thus, scientists must still fill big holes in both research and technology [6].

Electronic Versus Human Nose

How do electronic noses work? Let's compare your nose with the electronic version. Most substances contain volatile chemicals. Due to them, you can smell something. Sensors in your nose, which number about 10,000 in number and are nonspecific-task in nature, react to those complex chemical vapors (which may consist of 670 chemicals, as for coffee) and send the appropriate electric signals to your brain, which has about 10 million sensory neurons. The set of signals transmitted by these set of sensors create a pattern. Your brain records the pattern and, if it cannot match the pattern to any pattern already stored, the new one will be added to its already large library of patterns. Variation between this smell and the already stored pattern will highlight any difference in the constituent of vapor from the known pattern. The next time you encounter this smell, your brain will be able to recognize it [6].

The human nose is needed in many jobs; for example, in the coffee grading process, a human panel of smell experts will sniff out a batch of beans to determine its grade. However, this process is prone to give incorrect results, as the human olfactory system is sensitive to the environment, health, diet, and fatigue [6].

ENoses, however, are much simpler than the biological version, and able to detect only a small range of odors. An ENose utilizes a much smaller number of volatile chemical sensors, usually between 12 and 20, and a proportionate number of artificial neurons [6].

A conventional method for odor identification is both expensive and complicated. There must be a huge sensor array, in which each sensor is designed to respond to a specific odor. With this approach, the number of sensors must be at least as great as the number of odors being monitored. Apart from that, the quantity and complexity of the data collected by sensor arrays will cause trouble for this approach when it comes to being automated. As such, this method is not feasible [6].

The current trend seems to look to artificial neural networks (ANNs). When an ANN is combined with a sensor array, the number of detectable odors is generally greater than the number of sensors. Also, less selective sensors (thus, less expensive sensors) can be used for this approach. Once the ANN is trained for odor recognition, the operation will consist of propagating the sensor data through the network. With this approach, unknown odors can be rapidly identified in the field [6].

Due to limitations of current technology, many ANN-based ENoses have fewer than 20 sensors and fewer than 100 neurons. These systems are designed for odor-specific applications with a limited range of odors. Systems that mimic more of the functionality of the human olfactory system will require a larger set of sensing elements and a larger ANN [6].

Who Works with It?

Unfortunately, the state of the art in ENoses does not allow using these devices for such a perceptive task as human recognition. The work that is under development for person authentication is extremely expensive and, thus, not every laboratory can deal with it. However, there are at least two groups that have created devices for person recognition [6].

One company is the U.K. company Mastiff Electronic Systems. This company is said to be in development of Scentinel, a product that digitally sniffs the back of a computer user's hand to verify identity. This product is still very expensive ($48,600), but there is interest in its implementation from the British embassy in Argentina, Saudi Arabia's National Guard, and private Indian and Japanese companies [6].

The second group working to identify people by body odor through the use of artificial noses is the Pentagon's Defense Advanced Research Projects Agency. This agency gave out some $7.6 million in 2005, with the expectation that a people-sniffing electronic nose will be available in the next five to six years as specific milestones are met along the way. Figure 18-5 shows the prototype awaiting installation of its electronic nose at a laboratory at the University of Pennsylvania [6].

Figure 18-5 *Prototype of electronic nose, University of Pennsylvania. (Source: Reproduced with permission from the Lappeenranta University of Technology.)*

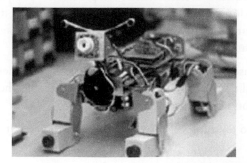

Current Applications

During the last decade, two dozen companies have developed over 200 electronic nose prototypes, and a number of commercial applications are expected in the next five to 10 years. According to industry analysts, a global market of 5,000 units annually is predicted by 2007 [6].

Inline electronic noses cost about $40,000–$50,000 each, while hand-held units are available for $5,000. As the gas sensor costs only about $5–$10, the major chunk of the cost lies in the odor recognition system. This is expected to reduce with improvements in pattern recognition software and advancement of ANN technology. Some uses are as follows:

- The most important application nowadays of ENoses is in medical diagnostics.
- Odors in the breath can indicate gastrointestinal problems, sinus problems, infections, diabetes, and liver problems. Infected wounds and tissues give off odors that can be detected by the electronic nose.
- Odors coming from body fluids can indicate liver and bladder problems.
- An electronic nose has been used to track glucose levels in diabetics, determine ion levels in body fluids, and detect pathological conditions such as tuberculosis.
- Environmental applications of electronic noses include identification of toxic and hazardous wastes, analysis of fuel mixtures, detection of oil leaks, and identification of household odors, monitoring factory emissions, and testing groundwater for odors.
- The biggest market for ENoses is the food industry. Application in this area includes quality assessment in food production, inspection of food quality by odor, control of food production, verifying if orange juice is really natural, grading whiskey, inspection of beverage containers, and classification of vintage of wines.
- ENoses are used in pharmaceuticals to determine whether stored drugs have reached their expiration date. This is necessary when companies are dealing on huge scales.
- They are also used in perfumery to identify counterfeit products [6].

Future Applications

Apart from human authentication as a specific computer's user, there are a number of prospective applications. As was previously mentioned, the real olfaction mechanism is still unknown to science. However, many scientists, research groups, and entrepreneurs are trying to understand it and even to approximate it. Some new applications follow.

Fight against Crime

The first application is the fight against crime: recognition of terrorists. There already are orders for the human recognition system from the British embassy in Argentina, Saudi Arabia's National Guard, and private Indian and Japanese companies [6].

Virtual Reality and Virtual Environments

The second new application is virtual reality and virtual environments. The main idea of this application is limited nowadays to 3D sound and stereo vision; thus, the users' immersion into virtual environments is restricted to two of five available senses. Virtual reality including smell is expected to aid training for perilous duties. Many real-life dangerous situations require managing more physical conditions than just visual and aural inputs. Among applications that rely strongly on smell are firefighting training and dangerous gas discharge [6].

Detection of Humans Buried in Rubble

The third important new application is detection of humans buried in rubble, for example, in earthquakes or from damage in coal mines. To detect human body odor, an electronic nose is applied. In principle, this ENose can be considered as an alternative to the dogs' work. Unfortunately, dogs only can go to a depth of 50 meters and work only with his or her handler. Another disadvantage is the long-term maintenance of the handler-dog team, because it is so expensive. An electronic nose for human detection will be a suitable replacement for the dogs. Of course, this nose will not be as sensitive as a dog's; but it will still be used perfectly for this specific application [6].

Telesurgery

Finally, a more futuristic application of ENose is telesurgery: The ENose would identify odors in the remote surgical environment. All the preceding applications are expected to appear in the next two to three years [6].

Messengers in Your Mouth

Scientists are creating super-sensors to pick up myriad molecules released at low concentrations that constitute human scents, including carbon dioxide, acetone, ethanol, and sulfur. To capture them, they poke tiny pores into glass (as many as 10,000 on a chip the width of a pencil eraser), each tailored to the size of the molecule. Excited by a laser, the chemicals trapped in the pores emit different colors, and computers then analyze the resulting pattern [1].

Dental researchers are attacking the challenges of identification and diagnosis from another vantage point—the mouth. They're studying whether saliva contains markers for various diseases. If the technology works, it has additional potential for biometric applications. Spit contains many of the proteins, nucleic acids, and other substances that are found in blood. While they are present in lesser quantities, they can also be sampled less intrusively [1].

Scientists have found that they can detect in human saliva some 3,000 messenger RNAs, molecules that carry genetic information within a human cell. These molecules might be able to serve as markers for disease, or perhaps for identity, just like DNA. And they are often easier to detect. About 180 RNA markers are common across all individuals, but the remainder can differ. Scientists don't know how constant these are to the individual on a Monday versus a Friday, but they could possibly serve as fingerprints for that individual. Recently, scientists have identified four RNA markers in saliva that may indicate the presence of oral cancer [1].

The use of bodily scents and secretions as biometrics presents an intriguing anti-terrorism weapon (see sidebar, "Beyond Hand Sniffing"). But if the science isn't rock-solid, it can lead to a nightmare of mistaken identities. That's a problem even with mature biometrics, such as fingerprints. For example, the fingerprints of Oregon lawyer Brandon Mayfield were erroneously matched to those of a suspect in the Madrid train bombing in 2004. This cast a cloud over the innocent man for weeks [1].

Beyond Hand Sniffing

The day may come when a computer can identify its user by body odor. Innovative sensing techniques being developed include recording fingerprint images using ultrasound.

Odor Detection

The premise of this technique is that chemicals called rolatiles make each person's distinctive smell. A number of sensors check the different compounds that make someone smell. This method is under development.

Mastiff Electronic Systems is said to be developing Scentinel, a product that digitally sniffs the back of a computer user's hand to verify identity. These prospective odor sniffers are the most exotic technology in a list that includes face and fingerprint readers, iris and retinal scans, finger and hand geometry, and signature and voice recognition.

Today, there are 500 applications in 70 markets. Some biometric measurements should be logically applied in certain markets. Law enforcement will use fingerprints, while voice recognition for telephone and face recognition with video are natural choices in those markets.

The use of biometric technology is expanding into voter registration, identification of students for testing, healthcare, and even for entry into Disney theme parks. Some industries are starting to use several biometric technologies, layering one on top of another [2].

Problems Now and Later

Biometrics can bring a host of troubles. As they become used more and more in office access, ATM passwords, passports, and ID cards, their value increases, and so do efforts to steal or spoof them. And because biometrics are cloaked in science, matches may acquire an unearned aura of dependability. Recently, cryptographers in Japan showed that common fingerprint-based systems can be easily duped using simple molds of melted Gummi Bear candies. In hopes of precluding such scams, Albuquerque's Lumidigm Inc. captures images of not only the fingerprint itself but also the terrain beneath the skin. This includes the swirling patterns of active capillaries, which help indicate that the finger is alive. Fujitsu Ltd. has just installed palm scanners that read vein patterns at Mitsubishi bank ATMs [1].

Despite the many failings of biometrics, the federal government is encouraging scientists to fashion them into covert surveillance tools. Face recognition (the most obvious way to track people) is still dogged by problems when matching images that may be distorted by a smile or ill-placed shadow. While scientists work out those glitches, others are improving iris-based technology for surveillance at a distance. Though computers can easily find eyes on a face, today's systems can't scan irises from afar as people rush through a crowd [1].

Another hope is that certain characteristic movements may be recognizable at a distance. Taking a page from Monty Python's Ministry of Silly Walks, the U.S. Defense Advanced Research Projects Agency, the research body credited with inventing the Internet, funds work on software that could identify individuals by their strides. Researchers measure the silhouette of the torso, the swinging of the shoulders and legs, and the time it takes to move through a single step. Right now, people can still trick the system by wearing Manolo Blahniks, but there may be signature rhythms that are harder to disguise. Such gait recognition systems may be three to eight years from commercialization [1].

Many people in building security welcome advances in surveillance. In New York, two-thirds of Class A residential and commercial buildings use some combination of biometrics and surveillance for access control or checking time and attendance. Incidents of mistaken identity are rare. Biometrics can vindicate an innocent person by establishing a correct ID. Highly accurate technology is a friend to privacy [1].

The most serious privacy breaches are almost all linked to the proliferation of fast and inexpensive data processing and storage systems [4]. The worst problems arise when each bit of information an individual gives up over the course of a day (from the E-Zpass vehicle scans on the morning commute to the credit card purchase at Starbucks to the logging of PC keystrokes at work) get tied across various databases to create a detailed dossier of an innocent Joe's daily activity. We're just a couple of generations away from technology that makes it possible for a computer to save everything you do [1].

But in information technology the generations can fly by at superhuman speeds. Ever since September 11, 2001, the U.S. government has been striving through the power of software to extend its investigatory net over an elusive enemy lurking among the populace. The idea is to riffle through multiple databases using algorithms that categorize and rank documents, ranging from airline manifests, car rental records, and hotel guest lists to credit, court, and housing records compiled and sold by private companies such as ChoicePoint. In this way, machines might recognize relationships among human beings that humans themselves can miss [1].

This is just one of many measures that trigger a Big Brother alert. One of the hot buttons is eavesdropping. An emerging wireless technology [5] called software-defined radio has the power to make cellular phones compatible with any network standard, but opens new frontiers of snooping. The commercial merits of the technology are self-evident: Say good-bye to dead zones and lack of interoperability between police and firefighter radios. But the technology also enables superscanners that can be tuned to pick up the images on your neighbor's computer. That's possible because all computers emit stray radiation. With software-defined radio, even amateurs could probably design equipment that could spot somebody surfing pornography in the next apartment. The technology can also make it easier to turn the cellphone of a spouse into a bug when it's not in normal use [1].

Pores and Wrinkles

Advances in many surveillance technologies piggyback on progress in fields such as wireless signal processing, nanotechnology, and genomics. Even plain old digital cameras are hotbeds of innovation. The imaging sensors in consumer cameras have been achieving ever-higher resolutions while plunging in price. Because the gadgets are so engaging, crowds end up participating in surveillance efforts. Witness spectators hold cameras and phones aloft whenever news breaks—an act that may aid investigations or hold police misbehavior in check. And in biometrics, today's high-res imaging chips are an answer to researchers' prayers. Now they can do things that they couldn't do five years ago [1].

Improved picture quality has given a boost to Identix Inc., allowing it to add in minute details of the skin to increase the accuracy of facial recognition. It divides a small area on the face into a 400-block grid, and then inspects each block for the size of skin pores, wrinkles, and spots. Using an infrared camera, researchers at A4Vision Inc., a Sunnyvale, California start-up funded in part by In-Q-Tel, the CIA's venture fund, cooked up a 3D approach. Its system creates a topographical map by projecting a grid pattern of infrared light onto a face and matching the features [1].

Strides in wireless signal processing are bringing the power of astronomical instruments to homeland security. Giant radio telescopes today listen to the faint energy waves emanating from stars billions of light years away. The first earthbound applications of this electronic wizardry will be airport scanners that scrutinize passengers' bags. The principle is simple: All matter gives off so-called background radiation, or millimeter-wave heat, whether it's a supernova or a switchblade. Brijot Imaging Systems Inc. recently unveiled a $60,000 system that, Brijot claims, can distinguish between the heat coming from a human body and that from a metal or plastic object—and can pull this off from distances of up to 45 feet. The company says its system doesn't capture anatomical details [1].

A kindred technology can "see" the molecular composition of matter using extremely short wavelengths of energy. When a machine made by Picometrix Inc. shines these terahertz waves on a target, its molecules resonate at a tell-tale frequency. One plastic explosive, for instance, vibrates at 800 gigahertz. T-rays pose no radiation hazard because they don't penetrate human skin. But people being scanned will appear naked on the monitor unless the system is programmed to cover up private parts [1].

Airport safety is just a small facet of the security challenge that lies ahead. Biological and chemical attacks can be instigated in any location, and spread with alarming speed. If we could put sensing devices everywhere, maybe we could stop such attacks. But the cost is now prohibitive [1].

Summary/Conclusion

The problem of personal authentication based on body odor was analyzed in this chapter. It is absolutely clear that people with differing immunity genes produce different body odors. Each human has a unique body odor that is a combination of approximately 30 different odorants. The main purpose of human body odor analysis technology is not just to define these entire components, but to estimate their concentration [6].

To identify people by their body odor, you must use a special device like an electronic/artificial nose: the so-called ENoses. The two main components of these noses, the sensing system and the pattern recognition system, were described in this chapter. The main task of ENoses is to try to repeat the process of the human olfactory model. The stumbling block here is the information-processing mechanisms of human olfaction, which are entirely still unknown because of the lack of knowledge about the overall dynamical properties of the brain. Thus, for the pattern recognition system of ENoses, any pattern recognition algorithm can be used. However, the sensing system represents a problem. It was emphasized in this chapter that state of the art in sensors' sensitivity does not allow you to estimate the concentration of the odorants within its mixture. All that is possible to do now is to detect whether a specific odorant is contained in this mixture [6].

Presently, there are no available commercial applications for human authentication through body odor. However, at least two research groups (Mastiff Electronic Systems and the Pentagon's Defense Advanced Research Projects

Agency) are working on the development of a device capable of catching a human's body odor. It must be emphasized that such research is extremely expensive and tedious. Unfortunately, there is no available information about either the accuracy of the methods used in the devices or the numerical algorithms [6].

Finally, although the human body odor recognition system is still under construction, the odor recognition technique is quite useful in real-life applications. There are a lot of current applications that, together with future ones, have been presented in this chapter. Among the current applications are medical diagnostics and the food and beverages industry. And, among future applications are computer user identification, virtual reality and virtual environments, and the recognition of terrorists [6].

References

1. "The State of Surveillance," Wasatch Venture Fund, 15 West South Temple Street, Suite 520, Salt Lake City, UT 84101, August 1, 2005. Copyright © 2006 Wasatch Venture Fund. All Rights Reserved.

2. Mo Krochmal, "Biometrics Makes Scents for Computer Users," TechWeb, 600 Community Dr., Manhasset, NY 11030, September 1, 1999. TechWeb, Copyright © 2006 CMP Media LLC.

3. John R. Vacca, *Net Privacy: A Guide to Developing and Implementing an Ironclad eBusiness Privacy Plan*, McGraw-Hill (2001).

4. John R. Vacca, *The Essentials Guide to Storage Area Networks*, Prentice Hall, Professional Technical Reference, Pearson Education (2001).

5. John R. Vacca, *Wireless Data Dymistified*, McGraw-Hill (2003).

6. Zhanna Korotkaya, "Biometric Person Authentication: Odor," Department of Information Technology, Laboratory of Applied Mathematics, Lappeenranta University of Technology, PL 20, 53851 Lappeenranta, Finland, 2003.

19

How DNA Measurement Technology Works

DNA has captured popular, government, and scientific attention as a unique and stable identifier that is more powerful than fingerprinting. Researchers and developers have sought to leverage respect for that technology in promoting more exotic—or merely more opportunistic—biometric proposals concerning attributes such as brainwaves, pulse, personal smell and salinity, or nailbeds as a human barcode [1].

Identification on the basis of an individual's unique, stable, and measurable genetic characteristics has gained fundamental judicial, administrative, and scientific recognition over the past two decades. DNA-based identification (based on examination of tissue, semen, or other samples) appears to be highly accurate when correctly conducted (most challenges in recent years have centered on the contamination or substitution of samples) and has thus resulted in proposals for large-scale DNA registers (see sidebar, "DNA Identification"). It has also resulted in proposals for DNA-based authenticity labeling of indigenous artworks [1].

DNA Identification

DNA is an abbreviation of "deoxyribonucleic acid." DNA is a unique and measurable human characteristic that is accepted by society as absolute evidence of one's identity. In reality, DNA identification is not absolute, but it has come to be considered as the best method of confirming someone's identity, with a near-perfect probability of 99.999% accuracy [2].

The chemical structure of everyone's DNA is the same. The only difference between people (or any animal) is the order of the base pairs. There are many millions of base pairs in each person's DNA. Using these sequences, every person can be identified based on the sequence of their base pairs [2].

However, because there are so many millions of base pairs, the task of analyzing them all would be extremely time-consuming. Hence scientists use a small number of sequences of DNA that are known to greatly vary among individuals in order to ascertain the probability of a match [2]. The major issues with DNA identification revolve around the realistic ability of capturing and processing the sample of a person in a controlled and lawful manner that does not violate civil rights [2].

How Is DNA Measurement Used as a Biometric Identifier?

DNA (deoxyribonucleic acid) is the complex substance that contains the genetic information of an individual. DNA has a double helix structure, discovered by James Watson and Francis Crick in 1953 at Cambridge University. Each helix is a linear arrangement of four types of nucleotides or bases: A adenine, C cytosine, G guanine, and T thymine. Between the four bases, only two pairings are chemically possible; A always pairs with T, and G with C. As the helixes are complementary, when the first helix contains the sequence AGTCCTAATGT, for instance, the second one contains the complementary sequence TCAGGATTACA. The sequence of the bases determines all the genetic attributes of a person. So, for this chapter and based on current knowledge about DNA, it is very important to observe the following points:

- Only 2%–3% of the DNA sequence represents the known genetic material.

- Almost 70% of the sequence is composed of noncoding regions.

- Almost 30% of the sequence is composed of noncoding repetitive DNA, and only one-third is tandemly repetitive. The rest (two-thirds) is randomly repetitive [3].

DNA identification is based on techniques using the noncoding tandemly repetitive DNA regions. Only 10% of the total DNA bears nonsensitive information [3].

In general, DNA identification is not considered by many as a biometric recognition technology, mainly because it is not yet an automated process. (It takes some hours to create a DNA fingerprint.) However, because of the accuracy level of the process and because it is considered as a possible future biometric trait, it is here analyzed together with the standard biometric technologies [3].

DNA Sample

DNA can be isolated from a sample, such as blood, semen, saliva, urine, hair, teeth, bone, or tissue. So, DNA counts several sources of biological evidence, which are especially easy to collect or to find (and consequently to steal) in every place that an individual has been [3].

DNA Template

In the case of DNA use as a biometric, it is necessary to transform the sample into a template, which is an irreversible process. A DNA fingerprint (see sidebar, "DNA Fingerprinting") or DNA profile does not enable analysis related to genetic or medical aspects, because the technique used for establishing a DNA template focuses on the noncoding regions of DNA, and more precisely only on a specific part of the noncoding regions characterized by a high polymorphic degree [3].

DNA Fingerprinting

DNA fingerprinting is a complex technique that uses genetic material to identify individuals. Biometrics relies on distinctive individual physiological traits. Within such traits, DNA—the hereditary material that determines what genetic traits you inherit—is supposed to be the most distinctive. Structurally, it is a long, double-helical chain with a phosphate backbone, to which are attached ribose sugars and nitrogenous bases. DNA is like a code that directs your cells to make or assemble things that are required for the functioning of your body. Although a first glance may not show any significant differences in the code, even a seemingly insignificant change in the sequence leads to the differences that you see. So much so for biology. Let's look at how this is utilized in identifying individuals [4].

A DNA molecule is divided into small functional units called genes, which determine your black hair and brown eyes. DNA samples can be taken from the body if the subject or his or her personal belongings are passed through chemical processes. This DNA fingerprint is in the form of a sequence of A's, T's, G's, and C's in random order. These letters refer to the nitrogenous bases [4].

Sounds easy? Now comes the difficult part. The length of this sequence is immense, almost beyond comprehension! Here is where matching algorithms come into the picture. Lots of schemes are followed for generating these algorithms. The most widely used is a kind of indexed algorithm, which is also probabilistic in nature. This means that these algorithms look for matches only in those places where they are most likely to find a match, and they ignore the rest of the database. From an existing database, the algorithm generates indices and then matches the sequence under consideration with them [4].

The next step in DNA sequence matching is matching the shape of molecules. Biologists indicate that much of what function a protein will have is dependent on the shape it takes. But researchers face a problem: A protein molecule can be in a variety of similar but not same orientations. Hence, researchers need to look at rigid elements of molecules and the way the nonrigid elements are connected to them.

Molecules with similar rigid elements form a group. Algorithms then analyze and match DNA sequences accordingly [4].

DNA matching has advantages over other means of biometric verification. DNA samples can be collected in many more forms than blood samples, retina scans, or fingerprints. Even people's personal belongings, like hairbrushes, toothbrushes, or clothes, carry their DNA from phenomenons like natural skin flaking. So it becomes close to impossible for an imposter to fake a DNA sample or avoid leaving a trace at a crime scene. The only drawback is that DNA testing takes longer than other methods, something like a couple of days to a week. But constant research will soon have faster, if not real-time, DNA analyzers that will ensure punishment for the guilty and justice to the innocent [4].

DNA Template: DNA Fingerprinting

DNA fingerprinting, discovered by Alec Jeffreys in 1984 at the University of Leicester, allows identifying DNA patterns at various loci (specific places within the DNA sequence) that are unique to each individual—except identical twins. Each pattern is a repeated DNA fragment section (known as variable number of tandem repeats [VNTR]), and its size depends on the number of repetitions. At a given locus, the number of repeated DNA fragments varies between individuals. The technique used to examine DNA patterns is based on restriction fragment

Figure 19-1   *DNA templates: DNA fingerprint image and DNA profile representation, respectively. (Source: Reproduced with permission from the Institute for Prospective Technological Studies.)*

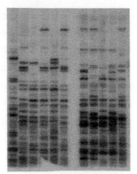

DNA fingerprint image

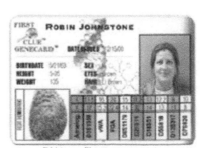

DNA profile representation

length polymorphism (RFLP). Due to the low quality and quantity of the DNA sample in a crime scene, the technique is based on the polymerase chain reaction (PCR) [3].

The procedure of creating a DNA fingerprint is composed of the following steps: isolation of DNA; denaturalization of DNA (cutting, sizing, and sorting); transfer; and probing (see sidebar, "Make a DNA Fingerprint: Detailed Procedure"). The DNA fingerprint is built by using several probes (five to 10 or more) simultaneously. The result resembles barcodes, as shown in Figure 19-1 [3].

Make a DNA Fingerprint: Detailed Procedure

Creating a DNA fingerprint is composed of the following steps:

1. **Isolation of DNA:** DNA must be recovered from a source of biological evidence. Avoiding any type of contamination is essential.

2. **Cutting, sizing, and sorting:** Special enzymes called restriction enzymes are used to cut the DNA at specific places. Thus, the repeated DNA fragments sections are recognized. Then they are separated and sorted by size through gel electrophoresis.

3. **Transfer:** The resulting distribution is transferred to a nylon or nitrocellulose sheet placed on the gel by blotting.

4. **Probing:** This step consists in adding radioactive or colored probes to the sheet in order to produce the DNA fingerprint.

5. **Final:** The DNA fingerprint is built by using five to 10 or more probes simultaneously. The result resembles barcodes [3].

DNA Template: DNA Profiling

From a DNA sample, it is possible to establish DNA profiles in order to represent the specific DNA patterns by numerical data. The numerical result is a string of values (such as 13.5, 17–16, 15.311, 9–10, 8); each pair of values is associated with a specific locus (D3S1358, VWA, FGA, etc.). In Figure 19-1 (right), the card shows the DNA profile (in fact just an extract) of this person, Robin Johnstone [3].

Nowadays, DNA identification is mainly used in forensics (see sidebar, "DNA Usage in Forensics"), or more precisely, in forensics investigation. The DNA fingerprint is a powerful tool to exclude an individual from a given DNA sample. Indeed, in criminal identification, it is necessary to contrast the DNA of suspects with the DNA evidence found in the crime scene; and, when a suspect has a different DNA pattern than the evidence, he or she is excluded. So, the DNA fingerprint is used to prove the suspect's innocence. It is easier to exclude an individual than it is to include an individual with the same certainty. This assessment is also made for the paternity proof enactment. In this field, there are only three legitimate conclusions from DNA to identity testing:

1.   Exclusion: The individual cannot be the source of the evidentiary sample;

2.   Non-exclusion: The individual cannot be excluded from being the source;

3.   No results: The analysis cannot be performed [3].

DNA Usage in Forensics

In 2005, Interpol launched an inquiry into the DNA database in order to obtain a global overview of DNA usage in forensics. The final objective was to gather DNA profiling information in order to facilitate the possible future exchange of DNA related to intelligence between the Interpol Member States [3].

Results for the European Region

- The European region consists of 46 countries and 1 sub-bureau;

- 96% of the European region replied to the DNA database in inquiry year 2005;

- 39 countries perform DNA analysis in criminal investigations;

- 29 countries have an implemented DNA database, including 12 CODIS software;

- 12 countries have an implemented DNA database, including 7 CODIS software;

- 24 countries have officially accredited laboratories and 10 countries are pending;

- The most prevalent category in the database is "stains," with the most quantitative being the "convicted" category;

- 27 out of the 38 countries with an implemented or planned DNA database allow the international exchange of profiles [3].

How Does It Work?

Unlike other biometrics identifiers (see sidebar, "Is DNA a Biometric?"), DNA enrollment is always possible; everyone has DNA. In addition, DNA allows enrollment at birth. The main advantage of DNA for this step is that DNA enrollment presents a no-failure case (no probability that a user will not be able to be enrolled). That means that the DNA FTE rate (Failure to Enroll) is 0%. However, DNA enrollment is neither direct (it needs a physical extraction and biochemical process; you cannot take a picture of DNA as for fingerprint or iris) nor automatic (it needs human intervention). Consequently, DNA is frequently considered to be a specific case of biometrics because of the nonautomatic enrollment [3].

Is DNA a Biometric?

DNA differs from standard biometrics in several ways:

1. DNA requires a tangible physical sample as opposed to an impression, image, or recording;

2. DNA matching is not done in real time, nor are all stages of comparison always automated (though this is not likely to still be the case fairly soon);

3. DNA matching does not employ templates or feature extraction, but rather represents the comparison of actual samples [5].

Regardless of these basic differences, DNA is a type of biometric inasmuch as it is the use of a physiological characteristic to verify or determine identity. Furthermore, it is one biometric that may become usable as a unique identifier, as consistent "templates" may eventually be generated from DNA. For this reason, as well as the theoretical ability to determine information about a user from DNA, its usage is highly problematic from a privacy perspective [5, 9].

Whether DNA will find use beyond its current use in forensic applications is uncertain. Intelligent discussion on how, when, and where it should and should not be used, who will control the data, and how it should be stored is necessary before its use begins to expand into potentially troubling areas. These definitions will vary by application: It's illogical to suggest that the usage of DNA in public benefits programs, which nearly all would view as highly problematic, should be viewed as equivalent to the use of DNA in a criminal investigation. Thinking about the dangers of DNA as a biometric is helpful as it underscores the tremendous variety of biometric technologies available, and makes clear that blanket statements about biometrics are generally misleading [5].

Acquiring Samples

DNA collection consists of performing an extraction of cells from all biological evidence in order to obtain a DNA sample. DNA collection is easy and takes some seconds. Several methods exist: a fingerprick for blood, a buccal swab for saliva, or a patch for skin [3].

DNA is subject to degradation and contamination, so the preservation of a DNA sample is of a particular concern, in order to not interfere with the analysis and the final result. There are various types or sources of degradation (temperature, humidity, light) and contamination (chemical, biological, and human). Therefore, it is necessary to dry the sample and freeze it; otherwise the integrity and the quality of the sample cannot be guaranteed [3].

Extracting Features

With regards to transforming the DNA sample into template, it previously was shown that a DNA sample is used to provide a DNA fingerprint whose representation is an image. A DNA fingerprint is a representation of the specific DNA patterns (black bands in the image) at various loci. However, it is also possible to represent them through numerical data by establishing a DNA profile, as previously described. In both cases, the transformation is a time-consuming process (several hours) and requires specific skills [3].

Digitalization and Storage

For the digitalization of a DNA fingerprint, it is necessary to capture images, by using a digital camera for instance. Hence, the database should be a bank of images. Some aspects, such as the number of probes and the quality of the image (resolution, format), should be normalized, especially if in the future you will apply software to store and compare the DNA fingerprint [3].

In the case of DNA profiles, the database stores numerical data; more precisely, strings of values—the direct representation of the DNA profile. The length of the string depends on the number of loci used to provide the DNA profile. It seems that this number is not fixed among the DNA profile databases. Moreover, the precision of the value seems to be variable; it is possible to find a precision with one or two decimals. If this type of digitalization is used, these two points should be subject to standardization in order to enable future comparisons [3].

In the United States, for forensic perspectives, a Combined DNA Index System (CODIS) database has been launched (see sidebar, "Combined DNA Index System (CODIS)"). DNA samples have been collected in all states in order to link serial crimes and unsolved cases with repeat offenders. This database stores DNA templates. The CODIS database allows law enforcement to cross-reference their DNA templates with those of other agencies across the country. Four loci have been established by Interpol as the European standard (Interpol Standard Set of Loci, or ISSOL); and, the European Network of Forensic Science Institute (ENFSI) recommends the use of the European Standard Set (ESS) (three additional loci than the Interpol list) in laboratories throughout Europe [3].

Combined DNA Index System (CODIS)

CODIS provides a system for automated information processing and telecommunications. The CODIS system helps law enforcement agencies develop investigative leads and generates statistical inference data about the frequency of occurrence of a particular profile in a selected population set. Using powerful data analyses and matching algorithms, forensic scientists and investigators in local crime laboratories create DNA profiles and search local, state, and national indexes for a match to a target profile of unknown origin. CODIS is currently implemented at more than 285 U.S. forensic laboratories and in 29 foreign countries [7].

An important contribution of the CODIS system to improving criminal justice is that it allows DNA evidence left at the crime scene to be linked to the perpetrator. In many sexual assault cases, DNA is the only evidence left at the crime scene, and there are rarely any third-party witnesses. The new and powerful analysis techniques of CODIS are helping solve crimes that have remained unsolved for decades due to the lack of proper analysis tools [7].

Note: CODIS aided in DNA identification of the missing after the 9/11 collapse of the World Trade Center.

Aside from the standardization issue that digitalization and the storage [10] of DNA templates raises, storing DNA templates in databases generates further security and privacy concerns with the public. This will be discussed later in this chapter [3].

Comparing Templates

DNA matching is not a trivial process and is expensive due to the complex transformation from the sample into template. The time required for the verification process is long; it is around three to four hours. Even with a forensics marker (see sidebar, "DNA Markers System"), some parts of this process are still manual [3].

DNA Markers System

The following are some of the DNA markers systems in forensics:

- CODIS software
- ESS
- SGM+
- Profiler+
- Power Plex 16

DNA Markers

Short tandem repeats (STRs) are the most widely used DNA markers in forensics. The CODIS database uses 13 STRs as the core loci. STR is a small base sequence (TCTA for the STR marker at the locus DYS391) that repeats itself several times (eight times), so the DNA pattern is $(TCTA)_8$.

An STR marker is a simple sequence (preceding example: TCTA). A contrasting example would be a polymorphic DNA marker, like $(TCCTGTCAAAC(TAACC)_2)_8$ [3].

The comparison does not take place in real time. In addition, this process must be performed by scientists (and it also depends on the kind of marker system used), so it requires a lot of knowledge and skills. The risk concerning matching is a DNA-based system that would create a potential false matching (see sidebar, "DNA Matching"), because of the impossibility of differentiating identical twins [3].

DNA Matching

DNA matching involves proving that a suspect's DNA matches a sample left at the scene of a crime. This type of biometrics technology requires two things:

1.   Creating a DNA profile using basic molecular biology protocols;

2.   Crunching numbers and applying the principles of population genetics to prove a match mathematically [6].

DNA: Your Own Personal Barcode?

Humans have 23 pairs of chromosomes containing the DNA blueprint that encodes all the materials needed to make up your body as well as the instructions for how to run it. One member of each chromosomal pair comes from your mother, and the other is contributed by your father [6].

Every cell in your body contains a copy of this DNA. While the majority of DNA doesn't differ from human to human, some 3 million base pairs of DNA (about 0.10% of your entire genome) vary from person to person. The key to DNA matching evidence lies in comparing the DNA left at the scene of a crime with a suspect's DNA in these chromosomal regions that do differ [6].

Declaring a Match

In the case of a DNA fingerprint and forensic framework, the matching is performed with some DNA templates, the evidence template from the scene and some templates from suspects. The process of declaring a match (see sidebar, "Argument Against DNA Matching") is a human process supported by a computer. First of all, the examiner or analyst must verify that the laboratory comparison conditions are fulfilled. For example, he or she proceeds to run some computerized measurements in order to ensure that the templates are comparable. Then, the examiner must establish whether two templates match in accordance with a match criterion; finally, he or she must determine the probability of the match (the probability that this match is not a random match)—the so-called probability random match (RMP). Some European countries may also carry out other examinations from a second sample, in order to verify an inclusion declaration. In the case of CODIS, computer software is used to automatically search its two indexes: the Convicted Offender index (felony sex offences and other violent crimes) and the Forensic index (from crime scene evidence) for matching DNA profiles [3].

Argument Against DNA Matching

Are there logical arguments against DNA matching? Yes—it could be regarded as invasive—and yes—there may be possible human rights implications. So, to put it simply—if you don't like it, don't apply for U.S. visa-less access. The same proposition applies in the United Kingdom with regard to proposals for benefit entitlement smart cards. If biometric analysis is to be used, why not take it to its currently viable logical extreme? If we're going to all this effort, why not collect a DNA sample at the same time?

You will have to take your eyeball and/or thumb to an authorized "capture point" to have your profile collected and encoded in order to qualify for a card. Why shouldn't you also deliver a saliva swab and/or a nail or hair trimming at the same time? One assumes similar data encoding can be carried out and the pattern added to your card. DNA encoding may be slow and costly today, but what is the current speed and cost of a non–DNA based investigation into a major crime? Both aspects will improve with volume requirement and availability, so even if the collected item is somehow corrupted or mis-matched later, subsequent analysis will prove or disprove allegations based solely upon a DNA match. DNA matching has been hugely successful in identifying criminals and, most importantly, eliminating suspects. So, is there a logical rather than emotional argument for not promoting its use [8]? Time will tell.

Technology: State of Development

DNA testing is a technique with a very high degree of accuracy. A statistical sampling shows a 1-in-6-billion chance of two people having the same profile. Nevertheless, by using a DNA technique, it is impossible to distinguish identical twins (the probability of identical twins in the United States is 1 in 250 or 0.4%). And the accuracy of DNA is considered lower than iris or retina recognition. Moreover, the possibility of sample contamination and degradation impacts the accuracy of the method [3].

Concerning DNA fingerprints, there are systems in various stages of research and development that will enable rapid interpretation for the matching. You can therefore expect more automation for the DNA verification process in the future [3].

Challenges and Limitations

DNA is present in all human beings (universality) and, with the exception of monozygotic twins, it is the most distinct biometric identifier available for human beings. DNA does not change throughout a person's life; therefore, the permanence of DNA is incontestable. It performs well for the applications where it is currently used (forensics, paternity tests, etc.), though it is not suitable for every application. DNA tests are difficult to circumvent under certain conditions (supervised sample collection with no possibility of data contamination). If a sample collection is not supervised, however, an impostor could submit anybody's DNA [3].

We leave DNA traces wherever we go (a single hair can provide a sample). It is impossible to keep DNA samples private [3].

DNA faces several other challenges. Several hours are required in order to obtain a DNA fingerprint. In addition, the collection methods (involving an extraction of a physical sample) generally raise privacy concerns; and DNA data encompass not only identification data, but also genetic data. The public is fairly hostile to DNA usage and storage. In other words, DNA performs well on the aspects of universality, distinctiveness, permanence, performance and resistance to circumvention, but it is weak on collectability and acceptability [3].

Privacy and Security Concerns

DNA collection is regarded as invasive sampling (i.e., fingerprick for blood). However, current DNA sampling methods have evolved and now allow less invasive sampling (collection with a buccal swab of saliva or of epidermal cells with a sticky patch on the forearm). Thus, the new sampling methods are considered not to violate the social expectations for privacy [3].

The main problem with DNA is that it includes sensitive information that relates to genetic and medical aspects of individuals. Any misuse of DNA information can disclose information about hereditary factors or medical disorders [3].

However, DNA profile representation is just a list of numbers, so it is noninformative and is regarded as neutral. In forensics, the selection of DNA markers is performed with the sole purpose of being neutral. This includes endeavors to locate DNA markers away from or between genes rather than being part of gene products. Hence, DNA markers are not established in order to be associated with any genetic disease [3].

Race and ethnicity are actually cultural, not biological or scientific, concepts. Nevertheless, DNA can tell a person what parts of the world some of their ancestors came from [3].

Privacy concerns really are linked with the DNA sample, because that is allowed to establish sensitive information related to genetic aspects. So, that point directly leads to the security of the DNA sample's database or to the certainty of the DNA sample's destruction after the DNA template elaboration [3].

The three main security concerns are about the security of the DNA system (access rights, use of information only for the overriding purpose);

the implementation of security mechanisms in order to ensure a high level of confidentiality; and the security of the DNA database (access rights, length of information retention). It seems essential to define the conditions under which the samples can be banked (anonymous/coded/identified storage) and to guarantee data protection. Therefore, a quality assurance plan and safety regulations of DNA banking (certification of authorized personnel, responsibilities listing, safety measures, etc.) are primary requirements [3].

Applications

Each person has a unique DNA fingerprint that is the same for every cell. A DNA fingerprint, unlike a conventional fingerprint, cannot be altered by surgery or any other known treatment. Apart from its use in medical applications (diagnosis of disorders), DNA is widely used for paternity tests, criminal identification, and forensics. It is also used in certain cases for personal identification, as the following two examples illustrate [3].

For example, in the United States, a pack known as DNA PAK (Personal Archival Kit) is sold with the aim of conserving a sample so that an individual can be identified in case of kidnaping, accidents, or natural disaster. Another U.S. company, Test Symptoms @Home, sells several products and services based on DNA. One such product is a personal identification card that shows general data, such as name, weight, sex, and so on; a fingerprint picture; and an extract of the DNA profile based on the same loci used by the CODIS database. Despite these examples, commercial applications for DNA are very limited; privacy fears and low user acceptance will undoubtedly be a bottleneck for the use of DNA in large-scale applications [3].

Future Trends

Progress in DNA testing will come in two areas: Current techniques will improve, offering more automation, precision, and faster processing times; and new techniques will be developed by exploiting the electronic proprieties of DNA. Nowadays, it is impossible to distinguish identical twins. In the future, however, it may be possible to do so, either through technical improvements in current DNA testing or through a different approach. One such alternative is to study the DNA of the microorganisms each person carries, such as viruses, bacteria, or other parasites [3].

A joint partnership between a U.S. and a Taiwanese company currently exploits DNA technology for security solutions. It also provides several products based on plant DNA technology for anti-counterfeiting or tracking purposes, such as DNA ink with a real-time authentication (DNA test pen) or a DNA marker integrated into textile materials. For the purposes of this chapter, an interesting application of DNA ink would be to use it to authenticate passports or visas. Though this is not a direct use of DNA to identify a human, it is a potentially interesting application [3].

It is important to understand that DNA from bacteria, plants, animals, and humans is the same at the chemical and structural levels; the differences lie in the length of the DNA (number of letters; 4 million for a simple bacteria DNA and at least 3 billion for human DNA) and the sequence. So studying DNA from bacteria is easier than studying DNA from plants, and by transitivity, easier than studying DNA from animals and ultimately humans. From this assessment, you can infer two future tendencies. The first is the use of another type of DNA to supplant the human DNA for individual identification (such as the parasites constellation that each person carries or the application of DNA ink); and the second is that current applications based on plant DNA or on animal DNA may in future exist for human DNA [3].

For example, the Canadian Royal Botanical Garden has presented its future view on botany in the field. The botanist of tomorrow is likely to use a DNA scanner, a small hand-held device enabling some complete analysis from the collected sample. In addition, new methods will emerge (DNA may be scanned in a contactless way based on Bluetooth technology). Thus, you can easily imagine this idea of a hand-held device for the analysis of sample found in a crime scene or disaster scene [3].

The current time required for DNA testing (from the extraction through the matching) is around three to four hours due to the time needed for the amplification process (which takes one to two hours). Recent tests, however, suggest that the time required will be reduced in the near future. Real-time PCR provide good results on plant DNA. Recently, the time needed to extract and amplify animal DNA was reduced to less than 14 minutes using Extract-N-Amp technique based on PCR. All tests have been performed with a tissue sample from a mouse. As a result, this technique provides a DNA ready for sequencing. It has been tested using saliva, hair, and human tissue samples and seems to operate well [3].

Summary/Conclusion

In practice, DNA identification is technically challenging, expensive, and not particularly quick (upwards of 14 minutes). Accordingly, its use centers on retrospective forensic applications (Who has been here?), rather than on-the-spot verification and screening [1].

Community perceptions differ, with studies suggesting that some people are unconcerned about DNA collection/use, and that others are worried about potential misuse of information in DNA registers (unsurprising given broader concerns about genetic privacy highlighted earlier in this chapter). Or, they may be uncomfortable with perceived invasive collection mechanisms (providing a swab of cells from inside their mouth or a blood specimen) [1].

These concerns are likely to increase, given recent media coverage about poor practice in the laboratory and the alleged ease of salting an innocent person's DNA at a crime scene. The sci-fi film *Gattaca* was supposedly the inspiration for a DNA-substitution scam to subvert a U.K. community to register [1].

References

1. "Caslon Analytics Note Biometrics," Caslon Analytics Pty Ltd, GPO Box 3239, Canberra ACT 2601, Australia, 2006. Copyright © Caslon Analytics. All Rights Reserved.

2. William J. Lawson, "Enhancing Assistive Technologies: Through the Theoretical Adaptation of Biometric Technologies to People of Variable Abilities," PhD Dissertation, School of Business, Kennedy-Western University, Tampa, Florida, 2003. Copyright © 2003 William J. Lawson. All Rights Reserved.

3. "Biometrics at the Frontiers: Assessing the Impact on Society." For the European Parliament Committee on Citizens' Freedoms and Rights, Justice and Home Affairs (LIBE), European Commission, Joint Research Center (DG JRC), Institute for Prospective Technological Studies, 2005. Copyright © 2005 European Communities.

4. Ashish Sharma, "DNA Finger Printing," Krify Software Technologies Private Limited, #67/2, South Avenue Plaza, DVG Road, Basavanagudi, Near NagSandra Circle, Bangalore — 560 004, India, 2005. Copyright © 2005. Krify Software Technologies Private Limited. All Rights Reserved.

5. "FAQ's," International Biometric Group, One Battery Park Plaza, New York, NY 10004, 2006. Copyright © 2000–2005 by International Biometric Group, LLC.

6. "DNA Matching: Comparing Personal Barcodes," Quest Biometrics, 2005. Copyright © 2005.

7. "DNA Matching," SAIC Corporate Headquarters, 10260 Campus Point Drive, San Diego, CA 92121, 2006. © SAIC, Science Applications International Corporation.

8. Scott Pollard, "Is There a Logical Argument Against DNA Matching?" NewsScan Inc., 631 East Shore Drive, Canton, GA 30114.

9. John R. Vacca, *Net Privacy: A Guide to Developing and Implementing an Ironclad eBusiness Privacy Plan*, McGraw-Hill (2001).

10. John R. Vacca, *The Essentials Guide to Storage Area Networks*, Prentice Hall, Professional Technical Reference, Pearson Education (2001).

Part 7: How Privacy-Enhanced Biometric-Based Verification/Authentication Works

20

How Fingerprint Verification/Authentication Technology Works

The use of fingerprints as a biometric is both the oldest mode of computer-aided, personal verification and the most prevalent in use today. However, this widespread use of fingerprints has been and still is largely for law enforcement applications. There is expectation that a recent combination of factors will favor the use of fingerprints for the much larger market of personal authentication. These factors include small and inexpensive fingerprint capture devices, fast computing hardware, recognition rate, and speed to meet the needs of many applications, the explosive growth of network and Internet transactions [9], and the heightened awareness of the need for ease of use as an essential component of reliable security [4].

Fingerprinting cannot be forged since each person has their own unique skin characteristics. However, the means for capturing fingerprints is not foolproof. The contact-based devices over time reduce the quality of the image, resulting in errors [1].

The current fingerprinting systems offer a wide variety of technologies in capturing, storing, and verifying searches by the use of large databases. Fingerprinting will remain the leading biometric technology through 2011. According to industry analysts, this biometric system is projected to grow from $431 million in 2006 to $4.736 billion in 2011 due to its versatility [1].

This chapter contains an overview of fingerprint verification methods and related issues. This chapter first describes fingerprint history and terminology. Digital image-processing methods are described that take the captured fingerprint from a raw image to match result. Systems issues are discussed, including procedures for enrollment, verification, spoof detection, and system security. Recognition statistics are discussed for the purpose of comparing and evaluating different systems. This chapter also describes different fingerprint capture device technologies. Here, fingerprints are considered in combination with other biometrics in a multimodal system. Finally, the chapter looks to the future of fingerprint verification [4].

It is necessary to state at the onset that there are many different approaches used for fingerprint verification. Some of these are published in the scientific literature, some published only as patents, and many are kept as trade secrets. This chapter attempts to cover what is publicly known and used in the field, and to cite both the scientific and patent literature. Furthermore, while the chapter attempts to be objective, some material is arguable and should be regarded that way [4].

History

There is archaeological evidence that fingerprints as a form of identification have been used at least since 7000 to 6000 BC by the ancient Assyrians and Chinese. Clay pottery from these times sometimes contain fingerprint impressions placed to mark the potter. Chinese documents bore a clay seal marked by the thumbprint of the originator. Bricks used in houses in the ancient city of Jericho were sometimes imprinted by pairs of thumbprints of the bricklayer. However, though fingerprint individuality was recognized, there is no evidence this was used on a universal basis in any of these societies [4].

In the mid-1800s scientific studies were begun that would establish two critical characteristics of fingerprints that are true to this day: No two fingerprints from different fingers have been found to have the same ridge pattern; and fingerprint ridge patterns are unchanging throughout life. These studies led to the use of fingerprints for criminal identification, first in Argentina in 1896, then at Scotland Yard in 1901, and in other countries in the early 1900s [4].

Computer processing of fingerprints began in the early 1960s with the introduction of computer hardware that could reasonably process these images. Since then, automated fingerprint verification systems (AFVS) have been deployed widely among law enforcement agencies throughout the world [4].

In the 1980s, innovations in two technology areas, personal computers and optical scanners [10], enabled the tools to make fingerprint capture practical in noncriminal applications, such as ID card programs. Now, in the early 2000s, the introduction of inexpensive fingerprint capture devices and the development of fast, reliable matching algorithms has set the stage for the expansion of fingerprint matching to personal use [4].

Why include a history of fingerprints in this chapter? This history of use is one that other types of biometrics do not come close to. There is the experience of a century of forensic [5] use and hundreds of millions of fingerprint matches, through which you can say with some authority that fingerprints are unique and their use in matching is extremely reliable [4].

Fingerprints

As previously mentioned, fingerprinting is the oldest method of successfully matching an identity. A person's fingerprints are a complex combination of patterns known as lines, arches, loops, and whorls. The most distinctive characteristics are the minutiae, the smallest details found in the ridge endings. There are several factors in favor of using fingerprinting for the purpose of a verification system. Fingerprints cannot be forged, and every individual has a unique print. Due to the several layers of skin that makes up the fingerprint, chemicals can not erase the uniqueness of the ridges, although some prints can be altered due to finger surgery or hand injury [1].

Obtaining a high-quality image where the fingerprint ridges and minutiae are recognizable is a complex procedure. The area from which to take measurements is extremely small; thus, sophisticated devices have been developed to capture an image with significant detail. A fingerprint capture system accomplishes the task in a variety of ways. The following devices are capture methods used today: reflected light optical, solid-state capacity inductance scanners, ultrasound scanners, micropad pressure sensors, and polarized multifrequency infrared illumination. These capture methods differ in terms of convenience, cost, and accuracy. The following are the advantages of fingerprinting:

- Prints remain the same throughout a person's lifetime;
- Fingerprinting is neither frightening nor emotionally disturbing;
- People's prints are unique [1].

However, the disadvantages are as follows;

- Searching through a huge database can be rather slow;
- Dirt on the finger or injury can blur the print;
- A fingerprint template is rather large compared to other biometric devices [1].

The fact that fingerprint files are extremely large (250 kB), compared to other biometrics, has created a problem for installing fingerprint biometric data on portable ID cards. The users must have databases at each verification site or create a way to download data to a central site of identity verification, which increases cost and slows down the matching process [1].

Fingerprint Verification

Fingerprints have been found on pottery and cave paintings from thousands of years ago, suggesting that the use of fingerprints to identify an individual dates back to ancient times. But the idea that no two individuals have the same fingerprints and that fingerprint patterns do not change significantly throughout life became accepted during the course of the 19th century. This gave rise to the law enforcement practice of using fingerprints for the identification of criminals. As a result, criminals found it harder to deny their identity, while innocent people were less likely to be wrongly identified as criminals. Moreover, by comparing fingerprints at a crime scene with the fingerprint record of suspected persons, proof of presence could be established [2].

Fingerprint verification, however, could only be done by highly trained and skilled people. Demands for fingerprint verification from law enforcement authorities began to outpace the laborious manual and visual approach to fingerprint indexing, searching, and matching. The advent of computing power led to the development of automatic fingerprint verification systems (AFVS). These systems have greatly improved the operational productivity of law enforcement agencies and reduced the cost of hiring and training human fingerprint experts. The rapid growth of automatic fingerprint verification technology for forensic use has paved the way for the application of fingerprint technology in other (civilian) domains. Fingerprint-based biometric systems have almost become synonymous with biometric systems as a whole. Fingerprint systems account for almost 53% of the biometrics market. Other biometric technologies may gain in popularity, but the use of the fingerprint still remains the oldest method of computer-aided personal verification [2].

What Is Fingerprint Verification?

Fingerprint verification consists of comparing a print of the characteristics of a fingertip or a template of that print with a stored template or print. Fingerprints become fully formed in the seventh month of fetal development, and they do not develop further throughout the life of an individual (though injury or skin conditions may cause changes). Not only are the fingerprints of different people different, there are so many variations during the formation of fingerprints that it is virtually impossible for two fingerprints to be exactly alike. Fingerprints from different fingers of the same individual are not entirely unrelated, as they originate from the same genes. This means, for instance, that the fingerprints of identical twins are similar but not identical. Under good conditions and with state-of-the-art technology, it seems that automatic fingerprint verification is able to distinguish identical twins but with a slightly lower accuracy than for non-twins. It is important to note that the uniqueness of fingerprints is not an established physiological fact but rather an empirical observation. Fingerprint formations are well studied, but the debate on the real uniqueness of fingerprints is not completely resolved [2].

How Does It Work?

A fingerprint consists of the features and details of a fingertip. As discussed previously, there are three major fingerprint features: the arch, the loop, and the whorl. Each finger has at least one major feature. Loops are lines that enter and exit on the same side of the print. Arches are lines that start on one side of the print, rise into hills, and then exit on the other side of the print. Whorls are circles that do not exit on either side of the print. The smaller or minor features (or minutiae) consist of the position of ridge ends (ridges are the lines that flow in various patterns across fingerprints) and of ridge bifurcations (the point where ridges split in two), as shown in Figure 20-1 [2]. There are between 50 and 200 of such minor features on every finger. Fingerprint verification done on the basis of the three major features is called pattern matching, while the more microscopic approach is called minutiae matching. Other features may be used for verification, but patterns and minutiae are the main ones.

Figure 20-1
Minutiae of a fingerprint. (Source: Reproduced with permission from the Institute for Prospective Technological Studies.)

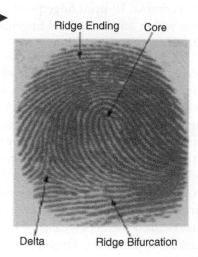

Ridge Ending   Core

Delta   Ridge Bifurcation

Acquiring a Sample

A fingerprint image can be captured voluntarily and/or consciously (with the person's consent and/or knowledge), or involuntarily or unconsciously. The latter typically occurs at the scene of crime where available fingerprints are investigated. People leave fingerprint trails on almost every surface they touch via the oil that coats the ridges of their print. The residue that is left behind is known as a latent fingerprint. For these to be used for identification or verification, they first need to be enhanced, for instance with special powders and brushes, and for matching they need to be photographed or lifted and placed on a fingerprint card [2].

Enrollment and acquisition can be done off-line or with a live sensor. An off-line image is typically obtained by smearing ink on the fingertip and creating an inked impression of the fingertip on a paper (or fingerprint card). This is the oldest and best-known acquisition technique, which is still used by law enforcement and other government agencies worldwide. Before the age of digitalization, these fingerprint cards were copied and sent to a centralized national verification office, where all cards were stored and where matching took place. Such a process is quite laborious and time-consuming. According to the FBI, a fingerprint check under this system would take usually three months to complete [2].

The off-line mode has been advanced during the last decade via digitization. The fingerprint cards are now scanned digitally, allowing the image data to be stored in databases and to be transferred via communication networks. This process is, of course, much faster compared to the physical fingerprint cards. In the United States, responses to criminal 10-print fingerprint submissions done electronically are now possible within one hour. Civil fingerprint submissions are done within 18 hours [2].

Live acquisition, on the other hand, is done by sensors reading the tip of the finger directly and in real time. A fingerprint scan contains a lot of information, but scanners normally focus only on getting an image of the information that is essential for matching. The quality of the sensed fingerprint image is of key importance for the performance of the system. Given the small area of the fingertip, its detailed minutiae, and its continuous use in everyday life (cuts, bruises, aging, weather conditions), poor image quality is a major concern in fingerprint applications. During the last few years, fingerprint scanners have considerably improved their performance and at the same time have become smaller and cheaper. This has enabled the deployment of fingerprint authentication beyond law enforcement applications. Fingerprint scanners are now being integrated in electronic devices such as laptops, keyboards, and PDAs. There are three types of live scanners:

1. Optical devices, which use a light source and lens to capture the fingerprint with a camera;

2. Solid-state sensors or silicon sensors, which appeared on the market in the mid-1990s to address the shortcomings of the early optical sensors;

3. Others, such as acoustic sensors that use acoustic signals to detect fingerprint details [2].

Note: Upcoming solid-state sensors are swiping sensors comparable to, for instance, swiping a credit card.

Important factors when describing and comparing fingerprint capture devices are cost, size, and performance (image resolution, bit depth, capture area, etc.), as well as their accompanying (usually proprietary) software containing the matching algorithms. There are standard requirements related to performance established by the FBI (resolution: 500 dots per inch; pixel depth: 8 bits). Commercial devices sometimes meet some of these requirements, but tradeoffs usually have to be made, especially between size and cost. Although solid-state sensors are currently small enough to be embedded in existing electronic devices (and even current optical sensors), another important tradeoff is the one between size and accuracy (both FAR and FRR): the smaller the fingerprint area rate, the worse the recognition rate (with the exception of "swiping sensors") [2].

Extracting Features

Getting a high-quality image of the fingerprint is very important for accurate fingerprint verification, but feature extraction also plays a crucial role. This consists of converting the fingerprint image into a usable and comparable format that does not require lots of storage space [8]. The format or template is a compressed version of the fingerprint characteristics. Several approaches to automatic minutiae extraction exist, but most of these methods transform fingerprint images into binary images. This means that only the coordinates of the minutiae (30 or 40) are stored, reducing it to a few hundreds of bytes. This is considerably less than the 10 Mbytes of storage per person needed for a 500 dpi image at 8 bits (FBI standards) for all 10 fingers. Central fingerprint databases would thus require terabits of storage [2].

Feature extraction is also needed because even a very precise fingerprint image will have distortions and false minutiae that need to be filtered out. An algorithm may search the image and eliminate one of two adjacent minutiae, as minutiae are very rarely adjacent. Anomalies can also be caused by scars, sweat, or dirt. The algorithms used for the feature extraction filter the image to eliminate the distortions and would-be minutiae [2].

Comparing Templates

The identification or verification process follows the same steps as the enrollment process with the addition of matching. It compares the template of the live image with a database of enrolled templates (verification), or with a single enrolled template (authentication) [2].

Declaring a Match

Matching can be separated into two categories: verification and identification. Verification is the topic of this chapter. It is the comparison of a claimant fingerprint against an enrollee fingerprint, where the intention is that the claimant fingerprint matches the enrollee fingerprint. To prepare for verification, a person initially enrolls his or her fingerprint into the verification system. A representation of that fingerprint is stored in some compressed format along with the person's name or other identity. Subsequently, each access is authenticated by the person identifying himself or herself, and then applying the fingerprint to the system such that the identity can be verified. Verification is also termed one-to-one matching [4].

Identification is the traditional domain of criminal fingerprint matching. A fingerprint of unknown ownership is matched against a database of known fingerprints to associate a crime with an identity. Identification is also termed one-to-many matching [4].

There is an informal third type of matching that is termed one-to-few matching. This is for practical applications where a fingerprint system is used by "a few" users, such as by family members to enter their house. A number that constitutes "few" is usually accepted as being somewhere between five and 20 [4].

The comparison between the sensed fingerprint image or template against records in a database or a chip usually yields a matching score quantifying the similarity between the two representations. If the score is higher than a certain threshold, a match is declared (belonging to the same finger(s)). The decision of a match or nonmatch can be automated, but it depends on whether matching is done for identification or verification purposes [2].

With verification applications, automated decision making is possible when conditions are ideal. In the case of the FBI, for instance, this means that fingerprint cards can be matched automatically when both enrollment and acquisition were done by law enforcement staff. But with latent prints (collected at a crime scene) and prints with a lower quality image, the automated process is less reliable. Automated systems imitate the way human fingerprint experts work, but the problem is that these systems cannot have observed the many underlying information-rich features an expert is able to detect visually. Automatic systems are reliable, rapid, consistent, and cost effective when matching conditions are good, but their level of sophistication can not rival that of a well-trained fingerprint expert. Therefore, for instance, a fingerprint expert can overrule an automated match [2].

Verification applications, especially mainstream commercial fingerprint verification, may be to a certain extent less accurate. This is because the issues at stake are different (identifying criminals), and also because verification consists of 1:1 matching. Verification may use less information from a fingerprint compared to forensic scientists identifying a fingerprint. The former seems to be more like a possible, "close-enough" correlation of similarities. Because of background interference (dirt, scratches, light, etc.) and no human supervision, the quality of fingerprint images is lower. The result is a "best" matching score that would not be feasible for law enforcement [2].

Feature Types

As discussed previously, the lines that flow in various patterns across fingerprints are called ridges and the spaces between ridges are called valleys. These ridges are compared between one fingerprint and another when matching. Fingerprints are commonly matched by one (or both) of two approaches. This part of the chapter describes the fingerprint features as associated with these approaches [4].

The more microscopic of the approaches is called minutia matching. The two minutia types that are shown in Figure 20-2 are a ridge ending and bifurcation. An ending is a feature where a ridge terminates [4]. A bifurcation is a feature where a ridge splits from a single path to two paths at a Y-junction. For matching purposes, a minutia is attributed with features of type, location (x, y), and direction. (Some approaches use additional features.)

Figure 20-2 *Fingerprint minutiae: ending and bifurcation. (Source: Reproduced with permission from Veridicom International Inc.)*

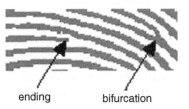

The more macroscopic approach to matching is called global pattern matching or simply pattern matching. In this approach, the flow of ridges is compared at all locations between a pair of fingerprint images. The ridge flow constitutes a global pattern of the fingerprint. Three fingerprint patterns are shown in Figure 20-3 [4].

Note: Different classification schemes can use up to 10 or so pattern classes, but these three are the basic patterns.

Figure 20-3 *Fingerprint patterns: arch, loop, and whorl. Fingerprint landmarks are also shown: core and delta. No delta locations fall within the captured area of the whorl here. (Source: Reproduced with permission from Veridicom International Inc.)*

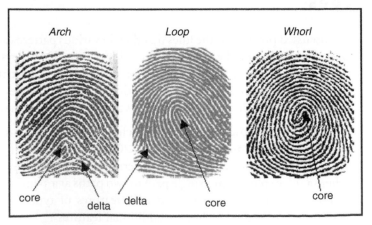

Two other features are sometimes used for matching: core and delta, as shown in Figure 20-3 [4]. The core can be thought of as the center of the fingerprint pattern. The delta is a singular point from which three patterns deviate. The core and delta locations can be used as landmark locations by which to orient two fingerprints for subsequent matching, although these features are not present on all fingerprints [4].

Other features of the fingerprint may be used in matching. For instance, pores can be resolved by some fingerprint sensors, and there is a body of work (mainly research at this time) to use the position of the pores for matching in the same manner as the minutiae are used. Size of the fingerprint and average ridge and valley widths can be used for matching; however, these are changeable over time. The positions of scars and creases can also be used, but are usually not used because they can be temporary or artificially introduced [4].

Image Processing and Verification

Following image capture to obtain the fingerprint image, image processing is performed. The ultimate objective of image processing is to achieve the best image by which to produce the correct match result. The image processing steps are the following: image noise reduction and enhancement, feature detection, and matching [4].

This part of the chapter is organized to describe first the sequence of processing and verification via a common minutia-based approach. This is described without variants and optional methods (of which there are many) for the sake of reading flow and simplicity. Furthermore, variations of this approach, both minutia-based and nonminutia-based, are also described [4].

Note: It is important to know that, though many researchers and product developers follow the preceding approach, all do not, and even the choice of what constitutes "common" may be contentious.

Image Specifications

Depending upon the fingerprint capture device, the image can have a range of specifications. Commonly, the pixels are eight-bit values, and this yields an intensity range from zero to 255. The image resolution is the number of pixels per unit length, and this ranges from 250 dots per inch (100 dots per centimeter) to 625 dots per inch (250 dots per centimeter), with 500 dots per inch (200 dots per centimeter) being a common standard. The image area is from 0.5 inches square (1.27 centimeter) to 1.25 inches (3.175 centimeter), with 1 inch (2.54 centimeter) being the standard [4].

Image Enhancement

A fingerprint image is one of the noisiest of image types. This is due predominantly to the fact that fingers are your direct form of contact for most of the manual tasks you perform: fingertips become dirty, cut, scarred, creased, dry, wet, worn, and so on. The image enhancement step is designed to reduce this noise and to enhance the definition of ridges against valleys. Two image processing operations designed for these purposes are the adaptive, matched filter and adaptive thresholding. The stages of image enhancement, feature detection, and matching are illustrated in Figure 20-4 [4].

There is a useful side to fingerprint characteristics as well. That is the "redundancy" of parallel ridges. Even though there may be discontinuities in particular ridges, one can always look at a small, local area of ridges and determine their flow. You can use this "redundancy of information" to design an adaptive, matched filter. This filter is applied to every pixel in the image (spatial convolution is the technical term for this operation). Based on the local orientation of the ridges around each pixel, the matched filter is applied to enhance ridges oriented in the same direction as those in the same locality, and to decrease anything oriented differently. The latter includes noise that may be joining adjacent ridges, thus flowing perpendicular to the local flow. These incorrect "bridges" can be eliminated by use of the matched filter. Figure 20-4(b) shows an orientation map where line sectors represent the orientation of ridges in each locality [4]. This filter is adaptive because it orients itself to local ridge flow. It is matched because it should enhance (or match) the ridges and not the noise [4].

After the image is enhanced and noise reduced, you are ready to extract the ridges. Though the ridges have gradations of intensity in the original gray-scale image, their true information is simply binary: ridges against background. Simplifying the image to this binary representation facilitates subsequent processing. The binarization operation takes as input a gray-scale image and returns a binary image as output. The image is reduced in intensity levels from the original 256 (eight-bit pixels) to two (one-bit pixels) [4].

The difficulty in performing binarization is that all the fingerprint images do not have the same contrast characteristics, so a single intensity threshold

Figure 20-4

Figure 20-4
*Sequence of
fingerprint
processing steps:
(a) original,
(b) orientation,
(c) binarized,
(d) thinned,
(e) minutiae,
(f) minutia
graph. (Source:
Reproduced with
permission from
Veridicom
International
Inc.)*

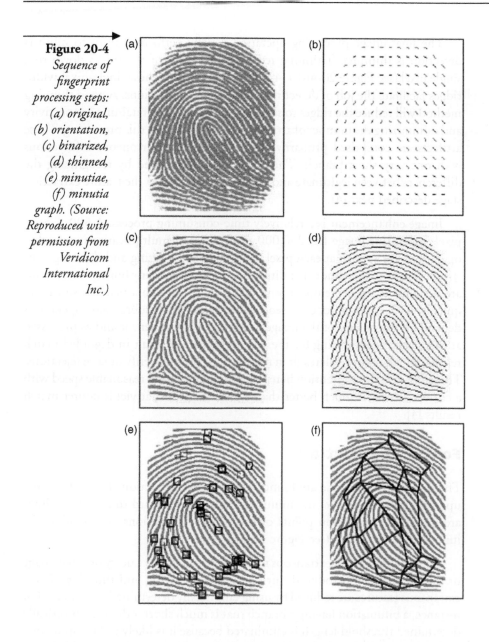

cannot be chosen. Furthermore, contrast may vary within a single image, for instance, if the finger is pressed more firmly at the center. Therefore, a common image processing tool is used called locally adaptive thresholding. This operation determines thresholds adaptively to the local image intensities. The binarization result is shown in Figure 20-4(c) [4].

The final image processing operation usually performed prior to minutia detection is thinning. Thinning reduces the widths of the ridges down to a single pixel (see Figure 20-4(d)) [4]. It will be seen how these single-pixel-width ridges facilitate the job of detecting endings and bifurcations. A good thinning method will reduce the ridges to single-pixel width while retaining connectivity and minimizing the number of artifacts introduced due to this processing. These artifacts are comprised primarily of spurs, which are erroneous bifurcations with one very short branch. These artifacts are removed by recognizing the differences between legitimate and erroneous minutiae in the feature extraction stage described next [4].

Image enhancement is a relatively time-consuming process. A 500 × 500-pixel fingerprint image has 250,000 pixels; several multiplications and other operations are applied at each pixel. Both matched filtering and thinning contribute largely to this time expenditure. Consequently, many fingerprint systems are designed to conserve operations at this stage to reach a match result more quickly. This is not a good tradeoff. The results of all subsequent operations depend on the quality of the image as captured by the sensor and as processed at this stage. Economizing for the sake of speed will result in degraded match results, which in turn will result in repeated attempts to verify or false rejections. Therefore, it is the contention here that a system offering reasonable speed with a correct answer is much better than a faster system that yields poorer match results [4].

Feature Extraction

The fingerprint minutiae are found at the feature extraction stage. Operating upon the thinned image, the minutiae are straightforward to detect. Endings are found at termination points of thin lines. Bifurcations are found at the junctions of three lines (see Figure 20-4(e)) [4].

There will always be extraneous minutiae found due to a noisy original image or due to artifacts introduced during matched filtering and thinning. These extraneous features are reduced by using empirically determined thresholds. For instance, a bifurcation having a branch that is much shorter than an empirically determined threshold length is eliminated because it is likely to be a spur. Two endings on a very short isolated line are eliminated because this line is likely due to noise. Two endings that are closely opposing are eliminated because these are likely to be on the same ridge that has been broken due to a scar or noise or a dry finger condition that results in discontinuous ridges. Endings at the boundary of the fingerprint are eliminated because they are not true endings but rather the extent of the fingerprint in contact with the capture device [4].

Feature attributes are determined for each valid minutia found. These consist of ridge ending or bifurcation type, the (x, y) location, and the direction of the ending or bifurcation. Although minutia type is usually determined and stored, many fingerprint matching systems do not use this information because discrimination of one from the other is often difficult [4].

The result of the feature extraction stage is what is called a minutia template. This is a list of minutiae with accompanying attribute values. An approximate range on the number of minutiae found at this stage is from 10 to 100. If each minutia is stored with type (one bit), location (nine bits each for x and y), and direction (eight bits), then each will require 27 bits (say, four bytes) and the template will require up to 400 bytes. It is not uncommon to see template lengths of 1,024 bytes [4].

Verification

At the verification stage, the template from the claimant fingerprint is compared against that of the enrollee fingerprint. This is done usually by comparing neighborhoods of nearby minutiae for similarity. A single neighborhood may consist of three or more nearby minutiae. Each of these is located at a certain distance and relative orientation from each other. Furthermore, each minutia has its own attributes of type (if it is used) and minutia direction, which are also compared. If comparison indicates only small differences between the neighborhood in the enrollee fingerprint and in the claimant fingerprint, then these neighborhoods are said to match. This is done exhaustively for all combinations of neighborhoods and, if enough similarities are found, then the fingerprints are said to match. Template matching can be visualized as graph matching, that is, comparing the shapes of graphs joining fingerprint minutiae. This is illustrated in Figure 20-4(f) [4].

Note: The word "similar" is used in the discussion of verification instead of "same."

Neighborhoods will rarely match exactly because of two factors. One is the noisy nature of a fingerprint image. The other is that the skin is an elastic surface, so distances and minutia directions will vary [4].

One result of the verification stage is a match score, usually a number between 0 and 1 (or 10 or 100). Higher values in the range indicate higher confidence in a match. This match score is then subject to a user-chosen threshold value. If the score is greater than the threshold, the match result is said to be true (or 1) indicating a correct verification; otherwise, the match is rejected and the match result is false (or 0). This threshold can be chosen to be higher to achieve greater confidence in a match result, but the price to pay for this is a greater number of false rejections. Conversely, the threshold can be chosen to be lower to reduce the number of false rejections, but the price to pay in this case is a greater number of false acceptances [4].

The user has control of only one parameter, the threshold, for most commercial verification products. This customization procedure is called back-end adjustment, because a match score is calculated first and the threshold can be chosen after to determine the match result. There are systems that, in addition to offering back-end adjustment, offer front-end adjustment as well. This enables the user to adjust some of the parameter values before the match score is calculated, and then to adjust the threshold after. Systems with front-end adjustment offer more versatility in obtaining the best results for different conditions, but are more complex for the user to adjust. This is why, for most systems, the vendor sets the optimum front-end parameter values and the user has control only of the matching threshold value via back-end adjustment [4].

Identification and One-to-Few Matching

Although the emphasis in this chapter is verification, it briefly will mention identification and one-to-few matching methods. For identification, the objective is to determine a match between a test fingerprint and one of a database of fingerprints whose size may range from 10,000 to tens of millions. One cannot simply apply the verification techniques just described to all potential matches because of the prohibitive computation time required. Therefore, identification is usually accomplished as a two-step process. Fingerprints in the database are first categorized by pattern type, or binned. The same is done for the test fingerprint. Pattern comparison is done between the test fingerprint and database fingerprints. This is a fast process that can be used to eliminate the bulk of nonmatches. For those fingerprints that closely match in pattern, the more time-consuming process of minutia-based verification is performed [4].

One-to-few matching is usually accomplished simply by performing multiple verifications of a single claimant fingerprint against the five to 20 potential matches. Thus, the execution time is linear in the number of potential matches. This time requirement becomes prohibitive if "few" becomes too large; in that case, an approach akin to identification must be used [4].

Variations on the Common Approach: Other Methods

Since one of the most vexing challenges of fingerprint processing is obtaining a clean image upon which to perform matching, various methods have been proposed to perform image enhancement. Most of these involve filtering that is adaptively matched to the local ridge orientations. The orientation map is first determined by dividing the image into windows (smaller regions) and calculating the local ridge orientations within these. The orientation can be determined in each window by spatial domain processing or by frequency domain processing after transformation by a two-dimensional fast Fourier transform [4].

After image enhancement and binarization of the fingerprint image, thinning is usually performed on the ridges. However, a different approach eliminates the binarization and thinning stages (both computationally expensive and noise-producing). This approach involves tracing ridges not from the binary or thinned image, but from the original gray-scale image. The result of the gray-scale ridge is the end point and bifurcation minutiae similar to the common approach [4].

Instead of using only a single-size window to determine the orientation map, multiple window sizes can be used via a multiresolution approach. Local orientation values are determined first throughout the image at a chosen, initial resolution level—that is, a chosen window size of pixels within which the orientation is calculated. A measure of consistency of the orientation in each window is calculated. If the consistency is less than a threshold, the window is divided into four smaller subwindows and the same process is repeated until consistency is above threshold for each window or subwindow. This multiresolution process is performed to avoid smoothing over small areas of local orientation, as will be the case, especially at the fingerprint core [4].

Because of the difficulty of aligning minutiae of two fingerprints, neighborhood matching was one of the earliest methods of facilitating a match. Groups of neighboring minutiae are identified in one fingerprint, usually two to four minutiae to a neighborhood, and each of these is compared against prospective neighborhoods of another fingerprint. There are two levels to matching. One is matching the configurations of minutiae within a neighborhood against another neighborhood. The other is matching the global configurations formed by the separate neighborhoods between enroll and verify fingerprints [4].

Because it is time-consuming to compare all neighborhood combinations between enroll and verify fingerprints, methods have been proposed to align the fingerprints to reduce the number of comparisons. A common method, and also a traditional method used for visual matching, is to locate a core and delta and align the fingerprints based on these landmarks. The core and delta are usually found on the basis of their position with respect to the ridge flow; therefore, the orientation map is determined and used for this. An elegant method to locate singular points in a flow field is the Poincaré index. For each point in the orientation map, the orientation angles are summed for a closed curve in a counterclockwise direction around that point. For nonsingular points, the sum is equal to 0 degrees; for the core, the sum is equal to 180 degrees; for a delta, the sum is equal to −180 degrees [4].

Other methods have been proposed to reduce the computational load of minutia matching. One approach is to sort the list of minutiae in some order conducive to efficient comparisons prior to matching [4].

Note: The preceding is especially appropriate for one-to-many matching, since sorting is done once per fingerprint, but matching many times.

A linearly sorted list of minutiae can be compiled by scanning the fingerprint from a selected center point outward by a predetermined scanning trajectory such as a spiral. In this way, one-dimensional vectors of minutiae, including their characteristics, can be compared between enroll and verify fingerprints. Another method to linearize the minutia comparison is the "hyperladder" matcher. This hyperladder is constructed sequentially by comparing minutia pairs in enroll and verify fingerprints, and adding more rungs as consecutive neighboring minutiae match. In another approach, an attributed graph can be constructed in which branches constitute nearest neighbor minutiae and emanate like "stars" on the graph. These stars are compared between fingerprint pairs—the number of matching branches constitutes the degree of confidence in the match [4].

Because there is so little discriminating information at a single minutia (even the type is unreliable), a different approach is to describe minutiae by more features. For instance, a minutia can be described by the length and curvature of the ridge it is on and of similar features on neighboring ridges [4].

Variations on the Common Approach: Correlation Matching

This discussion of matching has been minutia-focused to this point, to the exclusion of the global pattern matching approach mentioned earlier. Instead of using minutiae, some systems perform matches on the basis of the overall ridge pattern of the fingerprint. This is called global matching, correlation, or simply image multiplication or image subtraction [4].

It is visibly apparent that a pair of fingerprints of different pattern types, for instance whorl and arch, does not match. Global matching schemes go beyond the simple (and few) pattern categories to differentiate one whorl from a different whorl, for instance. Simplistically, this can be thought of as a process of aligning two fingerprints and subtracting them to see if the ridges correspond. There are four potential problems (corresponding to three degrees of freedom and another factor):

1. The fingerprints will likely have different locations in their respective images (translational freedom). You can establish a landmark such as a core or delta by which to register the pair; however, if these are missing or not found reliably, subsequent matching steps will fail.

2. The fingerprints may have different orientations (rotational freedom). If a proper landmark has been found in (1), the fingerprint can be rotated around this, but this is error-prone and/or computationally expensive. It is error-prone because the proper center of rotation depends on a single, reliably determined landmark. It is computationally expensive because performing correlation for many orientations involves repeatedly processing the full image.

3. Because of skin elasticity (nonlinear warping), even if matching fingerprints are registered in location and orientation, all subregions may not align.

4. Finally, there is the inevitable problem of noise. Two images of matching fingerprints will have different image quality, ridges will be thicker or thinner, discontinuities in ridges will be different depending on finger dryness, and the portion of the fingerprint captured in each image will be different [4].

The descriptions coming up next are more sophisticated modifications and extensions to the basic correlation approach to deal with the problems listed. Strictly speaking, the correlation between two images involves translating one image over another and performing multiplication of each corresponding pixel value at each translation increment. When the images correspond at each pixel, the sum of these multiplications is higher than if they do not correspond.

Therefore, a matching pair will have a higher correlation result than a non-matching pair. A threshold is chosen to determine whether a match is accepted, and this can be varied to adjust the false acceptance rate versus false rejection rate tradeoff, similar to the case for minutia matching [4].

Correlation matching can be performed in the spatial frequency domain instead of in the spatial domain as just described. The first step is to perform a two-dimensional fast Fourier transform (FFT) on both the enrollee and claimant images. This operation transforms the images to the spatial frequency domain. The two transformed images are multiplied pixel-by-pixel, and the sum of these multiplications is equivalent to the spatial domain correlation result. An advantage of performing frequency domain transformation is that the fingerprints become translation-independent; that is, they do not have to be aligned translationally because the origin of both transformed images is the zero-frequency location (0,0). There is a tradeoff to this advantage, however, in the cost of performing the two-dimensional FFT [4].

Frequency domain correlation matching can be performed optically instead of digitally. This is done using lenses and a laser light source. Consider that a glass prism separates projected light into a color spectrum; in other words, it performs frequency transformation. In a similar manner, the enrollee and claimant images are projected via laser light through a lens to produce their Fourier transform. Their superposition leads to a correlation peak whose magnitude is high for a matching pair and lower otherwise. An advantage of optical signal processing is that operations occur at the speed of light, much more quickly than for a digital processor. However, the optical processor is not as versatile (as programmable) as a digital computer, and because of this, few or no optical computers are used in commercial personal verification systems today [4].

One modification of spatial correlation is to perform the operation not upon image pixels but on grids of pixels or on local features determined within these grids. The enrollee and claimant fingerprint images are first aligned, and then (conceptually) segmented by a grid. Ridge attributes are determined in each grid square: average pixel intensity, ridge orientation, periodicity, or number of ridges per grid. Corresponding grid squares are compared for similar attributes. If enough of these are similar, this yields a high match score and the fingerprints are said to match [4].

The relative advantages and disadvantages between minutia matching and correlation matching differ between systems and algorithmic approaches. In general, minutia matching is considered by most to have a higher recognition accuracy. Correlation can be performed on some systems more quickly than minutia matching, especially on systems with vector-processing or

FFT hardware. Correlation matching is less tolerant of elastic, rotational, and translational variances of the fingerprint and of extra noise in the image [4].

Technology: State of Development

Since fingerprint technology is one of the oldest automated biometric identifiers (supported by strong demand from law enforcement), it has undergone extensive research and development. According to industry analysts, though, there is a popular misconception that automatic fingerprint verification technologies are without problems. They believe that fingerprint verification is still a challenging and important machine pattern recognition problem [2].

One of these challenges relates to the question of interoperability. Fingerprint verification normally consists of a closed system that uses the same sensors for enrollment and acquisition; the same algorithms for feature extraction and matching; and clear standards for the template (for instance, the enrollment procedure [FBI standard is nail-to-nail]). Take the example of fingerprint sensors. There are many different vendors on the market that all have proprietary feature extraction algorithms that are strongly protected, although there are some (proprietary) sensor independent verification algorithms on the market. Different sensors using the same technology (solid-state) produce different fingerprint raw image data, in the same way that sensors using different technologies (optical and solid-state) deliver raw images that are significantly different. Sensor interoperability is a problem that hitherto hardly has been studied and addressed, but it will become increasingly important as fingerprint scanners are more and more embedded in consumer electronics. In addition to image data, there is also the issue of interoperability of minutiae data that has been put forward recently [2].

Challenges and Limitations: Seven Pillars

Fingerprint verification has a good balance related to the so-called seven pillars of biometrics. Nearly every human being possesses fingerprints (universality) with the exception of those with hand-related disabilities. Fingerprints are also distinctive and the fingerprint details are permanent, although they may temporarily change due to cuts and bruises on the skin or external conditions (wet fingers). Live-scan fingerprint sensors can capture high-quality images (collectibility). The deployed fingerprint-based biometric systems offer good performance, and fingerprint sensors have become quite small and affordable. Fingerprints do have a stigma of criminality associated with them, but that is changing with the increased demand for automatic verification and authentication in a digitally interconnected society (acceptability). By combining the use of multiple fingers, cryptographic techniques, and liveness detection, fingerprint systems are becoming quite difficult to circumvent [2].

When only one finger is used, however, universal access and permanent availability may be problematic. Moreover, everyday life conditions can also cause deformations of the fingerprint, for instance, as a result of doing manual work or playing an instrument. Certain conditions, such as arthritis, affect the ease of use of fingerprint readers. Other conditions such as eczema may affect the fingerprint itself. It is estimated that about 5% of people would not be able to register and deliver a readable fingerprint. With large-scale applications that entail millions of people, an estimated 5% of people being temporarily or permanently unable to register amounts to a significant number. This will not only lead to serious delays (decrease in task performance) and annoyance (decrease in user satisfaction) but also makes fingerprinting not fully universally accessible [2].

Security

The security of the fingerprint verification system, as such, is dependent on two main areas: electronic security and liveness testing. Electronic security has to do with traditional digital security issues and is tackled with encryption [7] and other techniques to make it difficult to capture fingerprint information when it is being transmitted. For verification applications, one of the most secure systems (it is argued) consists of having the full system on a smart card (template, sensor, feature extraction, and matching). The output would then be a simple yes or no, or an encrypted message. Such a decentralized system (which is expected to become possible in the near future) would combine the biometric advantage of strong authentication with the user being in full control and without the biometric privacy risks [2, 6].

Apart from the cases where physical threats and force are used to get someone's fingerprint (or a dead finger), liveness testing also deals with spoofing the system with a fake, artificial fingerprint taken, for example, from fingerprint images people leave everywhere (latent fingerprints). There are some reports that fingerprints were relatively easy to reproduce with gelatine, but liveness detection procedures (three-dimensional imaging, temperature measuring) are increasingly being integrated in fingerprint readers. It is therefore argued that fingerprint verification is getting less vulnerable to artificial fingerprints [2].

Privacy

The privacy risks related to fingerprints are mainly the ones that are similar to most biometrics: the risk that unauthorized third parties get access to the biometric data as unique identifiers; the digital traces that biometric identification leave behind; and the traditional data protection issues related to storage (central or not), access (who has access), consent, transparency, and so on. There is also the issue of purpose creep or function creep, whereby the data collected for one purpose are used for other purposes. In addition, specific privacy concerns with fingerprints may come from its use by law enforcement agencies [2].

Applications

Fingerprint verification of criminals for law enforcement continues to be one of the major applications for this technology. Another large-scale application in Europe is EURODAC for asylum requests. In New York, fingerprints are used to prevent fraudulent enrollment for benefits. Using fingerprint verification to secure physical access is another popular application. Embedding of fingerprint readers in electronic devices opens up a whole range of digital applications that are based on online authentication. Finally, decisions have been taken for the future integration of fingerprints (with other biometrics) on travel documents and passports [2].

The Integrated Automated Fingerprint Verification System, more commonly known as IAFVS, is one of the largest biometric databases in the world. It is a U.S. national fingerprint and criminal history system maintained by the FBI that contains the fingerprints and corresponding criminal history information for more than 70 million subjects in the Criminal Master File. The fingerprints and corresponding criminal history information are submitted voluntarily by state, local, and federal law enforcement agencies. The IAFVS provides automated fingerprint search capabilities, electronic image storage, and electronic exchange of fingerprints and responses, 24 hours a day, 365 days a year. In Europe, there is no such database. Criminal fingerprint databases are under the control of national criminal authorities. The United Kingdom, for instance, has a national automated fingerprint verification system (NAFV) containing more than 7 million records [2].

There also exists a large central fingerprint database in the European Union. But this database exists for other purposes: It aims at preventing duplication of asylum requests in the EU member states. EURODAC is an EU-wide database (AFVS) set up to check the fingerprints of asylum seekers against the records of other EU countries. After four years of operation, an evaluation report on

EURODAC highlighted satisfactory results in terms of efficiency, quality of service, and cost-effectiveness. The EURODAC central unit has been operating continuously. Within four years, it processed almost 694,000 fingerprints of asylum seekers. It detected 51,621 cases of multiple applications (the same person having already made an asylum application in another country), which represents 7% of the total number of cases processed. In addition to asylum requests, illegal immigrants have been identified. Almost 51,000 fingerprints of illegal aliens were detected; and about 12,000 fingerprints were related to attempts to cross borders illegally. The evaluation report also states that there were no data protection problems raised by the member states' national data protection authorities regarding EURODAC operations [2].

The state of New York has over 1.3 million people enrolled in a system that tracks to entitlement to social services and protects against fraud known as "double-dipping" (enrolling for a benefit under multiple names). Fingerprint scanning is also being used to arrange secure access to schools and school premises such as cafeterias and libraries. Finally, with the embedding of fingerprint scanners in electronic devices, online authentication (replacement of passwords, PINs, etc.) becomes possible for a whole range of applications, including electronic payments [2].

In 2004, at the EU level, the Council of European Ministers adopted the regulation on mandatory facial images and fingerprints in EU passports at its meeting in Brussels. This regulation applies to passports and travel documents issued by member states (excluding Ireland, the United Kingdom, and Denmark). The Official Journal passports issued will have to contain a facial image within 18 months, and fingerprints within three years. Also, a committee has been set up by the European Commission with representatives from 22 member states to decide on the details such as how many fingerprints are to be taken, the equipment needed, and the costs [2].

Future Trends

Fingerprint verification scores well on the so-called seven pillars of biometrics. The quality of the acquired image at enrollment determines to a large extent the accuracy of the fingerprint matching. But the size of the sensor, its price and quality, and the required threshold for the recognition rate are also important factors to be taken into account. They relate to each other, so tradeoffs have to be made. But in general, the theoretical accuracy with fingerprint verification is said to be quite high. Also, the current embedding of fingerprint technology in consumer electronics might help to relieve fingerprinting from its criminal connotation [2].

However, a non-negligible part of the population faces difficulties in being enrolled and verified through fingerprints. For large-scale applications, this limiting factor needs to be taken into account. Fears related to hygiene and to physical attacks to get one's fingerprints have been reported. Some argue that all this calls for the availability of an alternative, be it a second biometric (face) or something else. Fingerprint verification is currently being used in conjunction with large-scale central databases for forensic purposes and for asylum requests. Other applications are related to checking entitlements and authorizing physical access. With the emerging trend of embedding fingerprint readers into electronic devices, fingerprint technology is losing its criminal stigma in favor of a wide range of online applications that require secure authentication. Finally, decentralized system-on-chip solutions are foreseen to address both privacy and security concerns [2].

Summary/Conclusion

The use of fingerprints for identification has been employed in law enforcement for about a century. A much broader application of fingerprints is for personal authentication, for instance, to access a computer, a network, a bank machine, a car, or a home. The topic of this chapter was fingerprint verification, where "verification" implies a user matching a fingerprint against a single fingerprint associated with the identity that the user claims. The following topics were covered: history, image processing methods, enrollment and verification procedures, system security considerations, recognition rate statistics, fingerprint capture devices, combination with other biometrics, and the future of fingerprint verification [4].

Fingerprint verification is seen as the most trusted and convenient method. The cost for such technology is becoming easily affordable as more industries enter the marketplace. Even with the prices of fingerprinting systems falling, it is difficult to forecast the overall cost and cost-effectiveness of the biometric security system [1].

Fingerprints is an area in which there have been many new and exciting developments in the past several years. Fingerprints constitute one of the most important categories of physical evidence and are among the few that can be truly individual (see Figure 20-5) [3].

Fingerprints were first recognized as unique in 1684. Recent advances in computing power have given us the ability to capture and compare one fingerprint against another at a user level. Efforts to develop automatic fingerprint verification systems were initiated in the early 1960s in at least three

Figure 20-5 *A fingerprint. (Source: Reproduced with permission from Ball State University.)*

Western countries: the United States, France, and Great Britain. The reason for this push was the availability of the digital computer. There was hope that this new technology could assist or even replace the labor-intensive processes of classifying, searching, and matching that are involved in using fingerprints for personal verification. The Federal Bureau of Investigation sponsored research in automatic fingerprint verification in the United States (see Figure 20-6) [3].

Figure 20-6 *Fingerprint verification device. (Source: Reproduced with permission from Ball State University.)*

Fingerprint verification compares a user's fingerprint to a previously stored template and determines validity or authenticity based on this comparison. The template is created from tiny points called minutiae (based on the position of end points and junctions of print ridges), which are extracted from the fingerprint during enrollment. The comparison of attributes are carried out using complex algorithms during verification [3].

References

1. Alex English, Christina Means, Kris Gordon and Kevin Goetz, "Biometrics: A Technology Assessment," Ball State University, 2000 W. University Ave., Muncie, IN 47306, 2006. Copyright © 2006.

2. "Biometrics at the Frontiers: Assessing the Impact on Society." For the European Parliament Committee on Citizens' Freedoms and Rights, Justice and Home Affairs (LIBE), European Commission, Joint Research Center (DG JRC), Institute for Prospective Technological Studies, 2005. Copyright © 2005 European Communities.

3. Barrett Key, Kelly Neal and Scott Frazier, "The Use of Biometrics in Education Technology Assessment," Ball State University, 2000 W. University Ave., Muncie, IN 47306, 2006. Copyright © 2006.

4. Lawrence O'Gorman, "Fingerprint Verification," [Michigan State University, 3115 Engineering Building, East Lansing, MI 48824-1226] Veridicom International Inc., Head Office, 5th Floor, 21 Water Street, Vancouver, British Columbia, Canada V6B 1A1, 2006.

5. John R. Vacca, *Computer Forensics: Computer Crime Scene Investigation, 2nd ed.*, Charles River Media (2005).

6. John R. Vacca, *Net Privacy: A Guide to Developing and Implementing an Ironclad eBusiness Privacy Plan*, McGraw-Hill (2001).

7. John R. Vacca, *Satellite Encryption*, Academic Press (1999).

8. John R. Vacca, *The Essentials Guide to Storage Area Networks*, Prentice Hall, Professional Technical Reference, Pearson Education (2001).

9. John R. Vacca, *Practical Internet Security*, Springer (2006).

10. John R. Vacca, *Optical Networking Best Practices Handbook*, John Wiley & Sons (2006).

21

Vulnerable Points of a Biometric Verification System

Because biometric-based verification systems offer several advantages over other verification methods, there has been a significant surge in the use of biometrics for user verification in recent years. It is important that such biometric-based verification systems be designed to withstand attacks when employed in security-critical applications, especially in unattended remote applications such as e-commerce. This chapter outlines the inherent vulnerability of biometric-based verification, identifies the weak links in systems employing biometric-based verification, and presents new solutions for eliminating some of these weak links. Although, for illustration purposes, fingerprint verification is used throughout, this analysis extends to other biometric-based methods as well [1].

Reliable user verification is becoming an increasingly important task in the Web-enabled world. The consequences of an insecure verification system in a corporate or enterprise environment can be catastrophic, and may include loss of confidential information, denial of service, and compromised data integrity. The value of reliable user verification is not limited to just computer or network access. Many other applications in everyday life also require user verification, such as banking, e-commerce, and physical access control to computer resources, and could benefit from enhanced security [1].

The prevailing techniques of user verification, which involves the use of either passwords and user IDs, or identification cards and PINs, suffer from several limitations. Passwords and PINs can be illicitly acquired by direct covert observation. Once an intruder acquires the user ID and the password, the intruder has total access to the user's resources. In addition, there is no way to positively link the usage of the system or service to the actual user; that is, there is no protection against repudiation by the user ID owner. For example, when a user ID and password is shared with a colleague, there is no way for the system to know who the actual user is. A similar situation arises when a transaction involving a credit card number is conducted on the Web. Even though the data are sent over the Web using secure encryption methods [3], current systems are not capable of assuring that the transaction was initiated by the rightful owner of the credit card. In the modern distributed systems environment, the traditional verification policy based on a simple combination of user ID and password has become inadequate [1].

Fortunately, automated biometrics in general, and fingerprint technology in particular, can provide a much more accurate and reliable user verification method. Biometrics is a rapidly advancing field that is concerned with identifying a person based on his or her physiological or behavioral characteristics. Examples of automated biometrics include fingerprint, face, iris, and speech recognition. User verification methods can be broadly classified into three categories, as shown in Table 21-1 [1]. Because a biometric property is an intrinsic property of an individual, it is difficult to surreptitiously duplicate and nearly impossible to share. Additionally, a biometric property of an individual can be lost only in case of a serious accident [1].

Biometric readings, which range from several hundred bytes to over a megabyte, have the advantage that their information content is usually higher than that of a password or a pass phrase. Simply extending the length of

Table 21-1   *Existing User Verification Techniques*

Method	Examples	Properties
What you know	User ID	Shared
	Password	Many passwords easy to guess
	PIN	Forgotten
What you have	Cards	Shared
	Badges	Can be duplicated
	Keys	Lost or stolen
What you know and what you have	ATM card + PIN	Shared
		PIN a weak link
		(Writing the PIN on the card)
Something unique about the user	Fingerprint	Not possible to share
	Face	Repudiation unlikely
	Iris	Forging difficult
	Voiceprint	Cannot be lost or stolen

passwords to get equivalent bit strength presents significant usability problems. It is nearly impossible to remember a 2K phrase, and it would take an annoyingly long time to type such a phrase (especially without errors). Fortunately, automated biometrics can provide the security advantages of long passwords while retaining the speed and characteristic simplicity of short passwords [1].

Even though automated biometrics can help alleviate the problems associated with the existing methods of user verification, hackers will still find there are weak points in the system, vulnerable to attack. Password systems are prone to brute force dictionary attacks. Biometric systems, on the other hand, require substantially more effort for mounting such an attack. Yet there are several new types of attacks possible in the biometric domain. This may not apply if biometrics is used as a supervised verification tool. But in remote, unattended applications, such as Web-based e-commerce applications, hackers may have the opportunity and enough time to make several attempts, or even physically violate the integrity of a remote client, before detection [1].

A problem with biometric verification systems arises when the data associated with a biometric feature has been compromised. For verification systems based on physical tokens such as keys and badges, a compromised token can be easily canceled and the user can be assigned a new token. Similarly, user IDs and passwords can be changed as often as required. Yet, the user only has a limited number of biometric features (one face, 10 fingers, two eyes). If the biometric data are compromised, the user may quickly run out of biometric features to be used for verification [1].

Fingerprint Verification

A brief introduction to fingerprint verification is presented here. Readers familiar with fingerprint verification may skip to the next section [1].

Fingerprints are a distinctive feature and remain invariant over the lifetime of a subject, except for cuts and bruises. As the first step in the verification process, a fingerprint impression is acquired, typically using an inkless scanner. Several such scanning technologies are available. Figure 21-1(a) shows a fingerprint obtained with a scanner using an optical sensor [1, 5]. A typical scanner digitizes the fingerprint impression at 500 dots per inch (dpi) with 256 gray levels per pixel. The digital image of the fingerprint includes several unique features in terms of ridge bifurcations and ridge endings, collectively referred to as minutiae [1].

Figure 21-1
*Fingerprint
recognition
(a) input image
(b) features.
(Source:
Reproduced with
permission from
the IBM
Corporation.)*

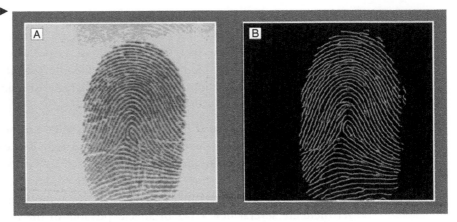

The next step is to locate these features in the fingerprint image, as shown in Figure 21-1(b), using an automatic feature extraction algorithm [1]. Each feature is commonly represented by its location (x, y) and the ridge direction at that location (theta). However, due to sensor noise and other variability in the imaging process, the feature extraction stage may miss some minutiae and may generate spurious minutiae. Furthermore, due to the elasticity of the human skin, the relationship between minutiae may be randomly distorted from one impression to the next [1].

In the final stage, the matcher sub-system attempts to arrive at a degree of similarity between the two sets of features after compensating for the rotation, translation, and scale. This similarity is often expressed as a score. Based on this score, a final decision of match or no-match is made. A decision threshold is first selected. If the score is below the threshold, the fingerprints are determined not to match; if the score is above the threshold, a correct match is declared. Often the score is simply a count of the number of the minutiae that are in correspondence. In a number of countries, 12–16 correspondences (performed by a human expert) are considered legally binding evidence of identity [1].

The operational issues in an automated fingerprint identification system (AFIS) are somewhat different from those in a more traditional password-based system. First, there is a system performance issue known as the fail-to-enroll rate to be considered. Some people have very faint fingerprints, or no fingers at all, which makes the system unusable for them. A related issue is a "Reject" option in the system based on input image quality. A poor-quality input is not accepted by the system during enrollment and verification [1].

Note: Poor-quality inputs can be caused by noncooperative users, improper usage, dirt on the finger, or bad input scanners. This has no analog in a password system. Then there is the fact that in a biometric system, that the matching decision is not clear-cut. A password system always provides a correct response—if the passwords match, it grants access, but otherwise it refuses access. However, in a biometric system, the overall accuracy depends on the quality of input and enrollment data along with the basic characteristics of the underlying feature extraction and matching algorithm [1].

For fingerprints, and for biometrics in general, there are two basic types of recognition errors: the false-accept rate (FAR) and the false-reject rate (FRR). If a nonmatching pair of fingerprints is accepted as a match, it is called a false accept. On the other hand, if a matching pair of fingerprints is rejected by the system, it is called a false reject. The error rates are a function of the threshold, as shown in Figure 21-2 [1]. Often the interplay between the two errors is presented by plotting FAR against FRR with the decision threshold as the free variable. This plot is called the receiver operating characteristic (ROC) curve. The two errors are complementary in the sense that if one makes an effort to lower one of the errors by varying the threshold, the other error rate automatically increases [1].

In a biometric verification system, the relative false-accept and false-reject rates can be set by choosing a particular operating point (a detection threshold). Very low (close to zero) error rates for both errors (FAR and FRR) at the same time are not possible. By setting a high threshold, the FAR error can be close to zero and, similarly by setting a significantly low threshold, the FRR rate can be close to zero. A meaningful operating point for the threshold is decided based on the application requirements, and the FAR versus FRR error rates at that operating point may be quite different. To provide high security, biometric systems operate at a low FAR instead of the commonly recommended equal error rate (EER) operating point where FAR = FRR. High-performance fingerprint recognition systems can support error rates in the range of 10^{-6}

Figure 21-2
Error tradeoff in a biometric system. (Source: Reproduced with permission from the IBM Corporation.)

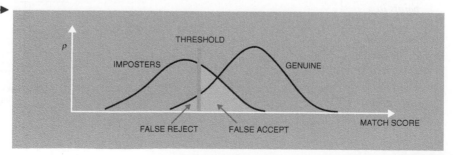

for false accept and 10^{-4} for false reject. The performance numbers reported by vendors are based on test results using private databases and, in general, tend to be much better than what can be achieved in practice. Nevertheless, the probability that the fingerprint signal is supplied by the right person, given a good matching score, is quite high. This confidence level generally provides better nonrepudiation support than passwords [1].

Vulnerable Points of a Biometric System

A generic biometric system can be cast in the framework of a pattern recognition system. The stages of such a generic system are shown in Figure 21-3 [1].

The first stage involves biometric signal acquisition from the user (the inkless fingerprint scan). The acquired signal typically varies significantly from presentation to presentation; hence, pure pixel-based matching techniques do not work reliably. For this reason, the second signal processing stage attempts to construct a more invariant representation of this basic input signal (in terms of fingerprint minutiae). The invariant representation is often a spatial domain characteristic or a transform (frequency) domain characteristic, depending on the particular biometric [1].

During enrollment of a subject in a biometric verification system, an invariant template is stored in a database that represents the particular individual. To verify the user against a given ID, the corresponding template is retrieved from the database and matched against the template derived from a newly acquired input signal. The matcher arrives at a decision based on the closeness of these two templates while taking into account geometry, lighting, and other signal acquisition variables [1].

Figure 21-3
Possible attack points in a generic biometric-based system. (Source: Adapted with permission from the IBM Corporation.)

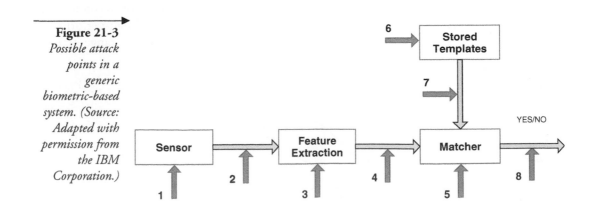

Note: Password-based verification systems can also be set in this framework. The keyboard becomes the input device. The password encryptor can be viewed as the feature extractor and the comparator as the matcher. The template database is equivalent to the encrypted password database.

Eight places in the generic biometric system of Figure 21-3 have been identified where attacks may occur [1]. The numbers in Figure 21-3 correspond to the items in the following list:

1.   **Presenting fake biometrics at the sensor:** In this mode of attack, a possible reproduction of the biometric feature is presented as input to the system. Examples include a fake finger, a copy of a signature, or a face mask.

2.   **Resubmitting previously stored digitized biometric signals:** In this mode of attack, a recorded signal is replayed to the system, bypassing the sensor. Examples include the presentation of an old copy of a fingerprint image or the presentation of a previously recorded audio signal.

3.   **Overriding the feature extraction process:** The feature extractor is attacked using a Trojan horse, so that it produces feature sets preselected by the intruder.

4.   **Tampering with the biometric feature representation:** The features extracted from the input signal are replaced with a different, fraudulent feature set (assuming the representation method is known). Often the two stages of feature extraction and matcher are inseparable and this mode of attack is extremely difficult. However, if minutiae are transmitted to a remote matcher (say, over the Internet [4]), this threat is very real. One could "snoop" on the Transmission Control Protocol/Internet Protocol (TCP/IP) stack and alter certain packets.

5.   **Corrupting the matcher:** The matcher is attacked and corrupted so that it produces preselected match scores.

6.   **Tampering with stored templates:** The database of stored templates could be either local or remote. The data might be distributed over several servers. Here the attacker could try to modify one or more templates in the database, which could result either in authorizing a fraudulent individual or denying service to the persons

associated with the corrupted template. A smart card-based verification system, where the template is stored in the smart card and presented to the verification system, is particularly vulnerable to this type of attack.

7.   **Attacking the channel between the stored templates and the matcher:** The stored templates are sent to the matcher through a communication channel. The data traveling through this channel could be intercepted and modified.

8.   **Overriding the final decision:** If the final match decision can be overridden by the hacker, then the verification system has been disabled. Even if the actual pattern recognition framework has excellent performance characteristics, it has been rendered useless by the simple exercise of overriding the match result [1].

There exist several security techniques to thwart attacks at these various points. For instance, finger conductivity or fingerprint pulse at the sensor can stop simple attacks at point 1. Encrypted communication channels can eliminate at least remote attacks at point 4. However, even if the hacker cannot penetrate the feature extraction module, the system is still vulnerable. The simplest way to stop attacks at points 5, 6, and 7 is to have the matcher and the database reside at a secure location. Of course, even this cannot prevent attacks in which there is collusion. Use of cryptography prevents attacks at point 8 [1].

The threats outlined in Figure 21-3 are quite similar to the threats to password-based verification systems [1]. For instance, all the channel attacks are similar. One difference is that there is no "fake password" equivalent to the fake biometric attack at point 1 (although, perhaps if the password was in some standard dictionary, it could be deemed "fake"). Furthermore, in a password- or token-based verification system, no attempt is made to thwart replay attacks (since there is no expected variation of the "signal" from one presentation to another). However, in an automated biometric-based verification system, one can check the liveness of the entity originating the input signal [1].

Brute Force Attack Directed at Matching Fingerprint Minutiae

This part of the chapter analyzes the probability that a brute force attack at point 4 of Figure 21-3, involving a set of fraudulent fingerprint minutiae, will succeed in matching a given stored template [1]. Figure 21-4 shows one such randomly

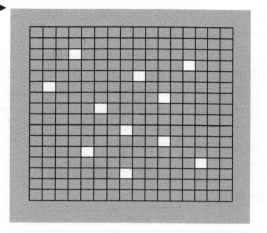

Figure 21-4
Example of a randomly generated minutiae set. (Source: Reproduced with permission from the IBM Corporation.)

generated minutiae set [1]. In a smart card system where the biometric template is stored in the card and presented to the verification system, a hacker could present these random sets to the verification system, assuming that the hacker has no information about the stored templates [1].

Note: An attack at point 2 of Figure 21-3, which involves generating all possible fingerprint images in order to match a valid fingerprint image, would have an even larger search space and consequently would be much more difficult.

A Naive Model

For the purpose of analyzing the "naive" matching minutiae attack, the following is assumed:

- The system uses a minutia-based matching method and the number of paired minutiae reflects the degree of match;

- The image size $S = 300$ pixels $\times$ 300 pixels;

- A ridge plus valley spread $T = 15$ pixels;

- The total number of possible minutiae sites $(K = S/(T^2)) = 20 \times 20 = 400$;

- The number of orientations allowed for the ridge angle at a minutia point $d = 4, 8, 16$;

- The minimum number of corresponding minutiae in query and reference template $m = 10, 12, 14, 16, 18$ [1].

These values are based on a standard fingerprint scanner with a 500 dpi scanning resolution covering an area 0.6×0.6 inches. A ridge and valley can span about 15 pixels on average at this scanning resolution. The other two variables d and m are being used as parameters to study the brute force attack. Let's start with 10 matching minutiae, since often a threshold of 12 minutiae is used in matching fingerprints in manual systems. Ridge angles in an automated system can be quantized depending on the tolerance supported in the matcher. A minimum of four quantization levels provides a 45 degree tolerance, while 16 levels provides roughly an 11 degree tolerance [1].

Note: It is assumed that the matcher will tolerate shifts between query and reference minutiae of at most a ridge and valley pixel width, and an angular difference of up to half a quantization bin (± 45 degrees for $d = 4$).

The $\log_2$ of the probability of randomly guessing a correct feature set through a brute force attack for different values of d and m is plotted in Figure 21-5 [1]. This measure (in bits) is referred to as strength, and it represents the equivalent

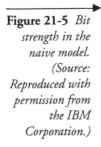

Figure 21-5 *Bit strength in the naive model. (Source: Reproduced with permission from the IBM Corporation.)*

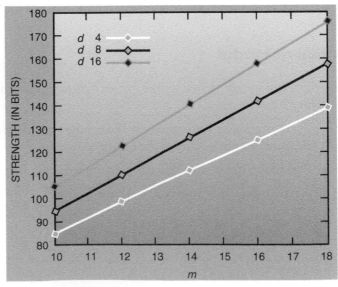

number of bits in a password verification system. This should convince the reader that a brute force attack in the form of a random image or a random template attempting to impersonate an authorized individual will, on average, require a very large number of attempts before succeeding [1].

The foregoing analysis assumes that each fingerprint has exactly m minutiae, that only m minutiae are generated, and that all of these minutiae have to match. A realistic strength is much lower because one can generate more than m query minutiae, say N_{total}, and only some fraction of these must match m minutiae of the reference fingerprint. This leads to a factor of about $(m^{Ntotal})^2$ or a loss of nearly 64 bits in strength for $m = 10$ with $N_{total} = 50$. The equivalent strength thus is closer to 20 bits for this parameter set. A more realistic model, which carefully incorporates this effect, is described next [1].

A More Realistic Model

In the naive approach, several simplifying assumptions were made. In this more realistic model, more realistic assumptions will be made. The brute force attack model will also be analyzed in more detail [1].

A more accurate model would require the consideration of the probability of a minutiae site being populated as a function of the distance to the center of the print (they are more likely in the middle). In addition, such a model would require that the directional proclivities depend on location (they tend to swirl around the core). In this model, however, you should ignore such dependencies and use the simpler formulation [1].

For example, the $\log_2$ of Pver (bit strength) is plotted in Figure 21-6 for $N = 40, d = 4, K = 400$ with m (the number of minutiae required to match) between 10 and 35 [1]. For a value of $m = 10$, you have about 22 bits of information (close to the prediction of the revised naive model). For the legal threshold of $m = 15$, you have around 40 bits of information (representing a number of distinct binary values equal to about 140 times the population of Earth). For a more typical value of $m = 25$, you have roughly 82 bits of information content in this representation (see Figure 21-6). This is equivalent to a 16-character nonsense password (such as "m4yus78xpmks3bc9") [1].

Other studies evaluate the individuality of a fingerprint based on the minutiae information. These analyses were based on the minutiae frequency data collected and interpreted by a human expert and involving a small set of fingers. Furthermore, these studies used all 10 types of Galton characteristics, whereas other studies are based on just one type of feature (with no

Figure 21-6 *Bit strength in the more realistic model. (Source: Reproduced with permission from the IBM Corporation.)*

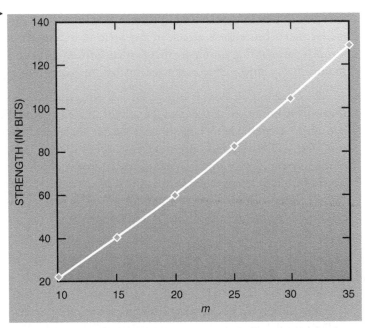

differentiation between ridge endings and bifurcations). The purpose of these studies was to quantify the information content of a fingerprint (similar to the naive method) rather than set thresholds for matching in the face of brute force attacks [1].

It should be pointed out that the brute force attack break-in probability is not dependent in any way on the FAR. That is, if the FAR is 10^{-6}, this does not mean that, on average, the system is broken into after 500,000 trials. The FAR is estimated by using actual human fingers and is typically attributable to errors in feature extraction (extra or missing features) and, to a lesser extent, to changes in geometry such as finger rolling or skin deformations due to twisting. The statistics governing the occurrence of these types of errors are different from those describing a brute force attack [1].

WSQ-Based Data Hiding

In both Web-based and other online transaction processing systems, it is undesirable to send uncompressed fingerprint images to the server due to bandwidth limitations. A typical fingerprint image is of the order of 512 × 512 pixels with 256 gray levels, resulting in a file size of 256 Kbytes. This would take nearly 40 seconds to transmit at 53 Kbaud. Unfortunately, many standard compression methods, such as Joint Photographic Experts Group (JPEG), have a tendency to distort the high-frequency spatial and structural ridge features of a fingerprint image. This has led to several research proposals regarding domain-specific compression methods. As a result, an open Wavelet Scalar Quantization (WSQ) image compression scheme proposed by the FBI has become the de facto standard in the industry, because of its low image distortion, even at high-compression ratios (over 10:1) [1].

Typically, the compressed image is transmitted over a standard encrypted channel as a replacement for (or in addition to) the user's PIN. Yet, because of the open compression standard, transmitting a WSQ-compressed image over the Internet is not particularly secure. If a compressed fingerprint image bit stream can be freely intercepted (and decrypted), it can be decompressed using readily available software. This potentially allows the signal to be saved and fraudulently reused (see attack point 2 in Figure 21-3) [1].

One way to enhance security is to use data-hiding techniques to embed additional information directly in compressed fingerprint images. For instance, if the embedding algorithm remains unknown, the service provider can look for the appropriate standard watermark to check that a submitted image was indeed generated by a trusted machine (or sensor). Most of the research, however, addresses issues involved in resolving piracy or copyright issues, not verification. An exception is the invisible watermarking technique for fingerprints. This involves examining the accuracy after an invisible watermark is inserted in the image domain. The proposed solution here is different because, first, it operates directly in the compressed domain and, second, it causes no performance degradation [1].

The approach is motivated by the desire to create online fingerprint verification systems for commercial transactions that are secure against replay attacks. To achieve this, the service provider issues a different verification string for each transaction. The string is mixed in with the fingerprint image before transmission. When the image is received by the service provider, it is decompressed and the image is checked for the presence of the correct one-time verification string. The method proposed here hides such messages with minimal impact on the appearance of the decompressed image. Moreover, the message is not hidden in a fixed location (which would make it more vulnerable to discovery) but is, instead, deposited in different places based on the structure of the image itself. Although this approach is presented in the framework of fingerprint image compression, it can be easily extended to other biometrics such as wavelet-based compression of facial images [1].

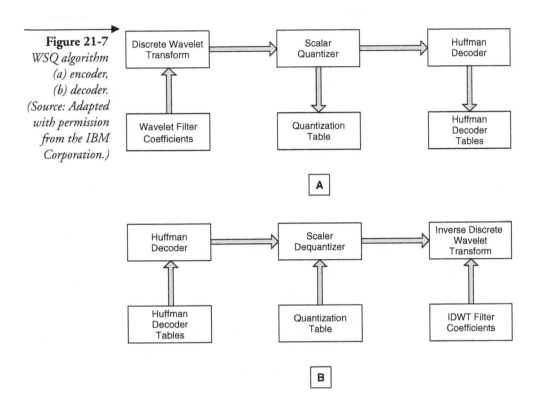

Figure 21-7
*WSQ algorithm
(a) encoder,
(b) decoder.
(Source: Adapted
with permission
from the IBM
Corporation.)*

The information hiding scheme proposed here works in conjunction with the WSQ fingerprint image encoder and decoder, which are shown in Figure 21-7, respectively [1]. In the first step of the WSQ compression, the input image is decomposed into 64 spatial frequency sub-bands using perfect reconstruction multirate filter banks based on discrete wavelet transformation filters. The filters are implemented as a pair of separable 1D filters. The two filters specified for encoder 1 of the FBI standard are plotted in Figures 21-8 and 21-9 [1]. The sub-bands are the filter outputs obtained after a desired level of cascading of the filters as described in the standard. For example, sub-band 25 corresponds to the cascading path of "00, 10, 00, 11" through the filter bank. The first digit in each binary pair represents the row operation index. A 0 specifies low-pass filtering using h0 on the row (column), while a 1 specifies high-pass filtering using h1 on the row (column). Thus, for the 25th sub-band, the image is first low-pass filtered in both row and column. This is followed by high-pass filtering in rows, then low-pass filtering in columns; the output of which is then low-pass filtered in rows and columns, ending with high-pass filtering in rows and columns [1].

Figure 21-8
Analysis filter h0.
(Source:
Reproduced with
permission from
the IBM
Corporation.)

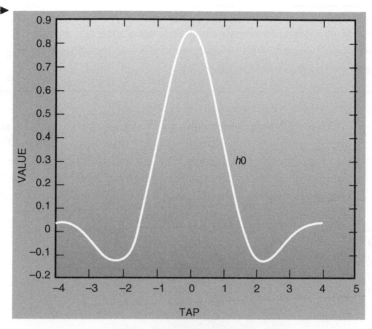

Figure 21-9
Analysis filter h1.
(Source:
Reproduced with
permission from
the IBM
Corporation.)

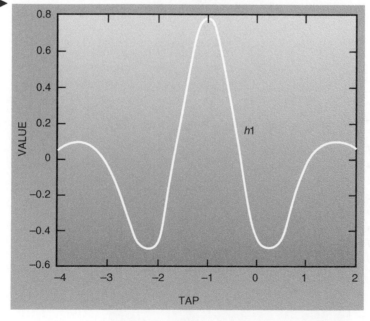

Note: There is appropriate down sampling and the symmetric extension transform is applied at every stage as specified in the standard. The 64 sub-bands of the gray-scale fingerprint image are shown in Figure 21-10, and are also shown in Figure 21-11 [1].

There are two more stages to WSQ compression. The second stage is a quantization process where the discrete wavelet transform (DWT) coefficients are transformed into integers with a small number of discrete values. This is accomplished by uniform scalar quantization for each sub-band. There are two characteristics for each band: the zero of the band (Z_k) and the width of

Figure 21-10
WSQ data-hiding results, original image. (Source: Reproduced with permission from the IBM Corporation.)

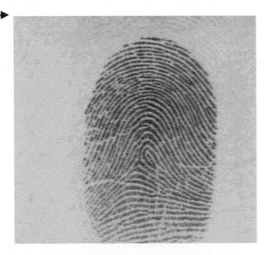

Figure 21-11
64 sub-bands of the image in Figure 21-9. (Source: Reproduced with permission from the IBM Corporation.)

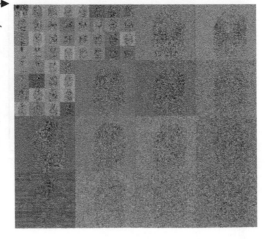

the bins (Q_k). These parameters must be chosen carefully to achieve a good compression ratio without introducing significant information loss that will result in distortions of the images. The Z_k and Q_k for each band are transmitted directly to the decoder. The third and final stage is the Huffman coding of the integer indices for the DWT coefficients. For this purpose, the bands are grouped into three blocks. In each block, the integer coefficients are remapped to numbers between 0–255, prescribed by the translation table described in the standard. This translation table encodes run lengths of zeros and large values. Negative coefficients are translated in a similar way by this table [1].

The data-hiding algorithm works on the quantized indices before this final translation (between stages 2 and 3). It is assumed that the message size is very small compared to the image size (or, equivalently, the number of DWT coefficients) [1].

Note: However, the Huffman coding characteristics and tables are not changed; the tables are computed as for the original coefficients, not after the coefficient altering steps.

As previously mentioned, the method here is intended for messages which are very small (in terms of bits), as compared to the number of pixels in the image. The basic principle is to find and slightly alter certain of the DWT coefficients. However, care must be taken to avoid corrupting the reconstructed image. To hide a message during the image encoding process (or, optionally, four), you should perform the following four basic steps:

1. Selecting a set of sites S;

2. Generating a seed for random number generation and then choosing sites for modification;

3. Hiding the message at selected sites by bit setting;

4. Appending the bits to the coded image [1].

Selecting a Set of Sites *S*

Given the partially converted quantized integer indices, this stage collects the indices of all possible coefficient sites where a change in the least significant bit is tolerable. Typically, all sites in the low-frequency bands are excluded.

Even small changes in these coefficients can affect large regions of the image because of the low frequencies. For the higher frequencies, candidate sites are selected if they have coefficients of large magnitude. Making small changes to the larger coefficients leads to a relatively small percentage change in the values and hence minimal degradation of the image [1].

Tip: Among the quantizer indices, there are special codes to represent run lengths of zeros, large integer values, and other control sequences. All coefficient sites incorporated into these values are avoided. During implementation, you should only select sites with translated indices ranging from 107–254, but excluding 180 (an invalid code).

Generating a Seed for Random Number Generation and then Choosing Sites for Modification

Sites from the candidate set S, which are modified, are selected in a pseudo-random fashion. To ensure that the encoder actions are invertible in the decoder, the seed for the random number generator is based on the sub-bands that are not considered for alteration. For example, in the selection process, the contents of sub-bands 0–6 are left unchanged in order to minimize distortion. Typically, fixed sites within these bands are selected, although in principle any statistic from these bands may be computed and used as the seed. Selecting the seed in this way ensures that the message is embedded at varying locations (based on the image content). It further ensures that the embedded message can only be read if the proper seed selection algorithm is known by the decoder [1].

Hiding the Message at Selected Sites by Bit Setting

The message to be hidden is translated into a sequence of bits. Each bit will be incorporated into a site chosen pseudorandomly by a random number generator seeded as described in the preceding. That is, for each bit, a site is selected from the set S based on the next output of the seeded pseudorandom number generator. If the selected site has already been used, the next randomly generated site is chosen instead. The low-order bit of the value at the selected site is changed to be identical to the current message bit. On average, half the time this results in no change at all of the coefficient value [1].

Appending the Bits to the Coded Image

Optionally, all the original low-order bits can be saved and appended to the compressed bit stream as a user comment field (an appendix). The appended bits are a product of randomly selected low-order coefficient bits, and hence these bits are uncorrelated with the hidden message [1].

The Decoder

The steps performed by the decoder correspond to the encoder steps in the preceding. The first two steps are identical to the first steps of the encoder. These steps construct the same set S and compute the same seed for the random number generator. The third step uses the pseudorandom number generator to select specific sites in S in the prescribed order. The least significant bits of the values at these sites are extracted and concatenated to recover the original message [1].

If the appendix restoration is to be included, the decoder can optionally restore the original low-order bits while reconstructing the message. This allows perfect reconstruction of the image (up to the original compression) despite the embedded message. Because the modification sites S are carefully selected, the decompressed image even with the message still embedded will be nearly the same as the restored decompressed image. In practice, the error due to the embedded message is not perceptually significant and does not affect subsequent processing and verification. Figures 21-10 and 21-12 show the original and the reconstructed images, respectively [1].

Figure 21-12
WSQ data-hiding results, reconstructed image. (Source: Reproduced with permission from the IBM Corporation.)

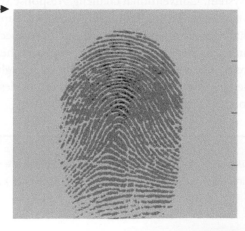

By using this process, only a specialized decoder can locate and extract the message from the compressed image during the decoding process. This message might be a fixed verification stamp, personal ID information that must match some other part of the record (which might have been sent in the clear), or some time stamp. If the bit stream does not contain an embedded message or the bit stream is improperly coded, the specialized decoder will fail to extract the expected message and will thus reject the image. If, instead, an unencoded WSQ-compressed fingerprint image is submitted to the special decoder, it will still extract a garbage message that can be rejected by the server [1].

Many implementations of the same algorithm are possible by using different random number generators or partial seeds. This means it is possible to make every implementation unique without much effort; the output of one encoder need not be compatible with another version of the decoder. This has the advantage that cracking one version will not compromise any other version [1].

This method can also be extended to other biometric signals using a wavelet compression scheme, such as facial images or speech. While the filters and the quantizer in the WSQ standard have been designed to suit the characteristics of fingerprint images, wavelet-based compression schemes for other signals are also available [1].

Image-Based Challenge/Response Method

Besides interception of network traffic, more insidious attacks might be perpetrated against an automated biometric verification system. One of these is a replay attack on the signal from the sensor (see attack point 2 in Figure 21-3) [1]. A new method is proposed here to thwart such attempts, based on a modified challenge/response system. Conventional challenge/response systems are based either on challenges to the user, such as requesting the user to supply the mother's maiden name, or challenges to a physical device, such as a special-purpose calculator that computes a numerical response. The approach here is based on a challenge to the sensor. The sensor is assumed to have enough intelligence to respond to the challenge. Silicon fingerprint scanners can be designed to exploit the proposed method using an embedded processor [1].

Note: Standard cryptographic techniques are not a suitable substitute. While these are mathematically strong, they are also computationally intensive and could require maintaining secret keys for a large number of sensors. Moreover, the encryption techniques cannot check for liveness of a signal. A stored image could be fed to the encryptor, which would happily encrypt it. Similarly, the digital signature of a submitted signal can be used to check only for its integrity, not its liveness [1].

The system here computes a response string, which depends not only on the challenge string but also on the content of the returned image. The changing challenges ensure that the image was acquired after the challenge was issued. The dependence on image pixel values guards against substitution of data after the response has been generated [1].

The proposed solution works as shown in Figure 21-13 [1]. A transaction is initiated at the user terminal or system. First, the server generates a pseudo-random challenge for the transaction and the sensor [1].

Note: It is assumed that the transaction server itself is secure. The client system then passes the challenge on to the intelligent sensor. Next, the sensor acquires a new signal and computes the response to the challenge that is based in part on the newly acquired signal. Because the response processor is tightly integrated with the sensor (preferably on the same chip), the signal channel into the response processor is assumed ironclad and inviolable. It is difficult to intercept the true image and to inject a fake image under such circumstances.

As an example of an image-based response, consider the function "x1+," which operates by appending pixel values of the image (in scan order) to the end of the challenge string. A typical challenge might be "3, 10, 50." In response to this, the integrated processor then selects the 3rd, 10th, and 50th pixel value from this sequence to generate an output response such as "133, 92, 176."

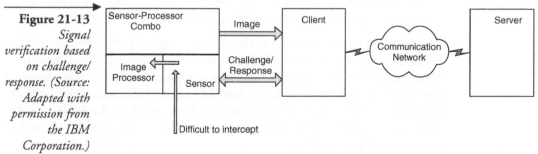

Figure 21-13
Signal verification based on challenge/ response. (Source: Adapted with permission from the IBM Corporation.)

The complete image as well as the response is then transmitted to the server, where the response can be verified and checked against the image [1].

Other examples of responder functions include computing a checksum of a segment of the signal; a set of pseudorandom samples; a block of contiguous samples starting at a specified location and with a given size; a hash of signal values; and a specified known function of selected samples of the signal. A combination of these functions can be used to achieve arbitrarily complex responder functions. The important point is that the response depends on the challenge and the image itself [1].

The responder can also incorporate several different response functions that the challenger could select among. For instance, the integrated processor might be able to compute either of two selectable functions, "x1+" and "x10+." The function "x10+" is similar to "x1+" except it multiplies the requested pixel values by 10 before appending them. Financial institution A might use function "x1+" in all its units, while institution B might use "x10+" in all of its units. Alternatively, for even-numbered transactions, function "x10+" might be used and for odd-numbered transactions, "x1+" might be used. This variability makes it even harder to reconstruct the structure and parameters of the response function. Large numbers of such response functions are possible because you have a large number of pixels, and many simple functions can be applied to these pixels [1].

Cancelable Biometrics

Deploying biometrics in a mass market, like credit card authorization or bank ATM access, raises additional concerns beyond the security of the transactions. One such concern is the public's perception of a possible invasion of privacy [2]. In addition to personal information such as name and date of birth, the user is asked to surrender images of body parts, such as fingers, face, and iris. These images, or other such biometric signals, are stored in digital form in various databases. This raises the concern of possible sharing of data among law enforcement agencies or commercial enterprises [1].

The public is concerned about the ever-growing body of information that is being collected about individuals in our society. The data collected encompass many applications and include medical records and biometric data. A related concern is the coordination and sharing of data from various databases (see sidebar, "Ordering Pizza in 2009"). In relation to biometric data, the public is, rightfully or not, worried about data collected by private companies being matched against databases used by law enforcement agencies. Fingerprint images, for example, can be matched against the FBI or Immigration and Naturalization Service (INS) databases, with ominous consequences [1].

Ordering Pizza in 2009

The coordination and sharing of data from various databases has gotten to the point of being ridiculous, and almost comical. The following is an e-mail that this author received recently from an anonymous friend. This is so close to what is probably going to be happening in 2009, with regards to data being collected by private companies and being matched against databases used by law enforcement agencies. Let's take a look:

Operator: Thank you for calling Pizza Hut. May I have your national ID number?

Customer: Hi, I'd like to place an order.

Operator: I must have your NIDN first, sir.

Customer: My National ID Number, yeah, hold on, eh, it's 6102049998-45-54666.

Operator: Thank you Mr. Damion. I see you live at 1666 Hades Drive, and the phone number is 494-2666. Your office number over at Omen Insurance is 745-2666 and your cell number is 266-2666. E-mail address is . . . Which number are you calling from, sir?

Customer: Huh? I'm at home. Where'd you get all this information?

Operator: We're wired into the HSS, sir.

Customer: The HSS, what's that?

Operator: We're wired into the Homeland Security System, sir. This will add only 15 seconds to your ordering time.

Customer: (sighs) Oh well, I'd like to order a couple of your All-Meat Special pizzas.

Operator: I don't think that's a good idea, sir.

Customer: Whaddya mean?

Operator: Sir, your medical records and commode sensors indicate that you've got very high blood pressure and extremely high cholesterol. Your National Health Care provider won't allow such an unhealthy choice.

Customer: What?!?! What do you recommend, then?

Operator: You might try our low-fat Soybean Pizza. I'm sure you'll like it.

Customer: What makes you think I'd like something like that?

Operator: Well, you checked out *Gourmet Soybean Recipes* from your local library last week, sir. That's why I made the suggestion.

Customer: All right, all right. Give me two family-sized ones, then.

Operator: That should be plenty for you, your wife and your four kids. Your two dogs can finish the crusts, sir. Your total is $46.66.

Customer: Lemme give you my credit card number.

Operator: I'm sorry sir, but I'm afraid you'll have to pay in cash. Your credit card balance is over its limit.

Customer: I'll run over to the ATM and get some cash before your driver gets here.

Operator: That won't work either, sir. Your checking account is overdrawn also.

Customer: Never mind! Just send the pizzas. I'll have the cash ready. How long will it take?

Operator: We're running a little behind, sir. It'll be about 45 minutes, sir. If you're in a hurry you might want to pick em up while you're out getting the cash, but then, carrying pizzas on a motorcycle can be a little awkward.

Customer: Wait! How do you know I ride a scooter?

Operator: It says here you're in arrears on your car payments, so your car got repo'ed. But your Harley's paid for and you just filled the tank yesterday at the discount price of $8.05 a gallon.

Customer: Well, I'll be a #%#^^&$%^$@#!

Operator: I'd advise watching your language, sir. You've already got a July 4, 2006 conviction for cussing out a cop and another one I see here in September for contempt at your hearing for cussing at a judge. Oh yes, I see here that you just got out from a 120-day stay in the State Correctional Facility. Is this your first pizza since your return to society?

Customer: (speechless)

Operator: Will there be anything else, sir?

Customer: Yes, I have a coupon for a free 2 liter of Coke.

Operator: I'm sorry sir, but our ad's exclusionary clause prevents us from offering free soda to diabetics. The New Constitution prohibits this. Thank you for calling Pizza Hut.

These concerns are aggravated by the fact that a person's biometric data are a given and cannot be changed. One of the properties that makes biometrics so attractive for verification purposes (their invariance over time) is also one of its liabilities. When a credit card number is compromised, the issuing bank can just assign the customer a new credit card number. When the biometric data are compromised, replacement is not possible [1].

In order to alleviate this problem, the concept of "cancelable biometrics" is introduced here. It consists of an intentional, repeatable distortion of a biometric signal based on a chosen transform. The biometric signal is distorted in the same fashion at each presentation, for enrollment and for every verification. With this approach, every instance of enrollment can use a different transform, thus rendering cross-matching impossible. Furthermore, if one variant of the transformed biometric data is compromised, then the transform function can simply be changed to create a new variant (transformed representation) for re-enrollment as, essentially, a new person. In general, the distortion transforms are selected to be noninvertible. So even if the transform function is known and the resulting transformed biometric data are known, the original (undistorted) biometrics cannot be recovered [1].

Example of Distortion Transforms

In the proposed method, distortion transforms can be applied in either the signal domain or the feature domain. That is, either the biometric signal can be transformed directly after acquisition, or the signal can be processed as usual and the extracted features can then be transformed. Moreover, extending a template to a larger representation space via a suitable transform can further increase the bit strength of the system. Ideally the transform should be noninvertible so that the true biometric of a user cannot be recovered from one or more of the distorted versions stored by various agencies [1].

Examples of transforms at the signal level include grid morphing and block permutation. The transformed images cannot be successfully matched against the original images, or against similar transforms of the same image using different parameters. While a deformable template method might be able to find such a match, the residual strain energy is likely to be as high as that of matching the template to an unrelated image. In Figure 21-14, the original image is shown with an overlaid grid aligned with the features of the face [1]. In the adjacent image, we show the morphed grid and the resulting distortion of the face. In Figure 21-15, a block structure is imposed on the image aligned with characteristic points [1]. The blocks in the original image are subsequently scrambled randomly but repeatably.

An example of a transform in the feature domain is a set of random, repeatable perturbations of feature points. This can be done within the same physical space as the original, or while increasing the range of the axes. The second case

Figure 21-14
Distortion transform based on image morphing. (Source: Reproduced with permission from the IBM Corporation.)

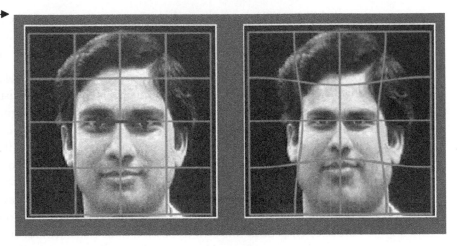

Figure 21-15
Distortion transform based on block scrambling. (Source: Reproduced with permission from the IBM Corporation.)

provides more brute force strength, as was noted earlier (this effectively increases the value of K). An example of such a transform is shown in Figure 21-16 [1]. Here, the blocks on the left are randomly mapped onto blocks on the right, where multiple blocks can be mapped onto the same block. Such transforms are noninvertible; hence, the original feature sets cannot be recovered from the distorted versions. For instance, it is impossible to tell which of the two blocks the points in composite block B, D, originally came from. Consequently, the owner of the biometrics cannot be identified except through the information associated with that particular enrollment [1].

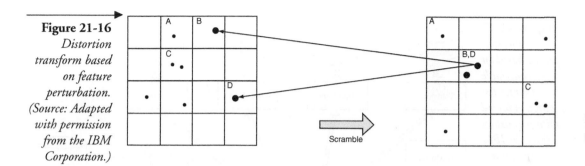

Figure 21-16
*Distortion
transform based
on feature
perturbation.
(Source: Adapted
with permission
from the IBM
Corporation.)*

Note: For the transform to be repeatable, you need to have the biometric signal properly registered before the transformation. Fortunately, this problem has been partially answered by a number of techniques available in the literature (such as finding the "core" and "delta" points in a fingerprint, or eye and nose detection in a face).

Encryption and Transform Management

The techniques presented here for transforming biometric signals differ from simple compression-using signal or image-processing techniques. While compression of the signal causes it to lose some of its spatial domain characteristics, it strives to preserve the overall geometry. That is, two points in a biometric signal before compression are likely to remain at a comparable distance when decompressed. This is usually not the case with distortion transforms. The technique also differs from encryption. The purpose of encryption is to allow a legitimate party to regenerate the original signal. In contrast, distortion transforms permanently obscure the signal in a noninvertible manner [1].

Finally, when employing cancelable biometrics, there are several places where the transform, its parameters, and identification templates could be stored. This leads to a possible distributed process model, as shown in Figure 21-17 [1]. The "merchant" is where the primary interaction starts in our model. Based on the customer ID, the relevant transform is first pulled from one of the transform databases and applied to the biometrics. The resulting distorted biometrics is then sent for verification to the "authorization" server. Once the user's identity has been confirmed, the transaction is finally passed on to the relevant commercial institution for processing [1].

Figure 21-17
Verification process based on cancelable biometrics. (Source: Adapted with permission from the IBM Corporation.)

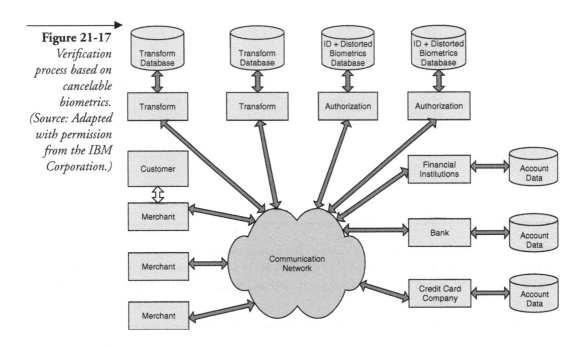

Note: An individual user may be subscribed to multiple services, such as e-commerce merchants or banks. The verification for each transaction might be performed either by the service provider itself or by an independent third party. Similarly, the distortion transform might be managed either by the verifier or by yet another independent agency. Alternatively, for the best privacy the transform might remain solely in the possession of the user, stored, say, on a smart card. If the card is lost or stolen, the stolen transform applied to another person's biometrics will have very little impact. However, if the transform is applied to a stored original biometric signal of the genuine user, it will match against the stored template of the person. Hence "liveness" detection techniques (such as described earlier) should be added to prevent such misuse.

Summary/Conclusion

Biometric-based verification has many usability advantages over traditional systems such as passwords. Specifically, users can never lose their biometrics, and the biometric signal is difficult to steal or forge. The intrinsic bit strength of a biometric signal can be quite good, especially for fingerprints, when compared to conventional passwords [1].

Yet any system, including a biometric system, is vulnerable when attacked by determined hackers. This chapter has highlighted eight points of vulnerability in a generic biometric system and has discussed possible attacks. The chapter has also suggested several ways to alleviate some of these security threats. Replay attacks have been addressed, using data-hiding techniques to secretly embed a tell-tale mark directly in the compressed fingerprint image. A challenge/response method has been proposed to check the liveliness of the signal acquired from an intelligent sensor [1].

Finally, the chapter has touched on the often neglected problems of privacy and revocation of biometrics. It is somewhat ironic that the greatest strength of biometrics, the fact that the biometrics does not change over time, is at the same time its greatest liability. Once a set of biometric data has been compromised, it is compromised forever. To address this issue, the chapter has proposed applying repeatable noninvertible distortions to the biometric signal. Cancellation simply requires the specification of a new distortion transform. Privacy is enhanced because different distortions can be used for different services and the true biometrics are never stored or revealed to the verification server. In addition, such intentionally distorted biometrics cannot be used for searching legacy databases and will thus alleviate some privacy violation concerns [1].

References

1. N. K. Ratha, J. H. Connell, and R. M. Bolle, "Enhancing Security and Privacy in Biometrics-Based Authentication Systems," *IBM Systems Journal*, 1133 Westchester Avenue, White Plains, New York 10604, Volume 40, Number 3, 2001. Copyright © 2006, IBM Corporation.

2. John R. Vacca, *Net Privacy: A Guide to Developing and Implementing an Ironclad eBusiness Privacy Plan*, McGraw-Hill (2001).

3. John R. Vacca, *Satellite Encryption*, Academic Press (1999).

4. John R. Vacca, *Practical Internet Security*, Springer (2006).

5. John R. Vacca, *Optical Networking Best Practices Handbook*, John Wiley & Sons (2006).

How Brute Force Attacks Work

As mentioned in the foregoing chapters, biometrics such as fingerprints, voiceprints, irises, and faces are becoming increasingly attractive tools for verification and access control. As replacements for passwords, biometrics have a number of advantages. First, biometrics are inherently linked to the user and cannot be forgotten, lost, or given away. Second, appropriately chosen biometrics have high entropy and are less susceptible to brute force attacks than poorly chosen passwords—but are susceptible to brute force attacks nonetheless. Finally, biometric verification requires very little user expertise and can be used for widespread deployment.

With the preceding in mind, let's first look at an approach for generating a cryptographic key from an individual's biometric for use in proven symmetric cipher algorithms to show how brute force attacks work. The proposed approach uses a method referred to as biometric aggregation. The encryption process begins with the acquisition of the required biometric samples [4]. Features and parameters are extracted from these samples and are used to derive a biometric key that can be used to encrypt a plaintext message and its header information. The decryption process starts with the acquisition of additional biometric samples from which the same features and parameters are extracted and used to produce a "noisy" key, as done in the encryption process. Next, a small set of permutations of the "noisy" key are computed. These keys are used to decrypt the header information and determine the validity of the key. If the header is determined to be valid, then the rest of the message is decrypted. The proposed approach eliminates the need for biometric matching algorithms, reduces the cost associated with lost keys, and addresses nonrepudiation issues. This chapter also very, very briefly reports on work in progress [1].

Biometric Cryptography: Key Generation Using Feature and Parametric Aggregation to Show How Brute Force Attacks Work

This part of the chapter proposes a technique for generating keys for symmetric cipher algorithms (such as the widely used Data Encryption Standard (DES) and 3-DES) to show how brute force attacks work and how they can be prevented. This technique can be extended to asymmetric algorithms as well. There are several problems that must be addressed in order to generate a useful biometric cryptographic key. We will consider those associated with:

1.   The entropy (strength) of the biometric key

2.   The uniqueness of the biometric key

3.   The stability of the biometric key [1]

Key Entropy (Strength)

Instead of simply developing longer cryptographic keys to resist brute force attacks, a more intelligent approach is to aggregate features and parameters from an individual in such a way that their correlation generates a key that is much stronger than the individual size of the actual key [1].

Key Uniqueness

The uniqueness of a biometric key will be determined by the uniqueness of the individual biometric characteristics used in the key. Instead of trying to find a single unique feature, a biometric key needs to find only a collection of somewhat unique features or parameters that, when assembled collectively, create a unique profile for an individual. The incorporation of a simple passphrase will improve the accuracy of the biometric key by incorporating "something you are" with "something you know" [1].

Key Stability

A major problem with biometric identification is that an individual's enrollment template and sample template can vary from session to session. This variation can occur for a number of reasons including different environments (lighting, orientation, emotional state) or physical changes (facial hair, glasses, cuts).

If a set of relatively stable features can be determined and the amount of variation can be reduced to an acceptable number of bits, then it might be possible for a valid user to search a limited key space to recover an encrypted transmission while making a brute force search by an attacker remain difficult, if not impossible [1].

The motivation for this approach comes from studying the improvements being made in biometrics and computer processing speeds as well as the limitations associated with existing cryptographic operational requirements. Biometric cryptography does not require that a complete solution to the biometrics problem (find a trait that will correctly identify a user under all conditions) be found [1].

Previous Work

There has been relatively little work done on generating keys using biometrics to date. This is primarily because biometrics does not produce the same matching templates every time a sample is collected and cryptography relies on a stable and unique key to encrypt and decrypt messages. There are two approaches, key release and key generation, that have been proposed to address incorporation of biometrics into cryptography [1]. Key release algorithms require that:

1.   The cryptographic key is stored as part of the user's database;

2.   There must be access to biometric templates for matching;

3.   User verification and key release are completely decoupled [1].

While this technique does work, several problems result. One problem is that there is no way to ensure who produced the key. The user could deliberately choose a known weak key. Second, if the key is stored in a database, the information could be hacked by spoofing the matcher and, third, an enrollment process is required to store the template [1].

Key generation approaches avoid some of the problems associated with key release approaches by binding the secret key to the biometric information and requiring access to a biometric template. Unfortunately, key generation approaches so far have required prealigned sample representations, intensive calculations, and more complicated systems than their key release counterparts [1].

Application Description

Figure 22-1 illustrates the system structure for biometric cryptography [1]. The encryption and decryption processes are laid out in this illustration. The encryption process takes as input a plaintext message and header and uses a biometric identifier as the cipher key. The decryption process is similar to the encryption process, except that it is responsible for computing a limited number of permutations of the sample key with the hope that one of the permutations will match the original encryption key [1].

Biometric Aggregation

A novel approach to defining a biometric key is currently being explored. Biometric aggregation is an extension of the AdaBoosting concept, in which aggregation of smaller, well-defined classifiers can provide a more accurate classification than any single classifier. Figure 22-2 provides a more detailed

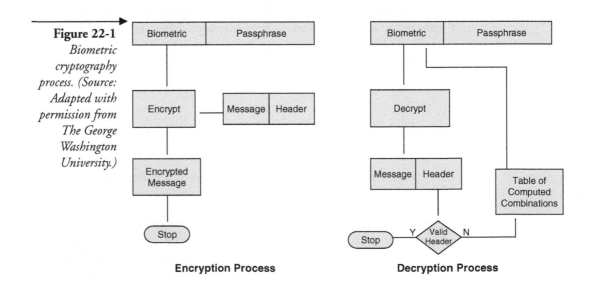

Figure 22-1 *Biometric cryptography process. (Source: Adapted with permission from The George Washington University.)*

Encryption Process   **Decryption Process**

Figure 22-2   *Biometric aggregate identifier. (Source: Adapted with permission from The George Washington University.)*

| Fingerprint Classifiers/Parameters (Each hand, X_1, bits) | Faceprint Classifiers/Parameters (Each hand, X_2, bits) | Palmprint Classifiers/Parameters (Each hand, X_3, bits) | Iris Classifiers/Parameters (Each hand, X_4, bits) | Retina Classifiers/Parameters (Each hand, X_5, bits) | Speechprint Classifiers/Parameters (Each hand, X_6, bits) |

illustration of the biometric identifier being used to encrypt and decrypt the message and header [1].

Note: A brute force attack could be attempted against each component of this key to find a weakness. However, if each part of the key could be considered to be strong enough (a 56-bit DES key can be cascaded) and since each part of the key is related, the overall key strength should be greater than the minimum bit length of a constituent [1].

Without further ado, let's look at brute force attacks by themselves in more detail. How do they really work, and are there many variations of them?

Brute Force Attacks in General

The most basic attack that can take several forms is brute force. First, the hacker can sit down at the terminal and perform an online attack. The adversary enters your username and tries to manually guess your password. Given unlimited opportunities, an adversary may be able to guess the password. Login guessing attacks are thwarted by configuring the operating system to limit the number of failed login attempts allowed for each user. After a threshold is reached, the account will be locked and that user will be unable to log in until the system administrator intervenes by resetting the password or unlocking the account in some other way [2].

Unfortunately, this defense policy allows for a denial-of-service attack. A malicious person could cycle through each user account on the system, exceed the failed login threshold, and lock out all of the accounts. The policy should be configured so that the root user, or another privileged user, is always able to access the system from a designated, physically secure console in order to correct the situation. To generalize, any verification system should be resistant to denial-of-service attacks that lock out the system administrator. On computers with an external media device, such as a tape or a CD-ROM drive, the administrator can always boot the system from backup media, enter a limited computing environment known as single-user mode in UNIX, and take corrective actions. One can further deter manual guessing by injecting some randomness into the login procedure. For example, if an incorrect password is entered, the system can delay for a time interval before presenting the next login prompt. The time interval can be computed as an increasing function of the number of failed logins from that terminal device, or it might be configured as a constant value such as always delaying for two minutes after three failed login attempts for a single user. If increasing interval delays are used, another configuration value is needed to determine when to reset the interval to its initial state. A variation on the interval delay is to change the input time between keystrokes used to read the password [2].

Another defense involves locking the terminal itself after a configured number of failed login attempts. As in the preceding case, the terminal can be reset only by an authorized administrator. Some operating systems provide an option for configuring which terminal devices individual users are able to use for login. A failed login attempt at an unauthorized terminal is usually not counted against the failed login threshold. With today's ubiquitous networked computer environments, some operating systems extend the notion of "authorized terminal" to include a range of valid network addresses [2].

Notice that these attempts to defend against manual guessing attacks are used only to deter the hacker. Failed login thresholds, delay intervals, and other defenses will not prevent a brute force attack from occurring. However, they will discourage someone from instigating a brute force attack and increase the difficulty of successfully executing an attack. Preventing a brute force attack from happening is nearly impossible. It is highly probable, though, that one can prevent manual guessing attacks from succeeding. To increase the sophistication of the online brute force attack, the hacker can write a program to carry out the manual guessing steps. A dictionary of passwords is compiled and perhaps even customized for the penetration attempt. The program consults the dictionary when attempting to crack an account. Heuristics can be used so that the program adapts to the targeted host. For example, guessing that the password is the same as the username is a plausible first attempt [2]. The Internet Worm [5] was able to crack a number of accounts throughout the Internet using the following heuristics for passwords:

1.   No password

2.   The username or the username concatenated with itself

3.   Variations on the user's first name or last name: uppercase, lowercase

4.   The username typed in backwards

5.   A dictionary of 432 words chosen by the worm's author

6.   The online UNIX dictionary, if available [2]

You might think that a programmed attack like this works only if the attacker is trying to log in via a network using one of the many built-in ways in which operating systems provide for network logins. However, it is just as easy to detach a terminal from its cable and directly attach the cable into the back of a personal computer programmed to emulate a terminal. From this configuration, the hacker can run an automated attack that appears like a user attempting to physically log in [2].

A more dangerous brute force attack occurs when the attacker is able to gain a copy of the stored password values. With this additional knowledge, an offline attack is possible, via one of two probable approaches. In the first, the attack is centered on repeatedly trying to guess the password on a victim machine. Your system will be unaware of any attempts by the user to guess passwords because all efforts are made on another system. If a UNIX password file is obtained, the attacker could use any number of collaborators to achieve offline guessing in parallel. Portions of the password could even be divided among the cohorts. The victims machine is configured by attackers so that the defense mechanisms, such as failed login thresholds, are disabled [2].

A second type of offline attack depends on a matching strategy. Instead of guessing passwords, the attacker computes a number of hashed password values in advance. The attack succeeds when a user's hashed password value matches one of the precalculated hash values. Tools to facilitate this approach are readily available for both UNIX and NT. Crack software is widely relied upon by professional penetration testing teams to probe the suitability of user passwords on UNIX systems. Crack has been very successful at breaking many passwords. The arrival of L0phtCrack showed that NT also was subject to this kind of offline brute force attack. You might find it interesting that the infamous Internet Worm (which does not seem so infamous, given the sensitivity of sites recently broken into) traveled with a mini-dictionary and password cracker [2].

You must assume that an attacker has more time to devote to hashing potential passwords than you have for managing your systems. Thus, this attack is particularly threatening if your users are choosing passwords from a small search space, such as passwords of only three alphabetic characters. Various statements about the difficulty of breaking the UNIX password hash have circulated in the community for years. Often one will encounter a statement describing the probability that a password can be cracked to be nearly impossible. Such statements are made from a theoretical basis and cite the length of the password, usually in bits, and how hard it would be to exhaustively search the entire space of these bit strings. Indeed, the usual phrase is that an exceedingly powerful computer running for the known life of the universe would not be able to complete the search. What these statements do not consider are the practical aspects of password guessing [2].

Users repeatedly choose weak or easily guessed passwords. Hackers and password crackers use this information to narrow the search space and substantially reduce the complexity of the problem. Although in theory, guessing an eight-character password from the universe of eight-character passwords is impossible; in practice, hackers are quite successful at it [2]!

A good defense against offline brute force guessing is to use stronger passwords. You also need to protect the password repository, whether it is a file on a local computer or a database on a central server. If the computing environment in which you are working will support encryption of the password file or database, you should take advantage of this feature. If the password file is stolen, mounting offline attacks against an encrypted file will be much more difficult [2].

If a user is required to remember different passwords for several systems, chances are that the passwords will be written down. Hackers who physically reconnoiter a site have favorite places they inspect for written passwords, such as under the keyboard, on a nearby filing cabinet, or on the back of the monitor. Needless to say, passwords that are written down are fairly easy to crack [2].

It is interesting that despite the many additional defense mechanisms in operating systems to deter brute force attacks, many computers today are cracked because of weaknesses in the password itself. Ignoring pleas from security experts, countless books, and trade magazines, users still continue to choose passwords that are easily guessed [2].

Social Engineering

Not all password threats are based on guessing or cryptographic techniques. Many hackers report that the easiest way to break into a system is social engineering. You would be amazed at how freely information is given over the phone without proper verification between the parties. Hacker lore is filled with tales of gullible users being conned into giving away their passwords, the passwords of their superiors, or other information that can be used to penetrate a network [2].

Sometimes, a social engineering attack requires physical surveillance of the work site. To accomplish this surveillance, an opponent impersonates someone from a maintenance company, courier service, or even a pizza delivery person to gain access to the site. Once inside, personal information about the target person can be gleaned from pictures on the desk, by sifting through the trash, or by listening to careless office gossip. Some even go so far as to dig through trash containers on the company's premises. Security guards who stumble upon these sifters are easily repelled when the hacker explains the activity as collecting aluminum cans, searching for a lost article such as a watch, or desperately trying to retrieve a lost report. Stories have been told of security guards helping a hacker find useful information in these situations [2].

The shoulder surfing technique is also a favorite. Try watching a friend type in a password. You will be surprised how easy it is to pick up at least a few characters. Remember, any information is useful. Knowing the password length and a few of its characters can help reduce the search space. If the password is particularly difficult to type, or if the user is unaccustomed to keying in the password, shoulder surfing is made easier by the slow keystroke pace. Another social engineering trick is to distract a user while the password is being entered. Verbal information processing can reduce the keystroke rate of a user. In other words, if the attacker is chatting in your ear about last night's football game, the time it takes you to enter your password will be increased [2].

When sufficient background material is obtained, the fun begins. The biggest problem the hacker faces is deciding which approach to use for social engineering. A particularly successful approach is for the attacker to call the target user and impersonate a superior. If the perpetrator can act convincingly, the hapless employee probably will respond automatically to any request. An alternative is to call a powerful network or system administrator over a period of time and build rapport by appealing to this person's ego [2].

Hackers have reported calling site experts with faked problems, only to gradually develop a "friendship" with the unsuspecting soul on the other end. Enough trust has been built up to trick the victim into divulging information useful for penetrating systems, even if passwords were not obtained. At one recent hacker conference, a successful social engineering attack was carried out via telephone as part of a keynote address. In order to avoid breaking any serious laws, the speaker disconnected the victim only seconds before some useful secrets were disclosed [2].

The purpose of a social engineering attack may be simply to gain additional information that makes password guessing easier. Almost any information is useful to an attacker. Names of children, favorite hobbies, project names, birthdays, and other personal data can help narrow the search space for a brute force password attack. A popular but predictable technique that some people use to "improve" the strength of passwords is to replace some characters with numerals. For example, the password "cocoon" would instead be "c0c00n." With respect to computer search speeds, this additional twist does not add significantly to the password combinations the cracker must test. Notice that the only defense, if you rely on reusable passwords, is to educate site users. Periodic reviews and trials can ensure that employees are complying [2].

Trojan Horses

Every computer science major has learned how to leave a login Trojan horse on a system. Before logging off the system, the perpetrator starts a problem that displays a login prompt and waits for a victim. The username and password entered into the Trojan horse are logged to a file or mailed to a collecting account. Usually, the Trojan horse fakes some type of problem and exits. The operating system then takes control and displays the true login prompt. Most users would assume that they had entered a password incorrectly or that some other glitch occurred in the system. Not surprisingly, this kind of attack can be very fruitful [2].

The temporary Trojan horse login succeeds because of a flaw in the login verification protocol. The user is required to authenticate to the computer, but the login program is assumed to be legitimate. To circumvent this problem, secure operating systems provide a secure attention key (SAK) sequence. The NT operating system instructs the user to enter Ctrl-Alt-Del to initiate a trusted path with the operating system. Most UNIX systems also provide a SAK [2].

When this special key sequence is pressed, the user is assured that a clean environment is made available for login. For example, the system will detach any processes that are attached to or running on that terminal. What happens to these processes depends on the operating system implementation. The net result is that there will not be a chance for the previous user's processes to act as a login impostor [2].

A more serious threat is replacement of the login program in the system itself. This attack depends on circumventing the system's access control mechanisms because login and other I&A routines are part of the TCB. A hacker who manages to install a permanent login Trojan horse can gain multiple username and password pairs. It is unlikely that only the login program would be replaced. Trojan horse versions of other security enforcing programs are certain to be found as well [2].

Network Sniffing

Many network protocols were designed with the assumption that users could be trusted or that the network was trustworthy. Precautions in protocol design for defending against network eavesdropping were not always taken. Network traffic monitoring is the electronic equivalent of shoulder surfing. A network sniffer is a program, or dedicated device, capable of capturing all traffic made available to one or more network adapters. Any data sent in the clear across the network is captured and inspected for usefulness. Countless network sniffers are running throughout the Internet today [2].

Network sniffers are freely available in the public domain or can be purchased as part of products such as RealSecure from Internet Security Systems. A user who has access to a personal computer connected to a network can easily install a sniffer program. Most sniffers are sophisticated enough to selectively find passwords used for network logins. The attacker does not need to monitor every packet traversing the network. Assuming that the communicating systems rely upon reusable passwords for verification, the person sniffing network traffic can effortlessly gather passwords to be used for later attacks. No evidence of this activity will be found on the attack targets, as is the case for online brute force attacks [2].

Network sniffing is not limited to watching for passwords used during the verification phase of a network login session. Because e-mail and other document delivery systems might contain lists of passwords, it is worth the effort to capture and scan these data forms as well. Remember that a new user must acquire the initial password from the security officer in an out-of-band manner. A common method chosen is e-mail, especially inside private corporate networks. Employees are often required to sign agreements declaring that they will not engage in network sniffing or scanning. Because many computer crimes include an insider, the threat of legal consequences clearly does not always outweigh the opportunity for financial reward [2].

Many private corporate networks also are accessed by contract vendors, who in turn may not adhere to the same restrictions. A successful social engineering attack could land a planted network sniffer on your network. The sniffer could periodically send passwords via e-mail to an external system. For these reasons, you should assume that passwords that are sent across a network in cleartext form have been compromised [2].

Electromagnetic Emissions Monitoring

Electromagnetic emissions also have been exploited as a means for sniffing passwords, albeit in a different wave spectrum than network traffic. Despite efforts by various standards agencies to limit emissions from monitors and even storage devices, surveillance of these data sources is a very serious threat. The U.S. TEMPEST standard is one guideline that manufacturers must follow to reduce electromagnetic emissions in an effort to eliminate this kind of attack. The general idea behind TEMPEST is to shield devices from emitting a strong signal. In some cases, an individual room or an entire building is built to the TEMPEST standard [2].

Software Bugs

Sometimes, the operating system does all the hard work for the hacker. Software bugs continue to be a major source of security problems. For example, a recent bug in the Solaris operating system made the hashed password values available to anyone on the system. One of the network application programs could be forced to end abnormally, and as a consequence, that program would dump its memory contents to disk in a core file (to aid in debugging the crash). Users with no special privileges could force the program to do this. The core file contained copies of the hashed password values that normally were stored in a shadowed file. The information could be used as input to Crack for an offline brute force attack [2].

Next, let's look at how biometric crypto algorithms are used to repel brute force attacks. In other words, this next part of the chapter examines symmetric cipher implementation techniques.

Biometric Crypto Algorithms to Repel Brute Force Attacks

The traditional areas of cryptography are symmetric ciphers and public key. Breaking cipher refers to finding a property (or fault) in the design or implementation of the cipher that reduces the number of keys required in a brute force attack (that is, simply trying every possible key until the correct one is found). For example, assume that a symmetric cipher implementation uses a key length of 128 bits. This means that a brute force attack needs to try up to all 2^{128} possible combinations to convert the ciphertext into plaintext, which is way too much for the current and foreseeable future computing abilities. However, a biometric cryptanalysis of the cipher with a 16-bit key allows the plaintext to be found in 2^{16} rounds, which is feasible [3].

There are numerous techniques for performing biometric cryptanalysis, depending on what access the biometric cryptanalyst has to the plaintext, ciphertext, secret key, or other aspects of the biometric cryptosystem. These are some of the most common types of attacks:

1. **Known-plaintext analysis:** With this procedure, the biometric cryptanalyst has knowledge of a portion of the plaintext from the ciphertext. Using this information, the biometric cryptanalyst attempts to deduce the key used to produce the ciphertext. The famous brute force is essentially a know-plaintext attack, although it can be extended to ciphertext-only attacks by using heuristics for detecting the plaintext.

2. **Chosen-plaintext analysis (also known as differential biometric cryptanalysis):** The biometric cryptanalyst is able to have any plaintext encrypted with a key and obtain the resulting ciphertext, but the key itself is not known. The biometric cryptanalyst attempts to deduce the key by comparing the entire ciphertext with the original plaintext. Older RSA algorithms have been shown to be somewhat vulnerable to this type of analysis.

3. **Ciphertext-only analysis:** Here, you need to work only from the ciphertext. This requires accurate guesswork about the style and how a message could be worded. It helps to know as much as possible about the subject of the message. At its extreme, such knowledge might bring this case closer to known plaintext analysis [3].

In addition to the preceding, other techniques are available. Even more important, the boundaries between a ciphertext-only attack, a known plaintext attack, and a chosen-plaintext attack are not necessarily rigid in practice. One of the techniques used to mount a ciphertext-only attack on a message is the probable word method, where a guess about possible plaintext is tried as a hypothesis. This was how Enigma messages were decrypted during World War II. Despite the fact that Enigma was a very secure stream cipher machine of its time, this negligence was successfully exploited by the British. This trick, in effect, at the cost of some wrong guesses and additional trials, turns the ciphertext-only case into the known plaintext case. In case of Enigma, the guessed plain text of the intercepted cipher message was called a crib. In 1939–40, Alan Turing and another Cambridge mathematician, Gordon Welchman, designed the most famous deciphering machine, the British Bombe. The basic property of the

Bombe was that it could break any Enigma-enciphered message, provided that the hardware of the Enigma was known and that a plaintext "crib" of about 20 letters could be guessed accurately [3].

Similarly, if a sufficient amount of known plaintext is available, that quantity will include plaintexts with different properties, including some that are desirable to the biometric cryptanalyst; hence, at least in extreme cases, the known plaintext case blurs into the chosen-plaintext case [3].

Successful biometric cryptanalysis is a combination of mathematics, inquisitiveness, intuition, persistence, powerful computing resources, and—more often than many would like to admit—luck. It's often a government's favorite game: Enormous resources are usually required. The breaking of the German Enigma code was probably the most famous case [3].

Today, biometric cryptanalysis is practiced by a wider range of organizations, which includes companies developing security products, and so forth. It is this constant battle between cryptographers trying to secure information and biometric cryptanalysts trying to break biometric cryptosystems and invent approaches that moves biometric cryptology forward. Cryptographers are now in a much better position than ever before to defeat attempts of biometric cryptanalysts to break their ciphers [3].

The reader needs to be very careful in judging the security of algorithms. The weak keys problem is especially important, much more so than resistance to differential biometric crypto attacks, although that is the currently fashionable area. One needs to understand that people are probably the weakest link, and many attacks have been possible because of poor choice of keys or other blunders [3].

Generally, attacks on well-studied biometric crypto algorithms are very difficult and, in this sense, research in this area often distort the truth by hinting that if algorithms contain a particular weakness it is easy to break it. Not true. The biometric cryptanalysis of single-key biometric cryptosystems depends on one simple fact—that some traces of the original structure of the plaintext may be visible in the ciphertext. For example, in a monoalphabetic substitution cipher (where each letter in the plaintext is replaced by a letter in the ciphertext that is the same each time), a simple analysis of a sizeable portion of the ciphertext can be used to retrieve most of the plaintext. This is due to the difference in frequencies of letters in the natural languages. That's why block ciphers are generally preferable [3].

Even a slight variation of the classic scheme of application of a cipher (one secret key to a monolithic plaintext) completely changes the rules of the game and may substantially increase the security of even a very simple algorithm. That means that restrictions of exporting a product that contains cryptographic algorithms is not as effective as one may think. For example, if the key is artificially limited to, say, 56 bits to make a brute force attack possible, there are still multiple cheap ways to foil such attempts. For example, if you think that DES is insecure, you can use Tripple DES without any problems [3].

With the advent of computers, nothing prevents the use of such methods as injection of random letters, striping, and compression. Striping involves splitting the text into stripes (for example, each nth character method produces n stripes) and then encoding each stripe separately, by using a different key. Striping reduces the effective length of the ciphertext and distorts statistical properties of the plaintext that are present, and thus makes biometric cryptanalysis a lot more difficult [3].

That means that with minor additional transformation, it's not that easy to break even very weak ciphers or algorithms that are often dismissed by the popular press as being insecure. For example, many consider DEC to be insecure. Let's assume that it is incorporated into some hardware of a black-box program, and that you need to enhance the security of the communication of information storage. There are several easy ways to do this:

- Use the two additional rounds of software DEC encryption, converting DES into Triple DES.

- Preprocess the plaintext with, say, bijective Huffman encoding. This is probably the most logical way to enhance the strength of stream ciphers. What you lose in speed you get back in transmission. Moreover, any type of compression effectively shortens the message, and this increases the ration length of plaintext/length of the password.

- A simple way that is the most logical for block ciphers like DEC is to add a random first byte to each block (salt) to invalidate "chosen plaintext" attacks. During the decryption, those first bytes can be easily discarded at almost no computational cost. The only penalty is a slight increase of the length of the ciphertext (1/8 or ∼12% for 64-bit blocks).

- Another possibility for clock ciphers is to use block-length-based striping (eight-way striping for DEC) with the simple transposition of each stripe as the extension of the secret key (random letters should be added to the stripes to make the length mod 8). In this case, the content of each block depends on the plaintext.

- Yet another method is to use for each message a key generated by pseudorandom generator with the actual key acting as the seed. This way there are no two messages with the same key. And, by knowing the initial key and pseudogenerator used, it's easy to recover the message even if some of the previous messages were lost. That also to a large extent invalidates "chosen plaintext" attacks [3].

Each of those measures makes attacks more difficult and negatively affects even the exhaustive search of the key space (brute force attack) [3].

Note: A brute force attack on plain vanilla DEC is not an easy task, and the cost of specialized computers is substantial ($1,000,000 in 1998; probably around $70,000 now).

No easy attack on DES has been discovered, despite the efforts of researchers over many years. The obvious method of attack is a brute force exhaustive search of the key space; this process takes 255 steps on average. Early on, it was suggested that a rich and powerful enemy could build a special purpose computer capable of breaking DES by exhaustive search in a reasonable amount of time. Later, Hellman showed a time-memory trade-off that allows improvement over exhaustive search if memory space is plentiful. These ideas fostered doubts about the security of DES. There were also accusations the NSA had intentionally weakened DES. Despite these suspicions, no feasible way to break DES faster than an exhaustive search has been discovered. The cost of a specialized computer to perform exhaustive search (requiring 3.5 hours on average) has been estimated by Wiener at one million dollars. This estimate was recently updated by Wiener to give an average time of 35 minutes for the same cost machine [3].

The first attack on DES that is better than an exhaustive search in terms of computational requirements was announced by Biham and Shamir, using a new technique known as differential cryptanalysis. This attack requires the encryption of 247 chosen plaintexts; that is, the plaintexts are chosen by the attacker. Although it is a theoretical breakthrough, this attack is not practical because of both the large data requirements and the difficulty of mounting a chosen plaintext attack [3].

More recently, Matsui has developed another attack, known as linear cryptanalysis. By means of this method, a DES key can be recovered by the analysis of 243 known plaintexts. The first experimental cryptanalysis of DES, based on Matsui's discovery, was successfully achieved in an attack requiring 50 days on 12 HP 9735 workstations. Clearly, this attack is still impractical [3].

Again, recently, a DES cracking machine was used to recover a DES key in 22 hours. The consensus of the cryptographic community is that DES is not secure, simply because 56-bit keys are vulnerable to exhaustive search. In fact, DES is no longer allowed for U.S. government use; triple-DES is the encryption standard, and AES is currently replacing that standard for general use [3].

A nontraditional method that might provide a substantial increase in security for weak ciphers is steganography. It is a very attractive method as the length of the transmitted text is not critical and can be at least doubled. The simplest example is imbedding the message into another "decoy" message before encryption. In this case, a chosen plaintext attack is useless, as for any given plaintext there is an infinite number of decoy texts. The simplest steganography-based defense is the injection of random characters using some formula that becomes the second key. For example, if only each second letter constitutes a plaintext and all other letters are decoy, the statistical properties of the text are distorted enough to consider the text a pseudorandom sequence of letters [3].

These are just random thoughts, and there are definitely better and still very simple ways to enhance any well-studied cipher like DEC or GOST. The area is definitely complex but, still, there is a level of hype, and distortions are enough to raise red flags [3].

The impression is that, even in the best case, brute force attacks are mostly impractical. "Other" methods are much more attractive, which is a very powerful argument against reusing keys and for using some kind of modern "one-time keys" technology (for example, any Secure-ID token is essentially a generator on one-time keys that can be used for encoding messages) [3].

All in all, the availability of powerful computers has radically changed this field. This is not in favor of biometric cryptanalysis specialists, but in favor of those who want to protect the text from decryption. Such methods include compression, steganography, new types of one-time pads, and code-length manipulation (with Huffman encoding, the length of the letter became a variable). All of these methods provide many interesting opportunities, and the IMHO can significantly enhance the level of security for any weak cipher. Unfortunately, these areas are usually ignored in the traditional cryptography textbooks [3].

With the current level of sophistication of biometric crypto algorithms, the chances that a particular message will be decrypted are very small. This also includes being outside of well-known three-letter agencies that are minimal, even with ciphers that are considered "weak" (DES). It might well be that further progress can be achieved, not by creating a stronger cipher, but by using new capabilities provided by computers. This includes but is not limited to striping (multiple pipelines encoding), compression, steganography (with adding noise as the simplest form), and alphabet substitution (each message actually can be encoded in its own alphabet). And, if each letter does not occupy a byte, then finding byte boundaries constitutes an additional solution. Consider, for example, an 11-bit key that contains three codes for each character that can be randomly chosen during the encoding phase [3].

It is important to understand that the cipher is kind of an envelope for the message and that nothing prevents you from sending the messages in parts. This would include several envelopes, possibly using not only different keys, but also different ciphers for each. This is different from striping, but achieves the same effect: Recovery of plaintext is more difficult for the same length of the message [3].

Outside of well-studied areas (like DES), ciphers still represent a little-known type of algorithm. And, like pseudorandom generators, more complex ciphers are not necessarily better or more secure. Unfortunately, multiple applications of the same cyphers increase security (like 3DES when the plaintext is encoded with one key and then with another) and represent a very little researched area. How two ciphers interact with each other in the scheme (cipher1-cipher2-cipher1), and to what extent they increase the strength of the original cipher in a general case is unclear. The answer may well be different depending on the type of cipher [3].

Note: While compression is a powerful method to increase the security of the messages, the usage of off-the-shelf compression utilities is a mixed blessing. While they reduce the overall redundancy of the text, they often put a predictable header at the beginning of the compressed stream, which can facilitate known-plaintext attacks. That problem can be solved on different levels, but even the simplest communication device now contains enough memory and CPU power to use bijective Huffman encoding, which was designed to nullify this disadvantage [3].

It has been argued that prevention of brute force attacks is actually one of the purposes of compression. As for chosen plaintext attacks, compression does not add much: You can always compress the chosen plaintext and compare results not with known plain text, but with the recompressed image. An even more aggressive approach to compression is to base it not on letters, but on diagrams or words. That might include replacing any frequently used word randomly chosen from a set of languages or (for articles) random letter combinations that can be automatically eliminated [3].

It has long been appreciated that there are advantages to eliminating regularities in the plaintext before encrypting. The primary advantages to doing this are that the opponents get less cyphertext to analyze; and what they do get has a corresponding plaintext with fewer redundancies and regularities [3].

The advantage of the first point should be obvious enough: The less data the enemy has to analyze, the fewer clues they have about the internal state of your cipher, and thus its key. The advantage of the second point is that it hinders cryptanalytic attacks. "Fewer redundancies and regularities" may be translated into more formal terms as "greater entropy per bit." The more closely the statistical properties of the file approach that of a random data stream, the fewer regularities the biometric cryptanalyst has to go on. All of this should be uncontroversial [3].

The fact that compression aids encryption was first realized by those who first employed "codewords" in their ciphers. By replacing frequently used words like "the" and "and" with otherwise little-used symbols before encrypting, they succeeded in reducing the volume of the text based on known regularities in the language. This type of cipher was employed, for example, by Mary Queen of Scots [3].

Finally, eliminating patterns in the frequency of occurrence of particular symbols in the text before enciphering is desirable. This was clearly realized by the time homophones were employed in conjunction with mono alphabetic-substitution ciphers [3].

Summary/Conclusion

This chapter dealt solely with the problem of verification and preventing brute force attacks. Depending on the application, biometrics can be used for identification or for verification. In verification, the biometrics is used to validate the claim made by the individual. The biometric of the user is compared with the biometrics of the claimed individual in the databases. The claim is rejected or accepted based on the match.

> **Note:** In essence, the verification system tries to answer the question, "Am I whom I claim to be?"

In identification, the system recognizes an individual. This is done by comparing his or her biometrics with every record in the database.

> **Note:** In essence, the identification system tries to answer the question, "Who am I?"

In general, biometric verification consists of two stages: enrollment and verification. During enrollment, the biometrics of the user are captured and the extracted features (template) are stored in the database. During verification, the biometrics of the user are captured again, and the extracted features are compared with the ones already existing in the database to determine a match. The specific record to fetch from the database is determined by using the claimed identity of the user. Furthermore, the database itself may be central or distributed, with each user carrying his or her template on a smart card. Biometrics offers several advantages over traditional security measures, including nonrepudiation, accuracy, and security.

Nonrepudiation

With token- and password-based approaches, the perpetrator can always deny committing the crime by pleading that his or her password or ID was stolen or compromised in some fashion. Therefore, a user can repudiate or deny the use of a service even when an electronic record exists. However, biometrics is indefinitely associated with a user and hence it cannot be lent or stolen, thus making repudiation infeasible.

Accuracy and Security

Password-based systems are prone to dictionary and brute force attacks. Furthermore, the system is as vulnerable as its weakest password. Biometric verification requires the physical presence of the user and therefore cannot be circumvented through a dictionary or brute force style attack.

References

1. Christopher Ralph Costanzo, "Biometric Cryptography: Key Generation Using Feature and Parametric Aggregation," School of Engineering and Applied Sciences, Department of Computer Science, George Washington University, October 14, 2004.

2. Paul Gurgul, "Exploits and Weaknesses in Password Security," TechTarget, 117 Kendrick Street, Suite 800, Needham, MA 02494, November 16, 2004.

3. Dr. Nikolai Bezroukov, "Softpanorama Slightly Skeptical Symmetric Crypto Algorithms Links," Softpanorama: (slightly skeptical) Open Source Software Educational Society. Copyright © 1996–2006 by Dr. Nikolai Bezroukov.

4. John R. Vacca, *Satellite Encryption*, Academic Press (1999).

5. John R. Vacca, *Practical Internet Security*, Springer (2006).

How Data-Hiding Technology Works

Biometric-based personal identification techniques that use physiological or behavioral characteristics are becoming increasingly popular compared to traditional token-based or knowledge-based techniques such as identification cards (ID), passwords, and so on. One of the main reasons for this popularity is the ability of the biometric technology to differentiate between an authorized person and an impostor who fraudulently acquires the access privilege of an authorized person [1]. Among various commercially available biometric techniques, fingerprint-based techniques are the most extensively studied and the most frequently deployed [1].

While biometric techniques have inherent advantages over traditional personal identification techniques, ensuring the security and integrity of the biometric data is critical. For example, if a person's biometric data (her or his fingerprint image) is stolen, it is not possible to replace it unlike replacing a stolen credit card, ID, or password (see sidebar, "Identity Documents"). A biometric-based verification system works properly only if the verifier system can guarantee that the biometric data came from the legitimate person at the time of enrollment. Furthermore, while biometric data provide uniqueness, they do not provide secrecy. For example, a person leaves fingerprints on every surface she or he touches, and face images can be surreptitiously observed anywhere that person looks. Figure 23-1 shows eight basic locations of attacks that are possible in a generic biometric system [1]. In the first type of attack, a fake biometric (such as a fake finger) is presented at the sensor. Resubmission of digitally stored biometric data constitutes the second type of attack. In the third type of attack, the feature detector could be forced to produce feature values chosen by the attacker instead of the actual values generated from the data obtained from the sensor. In the fourth type of attack, the features extracted using the data obtained from the sensor are replaced with a synthetic feature set. In the fifth type of attack, the matcher component could be attacked to produce high or low matching scores, regardless of the input feature set. Attack on the templates stored in databases is the sixth type of attack. In the seventh type of attack, the channel between the database and matcher could be compromised to alter transferred template information. The final type of attack involves altering

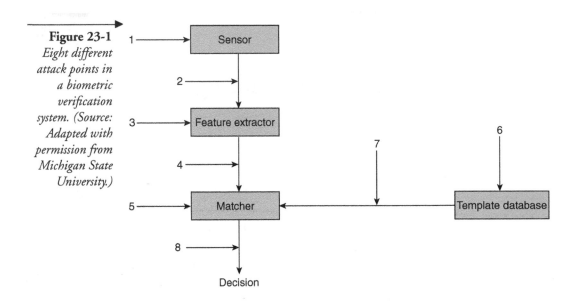

Figure 23-1
Eight different attack points in a biometric verification system. (Source: Adapted with permission from Michigan State University.)

the matching result itself. All of these attacks have the possibility of decreasing the credibility of a biometric system. As a solution to the second type of attack, called replay attacks, a challenge/response-based system is proposed here. In a related context, using biometric data in the generation of digital signatures in both symmetric and asymmetric systems is also proposed [1].

Identity Documents

Identity documents, such as ID cards, passports, and driver licenses, contain textual information, a portrait of the legitimate holder, and some other biometric characteristics (fingerprint, handwritten signature). As prices for digital imaging technologies (software, printers, and digital cameras) fall, making them more widely available, there has been an exponential increase in the ease with which counterfeiters can effectively forge documents. Today, with only limited knowledge of technology and a small amount of money, a counterfeiter can effortlessly replace a photo or modify identity information on a legitimate document to the extent that it is impossible to differentiate from the original [3].

This sidebar proposes a virtually fraud-proof ID document based on a combination of three different data hiding technologies: digital watermarking, 2D bar codes, and copy detection patterns, plus additional biometric protection. As will be shown, this combination of data-hiding technologies protects the document against any forgery, in principle without any requirement for other security features. To prevent a genuine document being used by an illegitimate user, this sidebar considers the additional use of biometric information to enhance the security by several more degrees. The biometrics data are also covertly stored in the document and, as will be discussed, these data can be linked in various ways to the other security features. Let's review the different security features and the way they are used to protect against forgeries or illicit use of the ID document [3].

Digital Watermark

A digital watermark is embedded in the digital image of the ID document (or portrait), which is then printed on a physical ID card. The watermark, which can be spread all over the document or just embedded in specific areas such as the portrait, can later be extracted from the physical ID. This can be done by scanning it and extracting the message with a watermark detector equipped with the encoding key (some specific processing of the document is required to cope with the geometrical transformation inherent to the print-scan process). The watermark on an ID card can serve three purposes: data hiding, verification of the portrait, and cross-verification with other information. But, because the digital analog transformation strongly damages the watermark, the data hiding capacity for reliable detection is quite low, in the order of 10–20 bytes, and a much higher capacity can be achieved with the 2D barcode. However, the watermark can be very useful in protecting the ID portrait against forgeries: Alterations to the portrait can be detected by locating the areas where the watermark is not present or significantly damaged. The watermark also can be used for cross-verification: By linking the watermark message/key to the 2D barcode, to the personal information printed in clear text, and/or to the biometric data, the different pieces of encrypted data are inseparably linked, and any replacement or modification can be detected. One way to link the encrypted data is to generate the watermark key as a hash value of the data in the barcode, itself containing an encrypted version of the personal information [3].

2D Barcode

A 2D barcode can be used to store several hundred bytes of payload. That payload is encrypted, and is used to contain a copy of the personal information or its most significant parts: birth date and ID number. The biometric data, which in the case of a compressed dynamic signature based on a two-dimensional pen position signal typically is in the order of 1kB, are also stored in an encrypted way [3].

Copy Detection Pattern (CDP)

At this point, a counterfeiter cannot forge the ID card (modify or replace the content, personal information, portrait, 2D barcode, etc.), but he or she may still be able to make a high-quality copy of it in which the watermark would survive. The CDP is a special, highly textured digital image that is inserted into the digital image of a document to be printed, and is maximally sensitive to digital-analog transformations. Automatic analysis of a scanned CDP can tell if the document is an original or a copy (by scanning and reprinting, photocopying, etc.). The secret key use to generate and detect the CDP can be made dependent on the information in the document, making each CDP unique [3].

Using the three data-hiding technologies just mentioned, the ID document is in principle cryptographically secure against any forgery or illicit copy. However, there remains the possibility that the document is used by an illegitimate holder. Illegitimate use of a valid ID card can be restricted by the use of biometrics for automatic verification of individuals. Indeed, biometric verification technologies recently have reached a great degree of maturity and a dramatic increase in the number of applications. A number of the behavioral or physical measures that are taken into account for biometric verification process have been used on ID cards for a long time. Examples for such biometric characteristics are the handwritten signature or fingerprint images, which are part of driver's licenses or passports in some countries. According to one possible classification of biometric systems, these biometric data can be referred to as overt biometric properties of the cardholder. Additionally, an encrypted 2D barcode allows for the inclusion of covert biometrics in ID cards for an automated cardholder verification. In a number of applications, it can be desirable to verify the ID card and the holder in the presence and by approval of the individual only. Cooperative biometric verification techniques such as signature verification include an explicit expression of intension and can be used for this purpose [3].

The majority of ID cards, passports, and other identity documents in use are poorly secured with archaic technologies, making them an easy target for counterfeiters. In the digital era, the security-by-obscurity principle cannot be seen as a persistent security strategy. However, by taking advantage of the great progress made in biometric verification, digital data-hiding, and cryptographic techniques, virtually fraudproof ID cards can be designed [3].

In order to promote the widespread utilization of biometric techniques, increased security of the biometric data, especially fingerprints, is necessary. Encryption [4], watermarking, and steganography are possible techniques to achieve this. Steganography, a word derived from Greek and meaning "secret communication," involves hiding critical information in unsuspected carrier data. While cryptography focuses on methods to make encrypted information meaningless to unauthorized parties, steganography is based on concealing the information itself. As a result, steganography-based techniques can be suitable for transferring critical biometric information, such as minutiae data, from a client to a server. Steganographic techniques reduce the chances of biometric data being intercepted by a pirate, hence reducing the chances of illegal modification of the biometric data. Digital watermarking techniques can be used to embed proprietary information, such as a company logo, in the host data to protect the intellectual property rights of that data [1]. They are also used for multimedia data verification. Encryption can be applied to the biometric templates for increasing security. The templates can reside in either:

1.   A central database;

2.   A token such as smart card;

3.   A biometric-enabled device such as a cellphone with fingerprint sensor [1].

They can also be encrypted after enrollment. Then, during verification, these encrypted templates can be decrypted and used for generating the matching result with the biometric data obtained online. As a result, the encrypted templates are secured, since they cannot be utilized or modified without decrypting them with the correct key, which is typically secret. One problem associated with this system is that encryption does not provide security once the data is decrypted. Namely, if there is a possibility that the decrypted data can be intercepted, encryption does not address the overall security of the biometric data. On the other hand, since watermarking involves embedding information into the host data itself, it can provide security even after decryption. The watermark, which resides in the biometric data itself and is not related to encryption-decryption operations, provides another line of defense against illegal utilization of the biometric data. For example, it can provide a tracking mechanism for identifying the origin of the biometric data. Searching for the correct decoded watermark information during verification can render the modification of the data by a pirate useless, assuming that the watermark embedding-decoding system is secure. Furthermore, encryption can be applied to the watermarked data, combining the advantages of watermarking and encryption into a single system. In the context of the work here, the security of the biometric data should be thought of as the means of eliminating at least some of the sources of attacks shown in Figure 23-1 [1].

Watermarking Techniques

Digital watermarking, or simply watermarking, which is defined as embedding information such as origin, destination, access level, and so on of multimedia data (image, video, audio, etc.) in the host data, has been a very active research area in recent years [1]. General image watermarking methods can be divided into two groups according to the domain of application of watermarking. In spatial domain methods [1], the pixel values in the image channel(s) are changed. In spectral-transform domain methods, a watermark signal is added to the host image in a transform domain such as the full-frame DCT domain [1].

There has been very little research on watermarking of fingerprint images. What is proposed is a data-hiding method that is applicable to fingerprint images compressed with a WSQ wavelet-based scheme. The discrete wavelet transform coefficients are changed during WSQ encoding, by taking into consideration possible image degradation. What is also proposed is a fragile watermarking method for fingerprint image verification. A spatial watermark image is embedded in the spatial domain of a fingerprint image by utilizing a verification key. The proposed method can localize any region of image that has been tampered with. Furthermore, this watermarking technique does not lead to a significant performance loss in fingerprint verification. A semiunique key based on local block averages is used to detect the tampering of host images, including fingerprints and faces. There are two spatial domain watermarking methods for fingerprint images. The first method utilizes gradient orientation analysis in watermark embedding, so that the watermarking process alters none of the features extracted using gradient information. The second method preserves the singular points in the fingerprint image, so that the classification of the watermarked fingerprint image (into arch, left loop, etc.) is not affected [1].

For more information on WSQ-based data hiding, see Chapter 21.

Hiding Biometric Data

This section considers two application scenarios. The basic data-hiding method is the same in both scenarios, but it differs in the characteristics of the embedded data, host image carrying that data, and the medium of data transfer. While you are using fingerprint and face feature vectors as the embedded data, other information such as user name or user identification number can also be hidden into the images. For example, you can use one type of biometric data to secure another type of biometric data to increase the overall security of the system [1].

Application Scenarios

The first scenario involves a steganography-based application (see Figure 23-2(a)) [1]. The biometric data (fingerprint minutiae) that need to be transmitted (possibly via a nonsecure communication channel) is hidden in a host (also called cover and carrier) image, whose only function is to carry the data. For example, the fingerprint minutiae may need to be transmitted from a law enforcement agency to a template database, or vice versa. In this scenario, the security of the system is based on the secrecy of the communication. The host image is not related to the hidden data in any way. As a result, the host image can be any image available to the encoder. In this application, you need to consider three different types of cover images: a synthetic fingerprint image,

Figure 23-2
Diagrams of application scenarios: (a) scenario 1 and (b) scenario 2. (Source: Adapted with permission from Michigan State University.)

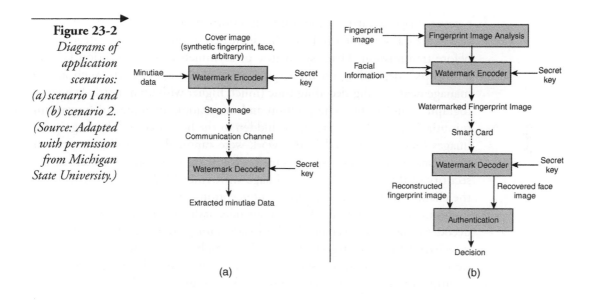

(a)

(b)

Figure 23-3
Sample cover images: (a) synthetic fingerprint, (b) face, and (c) "sailboat." (Source: Reproduced with permission from Michigan State University.)

(a)   (b)   (c)

a face image, and an arbitrary image (see Figure 23-3) [1]. The synthetic fingerprint image (360 × 280) is obtained after a post-processing of the image generated by using an algorithm. Using such a synthetic fingerprint image to carry actual fingerprint minutiae data provides an increased level of security, since the person who intercepts the communication channel and obtains the carrier image is likely to treat this synthetic image as a real fingerprint image [1]!

This application can be used to counter the seventh type of attack (on the communication channel between the database and the fingerprint matcher) depicted in Figure 23-1 [1]. An attacker will most likely not suspect that a cover image is carrying the minutiae information. Furthermore, the security of the transmission can be further increased by encrypting the stego image before transmission. Here, symmetric or asymmetric key encryption [1] can be utilized, depending on the requirements of the application, such as key management, coding-decoding time (much higher with asymmetric key cryptography), and so on. The position and orientation attributes of fingerprint minutiae constitute the data to be hidden in the host images. The fingerprint images (300 × 300) used in this work were captured by a solid-state sensor manufactured by Veridicom. A secret key is utilized in encoding to increase the security of the hidden data. The image with embedded data is sent through the channel that may be subject to interceptions. At the decoding site, using the same key that was used by the encoder (which can be delivered to the decoder using a secure channel prior to stego image transfer), the hidden data is recovered from the stego image. The keys can be different for every transmission, or several parameters such as receiver, sender, and fingerprinted subject identities can be used in determining the key assignment [1].

The second scenario is based on hiding facial information (eigenface coefficients) into fingerprint images. In this scenario, the marked fingerprint image of a person can be stored in a smart card issued to that person (see Figure 23-2(b)) [1]. At an access control site, for example, the fingerprint of the person possessing the card will be sensed and compared to the fingerprint stored on the smart card. Along with this fingerprint matching, the proposed scheme will extract the face information hidden in the fingerprint image. The recovered face will be used as a second source of verification, either automatically or by a human in a supervised biometric application. In this scenario, an additional biometric (face) is embedded into another biometric (fingerprint), in order to increase the security of the latter [1].

Data-Hiding Method

The amplitude modulation-based watermarking method described here is an extension of the blue-channel watermarking method [1]. The proposed method includes image adaptivity, watermark strength controller, and host image feature analysis along with a basic method in [1]. An earlier version of the method is the increase in data decoding accuracy related to these extensions. In the first step, the data to be hidden into the host image is converted to a binary stream. In the first scenario, where fingerprint minutiae data are hidden, every field of individual minutia is converted to a nine-bit binary representation. Such a representation can code integers between [0, 511] and this range is adequate for x-coordinate ([0, N-1]), y-coordinate ([0, M-1]), and orientation

([0, 359]) of a minutia, where N and M are the number of rows and number of columns in the fingerprint image, respectively. In the second scenario, eigenface coefficients are converted to a binary stream using four bytes per coefficient. A random number generator initialized with the secret key generates locations of the host image pixels to be watermarked. The details of this procedure are as follows: First, a sequence of random numbers between 0 and 1 is generated using uniform distribution. Then, every number with odd indices is linearly mapped to [0, X-1], and every number with even indices is linearly mapped to [0, Y-1], where X and Y are the number of rows and columns of the host image, respectively. Every pair comprised of one number with odd indices and one number with even indices indicates the location of a candidate pixel to be marked. During watermark embedding, a pixel is not changed more than once, as this can lead to incorrect bit decoding [1].

Experimental Results

In this part of the chapter, experimental results for the two application scenarios explained previously will be presented. Factors such as decoding accuracy and matching performance will be highlighted. For the first scenario, nearly 17% of the stego image pixels are changed during minutiae data hiding for all the three cover images shown in Figure 23-3 [1]. The key used in generating the locations of the pixels to be watermarked is selected as the integer 1,000. However, the exact value of the key does not affect the performance of the method. In this implementation, the key is used as the seed for the C++ random number generator. The generated random numbers are used as previously explained. Other random number generators can be used without affecting the performance of the proposed method. Remaining watermarking parameters are set to: $q = 0{:}1$, $A = 100$, $B = 1,000$. A higher q value increases the visibility of the hidden data. Increasing A or B decreases the effect of standard deviation and gradient magnitude in modulating watermark embedding strength, respectively. The size of the hidden data here is approximately 85 bytes. The extracted minutiae data from all of the three cover images is found to be exactly the same as the hidden data. Furthermore, the performance of the proposed algorithm was determined as follows: 15 images (five synthetic fingerprint, five face, five arbitrary) were watermarked with five different sets of minutiae data and by using five different keys. As a result, 375 different watermarked images were produced. Characteristics and sources of the host images and watermarking parameters are the same as given previously. Individual minutiae data sets contained between 23 to 28 points, with an average of 25 points. From all of these 375 watermarked images, researchers were able to extract the embedded minutiae information with 100% accuracy [1].

For the second application scenario, the fingerprint image (300×300) shown in Figure 23-4(a) is watermarked using the input face image (150×130) shown in Figure 23-4(b) [1]. The watermark information occupies 56 bytes, corresponding to the 14 eigenface coefficients (four bytes per coefficient). These 14 eigenface coefficients generate the 150×130 watermark face image of Figure 23-4(c) [1].

Note: The 14 eigenface coefficients are sufficient for a high-fidelity reconstruction of input face. A small face image database, which consists of 40 images, with four images for each of the 10 subjects, was used to generate the eigenfaces and coefficients.

Figures 23-4(d) and 23-4(e) correspond to minutiae-based data hiding [1]. The input image in Figure 23-4(a) is watermarked without changing the pixels shown in black (16% of the total image pixels) in Figure 23-4(d) [1]. This minutiae-based feature image is obtained by drawing 23×23 square blocks around every minutiae of the input fingerprint image. Figure 23-4(e) shows the image reconstructed during watermark decoding [1]. Nearly 15% of all the image pixels are modified during watermark encoding. This marking ratio is determined experimentally by requiring 100% correct decoding of the embedded data. Figures 23-4(f) and 23-4(g) correspond to ridge-based data hiding [1]. The input image in Figure 23-4(a) is watermarked without changing the pixels (31% of the total number of image pixels) in Figures 23-4(f) [1]. This ridge-based feature image is obtained from the thinned ridge image of the input fingerprint via dilation with a 3×3 square structuring element comprised of all nine pixels. Figure 23-4(g) shows the image reconstructed during watermark decoding [1]. Nearly 15% of all the image pixels are modified during watermark encoding. This embedding ratio is the same as the one used for minutiae-based embedding; fixing this parameter allows researchers to compare the two methods based on their mask characteristics. Effectively, the images in Figures 23-4(d) and 23-4(f) denote the binary maps [1].

In both cases, the key used in generating the locations of the pixels to be watermarked is selected as the integer 1,000. However, as mentioned earlier, the exact value of the key does not affect the performance of the method. Other watermarking parameters are set to the same values used previously, namely, $q = 0{:}1$, $A = 100$, $B = 1,000$. The watermark data are decoded correctly in

Figure 23-4 *Facial information embedding and decoding: (a) input fingerprint image with overlaid minutiae, (b) input face image, (c) watermark face image, (d) fingerprint feature image based on the minutiae, (e) reconstructed fingerprint image with overlaid minutiae, where watermarking did not change the pixels shown in black in (d), (f) fingerprint feature image based on the ridges, (g) reconstructed fingerprint image with overlaid minutiae, where watermarking did not change the pixels shown in black in (f). (Source: Reproduced with permission from Michigan State University.)*

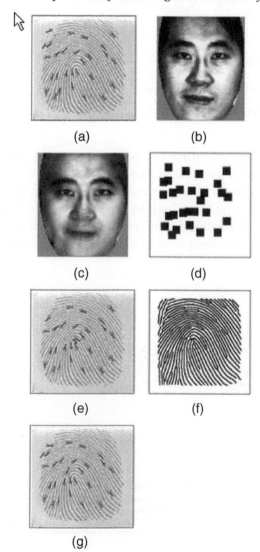

(a)   (b)

(c)   (d)

(e)   (f)

(g)

the decoding phase in both of the cases; the recovered faces are exactly the same as the watermark face image in Figure 23-4(c) [1].

In order to assess the effect of watermarking on fingerprint verification accuracy, receiver operating characteristics (ROC) curves for original images and images that are recovered after watermark decoding are computed. A total of 640 fingerprint images are used in these experiments. These images come from 160 users, with four impressions each of the right index finger captured using a Veridicom sensor. Three ROC curves given in Figure 23-5 correspond to fingerprint verification with:

1.   Data hiding

2.   Minutiae-based data hiding

3.   Ridge-based data hiding [1]

The proximity of the three curves in Figure 23-5 indicates that both minutiae-based and ridge-based watermarking methods do not introduce any significant degradation in fingerprint verification accuracy, though it is observed that ridge-based watermarking leads to less degradation [1]. Furthermore, in both of the cases, the embedded information (14 eigenface coefficients) was decoded with 100% accuracy from all of the 640 watermarked images [1].

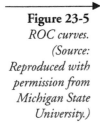

Figure 23-5
*ROC curves.
(Source:
Reproduced with
permission from
Michigan State
University.)*

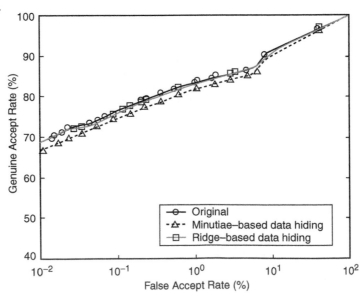

Now, let's take a very detailed look at an application of steganography and watermarking to enable secure biometric data (fingerprint) exchange. As previously mentioned, researchers hide fingerprint minutiae data in a host image, which can be a synthetic fingerprint image, a face image, or an arbitrary image. It is this carrier image that is transferred to the receiving party in this exchange, instead of the actual minutiae data. The hidden biometric data are extracted accurately from the carrier image using a secret key. Furthermore, when the host is a face image, the proposed method provides an additional cue in verifying the user. Data are hidden in the host image in an adaptive way to minimize possible degradations to that image. This method can tolerate several attacks on the carrier image [2].

Hiding a Fingerprint

A fingerprint-based biometric system has four stages: acquisition, representation, feature extraction, and matching. In the acquisition stage, a fingerprint image is captured via inked or live-scan methods. In most of the recent civilian applications, live-scan methods that directly produce the digital image of fingerprints are used. In the representation stage, the aim is to find invariant and discriminatory information inherent in the fingerprint image. In minutiae-based systems, the discontinuities in the regular ridge structure of fingerprint images, called ridge endings and ridge bifurcations, are identified in the feature extraction stage. During matching, a similarity value between the features extracted from the template and the input fingerprint images is calculated. This similarity value is used to arrive at an accept/reject decision [2].

Encryption does not provide security once the data is decrypted. On the other hand, since watermarking involves embedding information into the host data itself, it can provide security even after decryption. Encryption can also be applied to the watermarked data. Another option is to use steganography: By hiding fingerprint features in a carrier image, the security of fingerprint information can be increased [2].

Watermarking Techniques

To increase the security of the watermark data, the original watermark image is first transformed into another mixed image, and this mixed image is used as a new watermark image. The mixed image does not have a meaningful appearance, contrary to the original watermark image, which can contain specific logos or texts [2].

Pixel values at watermark-embedding locations are changed in a way to preserve the quantized gradient orientations around those pixels. As a result, the watermarking process alters none of the features extracted using gradient information [2].

Hiding Minutiae Data

In the following, host, carrier, and cover image terms will be used interchangeably. Figure 23-3 shows examples of these cover images [2]. The synthetic fingerprint image is obtained after a post-processing of the image, which is generated by using an algorithm [2]. The post-processing included increasing the image size from 320×240 to 360×280, eliminating the dominance of white background, and replacing this background with a uniform gray level distribution (between 225 and 235) with a mean of 230. This background transformation helps in hiding the minutiae data more invisibly. Using such a synthetic fingerprint image to carry actual fingerprint minutiae data provides an increased level of security, since a person who intercepts the communication channel and obtains the carrier image would treat this synthetic image as a real fingerprint image! The face image (384×256) was captured in a research laboratory. The "sailboat" image (512×512) is from the USC-SIPI database [2]. The luminance channel of this color image (although reproduced here in black and white) is used as the cover image. This application can be used to counter the attack on the communication channel between the database and the fingerprint matcher. An attacker will probably not suspect that a cover image is carrying the minutiae information [2].

Figure 23-6 shows an input fingerprint image, an overlaid minutiae image, and the attributes of the extracted minutiae [2]. These attributes constitute the data to be hidden in the host images. The minutiae data shown in Figure 23-6(c) contain three fields per minutiae, x-coordinate, y-coordinate, and orientation, for a total of 25 minutiae [2]. A secret key is utilized in encoding to increase the security of the hidden data. The image with embedded data (stego image) is sent through the channel that may be subject to interceptions. At the decoding site, using the same key that was used by the encoder (which can be delivered to the decoder using a secure channel prior to stego image transfer), the hidden data is recovered from the stego image [2].

The second scenario aims at increasing the security of face images. In this scenario, a person's face image, which also carries that person's fingerprint minutiae data, is encoded in a smart card (see Figure 23-7) [2]. At a controlled access site, this image will be read from the smart card and the original face image will be reconstructed. The extracted minutiae data will be compared to the minutiae

Figure 23-6
Minutiae data:
(a) input
fingerprint
image,
(b) overlaid
minutiae image,
(c) minutiae
point attributes.
(Source:
Reproduced with
permission from
Michigan State
University.)

(a)

(b)

x	y	θ
76	216	242
121	195	255
136	82	292
136	229	248
170	98	262
172	169	270
178	46	274
184	85	82
192	146	281
196	198	270
201	89	52
212	233	255
216	220	262
228	125	321
234	79	8
234	147	298
236	175	295
240	167	112
259	68	356
60	32	356
77	197	58
99	85	144
98	69	382
239	190	274
251	222	270

(c)

obtained from the user at the access site. These two minutiae data sets and the reconstructed face image will be used to accept or reject the user. This application can be used to eliminate several types of biometric system attacks. Fake biometric submission via a smart card that contains an inauthentic image will be useless, since that image will not contain the true minutiae data. Resubmission of digitally stored biometrics data (via a stolen but authentic smart card) will not be feasible, since the system verifies every user by using this data along with the minutiae data obtained online at a controlled access site. A user who succeeds in inserting a new template into the database will not be verified at the access site, since this new template will not contain the minutiae data [2].

Experimental Results

Figure 23-8(a)–(c) shows the stego images, which carry the minutiae data shown in Figure 23-6(c) for the cover images in Figure 23-3 [2]. Nearly 17% of the stego image pixels are changed during data hiding, for all three cover images.

Figure 23-7
*Minutiae hiding
for face
verification.
(Source: Adapted
with permission
from Michigan
State University.)*

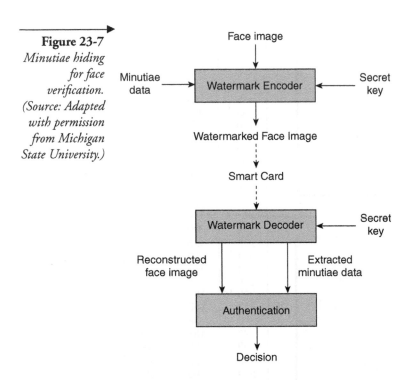

Face image

Minutiae data → Watermark Encoder ← Secret key

Watermarked Face Image

Smart Card

Watermark Decoder ← Secret key

Reconstructed face image   Extracted minutiae data

Authentication

Decision

Figure 23-8
*Stego images and
decoded data:
(a) synthetic
fingerprint,
(b) face,
(c) "sailboat," (d)
extracted
minutiae data.
(Source:
Reproduced with
permission from
Michigan State
University.)*

(a)   (b)

(c)

x	y	θ
76	216	242
121	195	255
136	82	292
136	229	248
170	90	262
172	169	270
178	46	274
184	85	82
192	146	281
196	198	270
201	89	52
212	233	255
216	220	262
228	125	321
234	79	8
234	147	298
236	175	295
240	167	112
259	68	356
60	32	356
77	197	58
99	85	144
98	69	382
239	190	274
251	222	270

(d)

Other watermarking parameters are set to: $q = 0.1$, $A = 100$, $B = 1,000$. A higher q value increases the visibility of the hidden data. Increasing A or B decreases the effect of standard deviation and gradient magnitude in modulating watermark embedding strength, respectively. The hidden data size is approximately 85 bytes. The extracted minutiae data from all the three cover images is shown in Figure 23-8(d) [2]; it is exactly the same as the hidden data shown in Figure 23-6(c) [2].

For the second application scenario, the face image shown in Figure 23-9(a) is watermarked by using the same minutiae data (see Figure 23-6(c)) [2]. Other parameters of the watermarking algorithm are the same as the ones used for the first application scenario. The extracted minutiae data are identical to the embedded data. Figure 23-9(c) shows the negative image of the difference between original image and the watermarked image [2]. The reconstructed face image is given in 23-9(d) [2].

In order to assess the average magnitude of changes in pixel values, several image and watermarking characteristics for the four host images are computed (see Table 23-1) [2]. The first column is the average value for all the image pixels. The second column is the average value for the changed (watermarked) pixels. The last column shows the average change in values (due to watermarking) of the watermarked pixels [2].

Additional tests to determine the performance of the proposed algorithm were conducted as follows: 15 images (five synthetic fingerprint, five face, five arbitrary) were watermarked with five different minutiae data, by using five different keys. As a result, 375 different watermarked images are produced. Characteristics and sources of the host images and watermarking parameters are the same as given previously. Individual minutiae data sets contained between

Figure 23-9 *Watermarking for user verification: (a) input face image, (b) watermarked face image, (c) negative of difference image, (d) reconstructed face image. (Source: Reproduced with permission from Michigan State University.)*

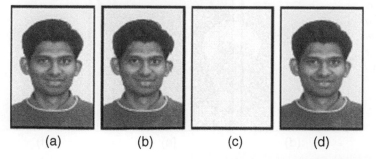

(a)   (b)   (c)   (d)

Table 23-1   *Host Image and Watermarking Characteristics*

Host Image	Overall Pixel Average	Changed Pixel Average	Avg. Change in Pixel Values
Figure 23-3(a)	221.8	222	24.9
Figure 23-3(b)	150.2	150.6	15.6
Figure 23-3(c)	124.8	124.9	15.4
Figure 23-9(a)	159.6	160.2	14.5

23 to 28 points, with an average of 25 points. From all of these 375 watermarked images, the researchers were able to extract the embedded minutiae information with 100% accuracy [2].

Finally, the proposed data hiding method is robust and can tolerate certain types of attacks, namely image cropping and JPEG compression. The hidden minutiae data are extracted correctly from (1) 40% cropped and (2) JPEG compressed (quality factor 90) versions of all four watermarked images (Figure 23-8(a)–(c) and Figure 23-9(b)). Figure 23-10 shows the attacked images for three of these host images [2].

Figure 23-10
Robustness to attacks: (a), (b), and (c) are 40% cropped versions of watermarked images; (d), (e), and (f) are JPEG compressed watermarked images. (Source: Reproduced with permission from Michigan State University.)

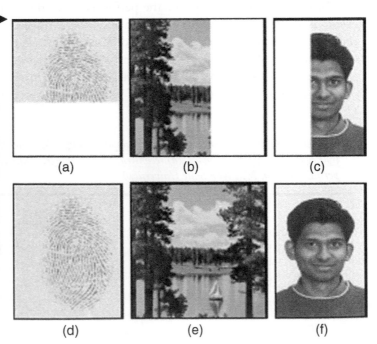

Summary/Conclusion

With the widespread utilization of biometric identification systems, establishing the verification of biometric data has emerged as an important research issue. The fact that biometric data is not replaceable and is not secret, combined with the existence of several types of attacks that are possible in a biometric system, make the issue of security/integrity of biometric data extremely critical. This chapter introduced two applications of an amplitude modulation-based watermarking method, in which the researchers hid a user's biometric data in a variety of images. This method has the ability to increase the security of both the hidden biometric data (eigenface coefficients) and host images (fingerprints) [1, 2].

Image-adaptive data-embedding methods used in the researchers' scheme lead to low visibility of the embedded signal. Feature analysis of host images guarantees high verification accuracy on watermarked (fingerprint) images [1, 2].

The ability of biometric-based personal identification techniques to differentiate between an authorized person and an impostor who fraudulently acquires the access privilege of an authorized person, is one of the main reasons for their popularity compared to traditional identification techniques. However, the security and integrity of the biometric data itself are important issues [1, 2].

Encryption, watermarking, and steganography are possible techniques to secure biometric data. In this chapter, two applications of watermarking to secure that data were presented. In addition to watermarking, encryption can be used to further increase the security of biometric data. The first application is related to increasing the security of biometric data exchange, which is based on steganography. In the second application, the researchers embedded facial information in fingerprint images. In this application, the data is hidden in such a way that the features that are used in fingerprint matching are not significantly changed during encoding/decoding [1, 2].

Finally, the verification accuracy based on decoded watermarked images is very similar to that with original images. The proposed method utilizes several properties of the human visual system to keep the visibility of the changes made to the host image low. Researchers are currently working on increasing the data-hiding capacity of the host images. Another topic for future research is to investigate how different (robust and fragile) watermarking schemes can be combined [1, 2].

References

1. Anil K. Jain and Umut Uludag, "Hiding Biometric Data," IEEE, Vol. 25, No. 11, November 2003 [Department of Computer Science and Engineering, Michigan State University, 3115 Engineering Building, East Lansing, MI 48824] 2003. IEEE, Published by the IEEE Computer Society, IEEE Transactions on Pattern Analysis and Machine Intelligence, Copyright © 2003.

2. Anil K. Jain and Umut Uludag, "Hiding Fingerprint Minutiae in Images," Computer Science and Engineering Department, Michigan State University, 3115 Engineering Building, East Lansing, MI, 48824, 2003.

3. Justin Picard, Claus Vielhauer, Otto-von-Guericke and Niels Thorwirth, "Towards Fraud-Proof ID Documents Using Multiple Data Hiding Technologies and Biometrics," [*Security, Steganography, and Watermarking of Multimedia Contents VI*, Edward J. Delp III, Purdue Univ.; Ping W. Wong, Consultant, San Jose CA, June 22, 2004, Volume 5306, No. 5306], 2004. © 2004; SPIE The International Society for Optical Engineering and IS&T (The Society for Imaging Science and Technology).

4. John R. Vacca, *Satellite Encryption*, Academic Press (1999).

24

Image-Based Challenges/Response Methods

Over the past decade, there has been a significant surge in the use of image-based biometric user verification applications. Image-based biometric user verification systems offer several useful advantages over knowledge- and possession-based methods such as password/PIN-based systems. When employed in security-critical applications, and more so in unattended remote applications, the image-based biometric user verification systems should be designed to resist different sources of security attacks on the system. This chapter covers the inherent strengths of an image-based biometric user verification scheme and also describes the security holes in such systems. A new solution is presented to alleviate one of the weak links in the system [1].

Image-Based Biometric User Verification Systems

Many applications in everyday life require user verification. The prevailing techniques of user verification, which involve passwords and user IDs or identification cards with PINs, suffer from several limitations. The main problems with such systems is that the verification subsystem can be fooled very easily and there is no way to link the user to the usage of the system. For example, the user ID and password can be shared with a colleague. Thus, the security of the system is compromised severely. There are many applications where such security lapses cannot be tolerated. It is more difficult to share a biometric of a person with another [1].

But, when biometrics is employed in security-critical applications, hackers will find the weak points in the system and attack the systems at those points. Unlike password systems, which are prone to password dictionary attacks, biometric systems require much more effort to hack into. In supervised use of biometrics as a verification tool, this may not be a concern. But in remote unattended application such as Web-based e-commerce applications [3], hackers will have enough time to make several attempts before giving up and remaining unnoticed. Standard crypto techniques will be useful in many ways to prevent a breach of security. But several new types of attacks are possible in the biometric domain [1].

Brute Force Attack: Possible Attack Points

As discussed in Chapter 22, the relationship between the number of brute force attack attempts is a function of the number of minutiae that are expected to match in the matcher subsystem. By generating all possible images to guess the matching fingerprint image, a much larger search space is required [1].

Possible Attack Points

A generic biometric system can be cast in the framework of a pattern recognition system. The stages of such a generic system are shown in Figure 24-1 [1]. There are in total eight possible sources of attack on such systems, as described below:

1. **Fake biometric at the sensor:** In this mode of attack, a possible reproduction of the biometric being used will be presented to the system. Examples include a fake finger, a copy of a signature, or a face mask.

2. **Resubmission of an old digitally stored biometric signal:** In this mode of attack, an old recorded signal is replayed into the system, bypassing the sensor. Examples include presentation of an old copy of fingerprint image or recorded audio signal of a speaker.

3. **Override feature extract:** The feature extractor could be attacked with a Trojan horse to change it to produce feature sets of choice.

4. **Tampering with the feature representation:** After the features have been extracted from the input signal in this mode, they are replaced with a synthesized feature set of choice, assuming the

Figure 24-1   *Possible attack points in a generic biometric-based system. (Source: Adapted with permission from the IBM Corporation.)*

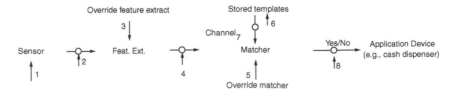

representation is known. Often the two stages of feature extraction and matcher are inseparable, and this mode of attack is extremely difficult. However, if minutiae are transmitted to a remote matcher (say, over the Internet [8]), then this threat is very real. One could snoop on the TCP/IP stack inside the computer and alter certain packets.

5.   **Override matcher:** The matcher is attacked to produce the desired result.

6.   **Tampering with stored templates:** The database of enrolled templates is available locally or remotely. This database can be distributed over several servers. The stored template attacker tries to modify one or more templates in the database, which could result in at least a denial of service for the corrupted template.

7.   **Channel attack between stored templates and the matcher:** The templates from the stored database are sent to the matcher through a channel that could be attacked to change the contents of the templates before they reach the matcher.

8.   **Decision override:** If the final result can be overridden with the choice of result from the hacker, the final outcome is very dangerous. Even if the actual pattern recognition system had an excellent performance characteristic, it will have been rendered useless by a simple exercise of overriding the result [1].

There exists several techniques to thwart attacks at various points. For instance, sensing finger conductivity or pulse can stop simple attacks at point 1. Encrypted communication channels [1] can eliminate at least remote attacks at point 4. However, even if the hacker cannot get inside the feature extract machine, the system is still vulnerable. The simplest way to stop attacks at points 5, 6, and 7 is to have the matcher and database reside in a secure location. Of course, even this cannot prevent attacks in which there is collusion. Cryptography can help at point 8 [1].

Challenge/Response Method

A new method is proposed here that can handle the attacks of type 2 on the input signal. The motivation of this approach is based on challenge/response systems. Conventional challenge/response systems are based on challenges to the user. This approach is based on challenges to the sensor. The sensor is assumed to

Figure 24-2
Signal verification based on challenge/ response. (Source: Adapted with permission from the IBM Corporation.)

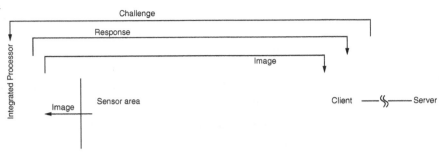

have enough intelligence to respond to the challenges. Standard cryptographic techniques, though mathematically strong, are computationally very intensive and would require maintaining a secret key [5] base for a large number of sensors. Moreover, the encryption techniques cannot check for liveliness of a signal. An old stored image can be given to the encryptor. Similarly, a digital signature of a signal does not check for its liveliness. The availability of a large number of image pixels is exploited here, as well as simple image-related challenges that can be posed to the sensor. Liveliness is assured by challenging the sensor with a pseudo random challenge. The sensor, being intelligent, responds to this challenge as per the scheme shown in Figure 24-2 [1].

The proposed solution works as follows. At the user terminal or system, the transaction gets initiated. The transaction server generates a pseudo random challenge for the transaction and the sensor [1].

Note: The transaction server is assumed to be secure. The user system passes the challenge to the intelligent sensor. The sensor acquires a signal at this point and computes the response to the challenge. For instance, the integrated processor might be able to compute either of two selectable functions: "x1+" and "x10+." Bank A might use function "x1+" in all its units, while Bank B might use "x10+." Alternatively, for even-numbered transactions, function "x10+" might be used, and for odd-numbered transactions "x1+" would be used. Hence, the challenge augmenter modifies the challenge through one or more functions.

The important point is that the response depends on the challenge and the image itself. A typical challenge might be "3, 10, 50." This would be augmented by function "x1+," by appending all the pixel values of the image (in scan order) to the end of the challenge string. The integrated processor then selects the 3rd, 10th, and 50th pixel value from this sequence to generate an output response such as "133, 92, 176." Other examples of responder functions include computing a checksum of a segment of the signal; a set of pseudorandom samples; a block of contiguous samples starting at a specified location; and, with a given size, a hash of signal values, and a specified known function of selected samples of the signal. A combination of these functions can be used to achieve arbitrarily complex responder functions. The signal as well as the response is transmitted to the server where the response can be verified [1].

By integrating the responder onto the same chip as the sensor, it is just about impossible to inject a fake image (point 2 attack). When the computed power is significant, you can carry out many novel solutions. For example, data hiding in a compressed domain can enhance the performance of the solution significantly. Many silicon fingerprint scanners [1] will be able to exploit the proposed method, as they can integrate a processor without much effort.

Keeping the preceding in mind, let's look at how image-based biometric verification systems for credit cards could put identity thieves out of business. Is it all being done with smoke and mirrors? Let's take a look.

Image-Based Biometric Verification Systems for Credit Cards

He stole the identities of the world's rich and famous—Paul Allen, Oprah Winfrey, Steven Spielberg, Warren Buffett, and Larry Ellison, to name a few. Until the New York City police busted 32-year-old Abraham Abdallah, it seemed that a diabolically gifted hacker, not a bus boy at a Brooklyn restaurant, had masterminded this multimillion-dollar caper [2].

However, a tattered copy of a *Forbes* magazine featuring America's 500 richest people found in Abdallah's possession (along with 900 credit cards) exposed the thief's simple modus operandi. Here were his targets, listed in order of their net worth, some with Social Security numbers and credit card information scrawled right next to their names. Investigators soon discovered that Abdallah had obtained most of this information from the Internet, as well as from credit bureaus Equifax, Experian, and TransUnion, by sending queries on the forged letterhead of several top investment banks [2].

With birth dates, addresses, and Social Security and credit card numbers in hand, Abdallah would use a computer at a public library to order merchandise online, withdraw money from brokerage accounts, and apply for credit cards in other people's names. Things started to unravel when he tried to transfer $20 million from the Merrill Lynch account of software entrepreneur Thomas Siebel. Someone at Merrill Lynch noticed that the same two Yahoo e-mail addresses, both Abdallah's, had been used in connection with five other clients. Soon after, on March 19, 2001, two New York City detectives wrestled Abdallah out of his car, ending one of the most sensational identity theft [6] sprees in history [2].

Catching identity thieves is like spearfishing during a salmon run: skewering one big fish barely registers when the vast majority just keep on going. According to industry analysts, the cumulative losses suffered by tens of millions of individuals and businesses worldwide registered at an estimated $554 billion in 2006. Industry analysts, which assumed an enormous 600% compound annual growth rate, projected that losses would rise to an almost unfathomable $5 trillion in 2008. More recent numbers indicate a much lower growth rate, at least in the United States, where total losses rose from about $48 billion in 2003 to $67.7 billion in 2006 [2].

Clearly, it is far too easy to steal personal information these days—especially credit card numbers, which are involved in more than 78% of identity thefts, according to a U.S. Federal Trade Commission study. It's also relatively easy to fake someone's signature or guess a password; thieves can often just look at the back of an ATM card, where some 40% of people actually write down their personal identification number (PIN) and give the thief all that's needed to raid the account. But, what if you had to present your fingers or eyes to a scanner built into your credit cards to verify your identity before completing a transaction? Faking fingerprints or iris scans would prove challenging to even the most technologically sophisticated identity thief [2].

The sensors, processors, and software needed to make secure credit cards that authenticate users on the basis of their physical, or biometric, attributes are already on the market. But so far, the credit card industry hasn't seen fit to integrate even basic fingerprint-sensing technology with their enormous IT systems. Concerned about biometric system performance, customer acceptance, and the cost of making changes to their existing infrastructure, the credit card issuers apparently would rather go on eating an expense equal to 0.25% of Internet transaction revenues and 0.08% of off-line revenues that now come from stolen credit card numbers [2].

Indeed, only a few companies worldwide have even experimented with biometric credit cards. The best known is the Bank of Tokyo–Mitsubishi. Since 2004, it has issued Visa cards embedded with chips that identify a customer according to vein patterns in the palm. All of the bank's ATMs have palm scanners that match the imaged vein patterns to a digitized copy of the customer's vein patterns (biometric template) that is stored in the card. But, because merchants lack the requisite palm scanners to go with this technology, customers still sign receipts or enter PINs when making purchases with the card [2].

All biometric systems recognize patterns, such as the veins in your palms, the texture of your iris, or the minutiae of your fingerprints. Researchers have recently proposed the broad outlines of a new verification system for credit cards based on biometric sensors that could dramatically curtail identity theft. The proposed system uses fingerprint sensors, though other biometric technologies, either alone or in combination, could be incorporated. The system could be economical, protect privacy [4], and guarantee the validity of all kinds of credit card transactions, including ones that take place at a store, over the telephone, or with an Internet-based retailer. By preventing identity thieves from entering the transaction loop, credit card companies could quickly recoup their infrastructure investments and save businesses, consumers, and themselves billions of dollars every year [2].

If credit card issuers don't act soon, customers, many of whom are becoming increasingly comfortable with biometric technologies, might just force the issue. In the United States, millions of people at hundreds of supermarkets have already given the thumbs-up to services offered by BioPay LLC, Herndon, Va., and Pay By Touch (see Figure 24-3) [2], San Francisco, which let shoppers

Figure 24-3
Scanners galore: Biometric ID systems are proliferating everywhere except in credit cards. A Pay By Touch fingerprint scanner for buying groceries. (Source: Reproduced with permission from IEEE.)

Figure 24-4
An ATM with a Fujitsu PalmSecure palm vein scanner. (Source: Reproduced with permission from IEEE.)

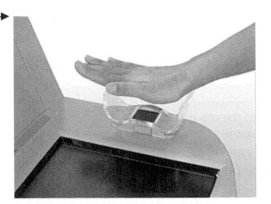

pay for their groceries by pressing a finger on a sensor mounted near the cash register—no card necessary. Millions more, mostly in Asia, have fingerprint sensors built into their cellphones to act as locks and into their laptops to replace text-based logins. All of this activity translates to 30% annual growth for a worldwide biometrics market that's expected to reach $4.5 billion in 2008, according to industry analysts. Finger-scanning technology made by

Figure 24-5
A fingerprint sensor from AuthenTec in an LG Electronics phone. (Source: Reproduced with permission from IEEE.)

companies like Atmel, AuthenTec (see Figure 24-5) [2], Digital Persona, Fujitsu (see Figure 24-4) [2], and Identix will account for almost 70% of the total market, industry analysts estimate. And that market will greatly expand if and when credit card companies get serious about combating ID theft [2].

Current Credit Card Verification Systems

Current credit card verification systems validate anyone (including impostors) who can reproduce the exclusive possessions or knowledge of legitimate card-holders. Presenting a physical card at a cash register proves only that you have a credit card in your possession, not that you are who the card says you are. Similarly, passwords or PINs do not authenticate your identity but rather your knowledge. Most passwords or PINs can be guessed with just a little information: an address, license plate number, birthdate, or pet's name. Patient thieves can and do take pieces of information gleaned from the Internet or from mail found in the trash and eventually associate enough bits to bring a victim to financial grief [2].

Besides trawling the Internet and diving into dumpsters for personal data, thieves exploit people through various cons known collectively as social engineering. A smooth-talking drifter can sometimes get a customer service representative to part with a PIN or reveal other things about an account, such as a mailing address or a phone number. The bank makes it easier for thieves if its verification protocol is riddled with exceptions. For instance, if you don't know the PIN, you might be able to provide a mailing address, mother's maiden name, phone number, or Social Security number (see sidebar, "27 Million Lost/Stolen VA Social Security Numbers") to get access to—or at least information about—a particular account. Sometimes those bits of data can be harvested from other sources.

27 Million Lost/Stolen VA Social Security Numbers

On May 25, 2006, the Veterans Affairs Office of Inspector General (VA OIG) and the FBI announced a $50,000 reward through the Montgomery County (Maryland) Crime Solvers organization for information that leads to the recovery of a laptop computer and external hard drive that contained personal information (including Social Security numbers) for 27 million U.S. veterans. The laptop was eventually recovered on July 5, 2006; but the damage had already occurred. The question here is: Were 27 million VA records compromised (downloaded) by identity thieves prior to the recovery of the stolen laptop? Inquiring minds want to know [10]!

Further questions remain. What happened to the external hard drive? Where are the records now? Will illegal aliens and terrorists have access to the stolen Social Security numbers? Was the robbery staged by our own government, so as to release the Social Security numbers to illegal aliens (and make it easy for illegal aliens to obtain stolen Social Security numbers), since the policy of the current Bush administration is to encourage entry by the illegals into the United States? Something here really smells to high heaven (see sidebar, "Identity Theft Victim: The Audra Schmierer Story"). But you be the judge; this author will just present the facts (such as they are) to date [10].

What Really Happened?

It all started when the Montgomery County Police started working with the FBI and the VA OIG in the investigation of a residential burglary that occurred on May 3, 2006, in the Aspen Hill community of Montgomery County, Maryland. Federal investigators discovered and removed other sensitive VA data the worker was not authorized to have at home. The primary objective of the investigation was the recovery of the laptop and external hard drive. As previously mentioned, the laptop was recovered, but not the external hard drive that contained the veterans' data [10].

In the meantime, U.S. Senator Mark Pryor (D-AR) made a plea to his chamber's leadership on May 25, 2006 to schedule floor debate on a bill that could help veterans better protect their stolen personal information. He is hopeful that the Senate will take some time on the floor in the near future to allow consideration of the Identify Theft Protection Act (5.1408) which he cosponsored with a bipartisan group of seven lawmakers. Pryor made the comment during a joint hearing hosted by the Veterans Affairs and Homeland Security committees on the recent data breach at the Veterans' Affairs Department. This bill was first introduced in July of 2005, but up to this point has not received much interest. A Senate aide who has worked on the bill said that there has been increased activity and discussion in recent days toward getting a data breach bill passed due to the Veterans' Affairs Department breach. Other provisions of the legislation are designed to [10]:

1.   Allow consumers to restrict access to their credit reports;

2.   Frustrate potential thieves from conducting transactions under other people's names;

3.   Require commercial entities and nonprofit institutions to implement security measures.

The committee needs to enact language on relevant areas that fall under its jurisdiction so that committee staffers can finalize and merge the various legislative proposals in their chamber. It is committee chairman Richard Shelby's (R-AL) intention to pass a bill that would cover the financial institutions covered by the Fair Credit Reporting Act and the Gramm-Leach-Bliley Act, which both address financial privacy issues. The House Energy and Commerce, Financial Services, and Judiciary committees all approved separate proposals for data protection legislation in late May 2006. The consumer groups generally support the Energy and Commerce bill, while business interests support the Financial Services legislation. Only the Financial Services legislation includes a provision that allows consumers to restrict third-party access to their credit reports. However, the bill would only permit them to do so after being victims of fraud [10].

U.S. Senator Max Baucus (D-MT), Ranking Member of the Senate Finance Committee, asked the Treasury Department via letter on May 26, 2006, for details on a plan to use IRS databases to notify U.S. veterans whose personal information was stolen. The IRS has an up-to-date database of most Americans' last known addresses, and that database will be used to contact veterans whose information was compromised. In a letter to Treasury Secretary John Snow, Baucus asked for detailed information on IRS plans to use private contractors to produce and send letters to veterans. In closing, he said: "Treasury must exercise the utmost care to ensure that the privacy of these veterans is completely protected and not further compromised. Veterans deserve assurances that the IRS's notification process will not result in their further victimization. Already, this incident has proved upsetting to many of our nation's veterans, who are now legitimately concerned that the security of their identities is at risk. Notification letters from the IRS may themselves add confusion. Letters from the IRS may cause veterans to wonder whether their tax or financial information has been compromised or whether they have a tax matter that needs to be addressed. I urge Treasury to ensure that these IRS letters are clear in their purpose and won't further complicate what could be an alarming situation for many veterans." See sidebar, "VA Answers to Frequently Asked Questions by Veterans [10]."

In addition, on May 22, 2006, Secretary of Veterans Affairs: R. James Nicholson, released the following memorandum (see Figure 24-6) to all VA employees:

"Today (May 22, 2006), I made an announcement regarding an incident in which an employee took home VA data without the authorization to do so. The employee's home was burglarized and the data was stolen. This memorandum is to remind you as a VA employee of your duty and responsibility in protecting sensitive and confidential information.

Each year, VA employees are required to complete Privacy and Cyber Security training. Those training courses are provided and required to serve as important reminders to all staff that public service is a public trust. The public trust requires us to be vigilant in safeguarding the personal information that we collect on the veterans and families as part of our service to them. Having access to such sensitive information requires that we protect Federal property and information, and that it shall not be used for other than authorized activities.

Because of the serious breach that has occurred by the actions of this VA employee in removing Federal property to his home without authorization, you will be asked to complete your annual General Privacy Training and VA Cyber Security Awareness for 2006 by June 30. All employees will then be required to sign a Statement of Commitment and Understanding. By signing this statement, you will confirm your understanding of the training, and the consequences for noncompliance, and your commitment to protecting sensitive and confidential information in the Department of Veterans Affairs.

In addition I have convened a task force of senior VA leaders to review all aspects of information security and make recommendations, if appropriate, to strengthen our protection of sensitive information. VA's mission is to honor and serve our nation's veterans. We must take very seriously the impact of this incident on the confidence veterans will have in our ability to handle their sensitive information [10]."

Figure 24-6
The VA stolen records letter to veterans. (Source: Department of Veterans Affairs.)

THE SECRETARY OF VETERANS AFFAIRS
WASHINGTON

Dear Veteran:

The Department of Veterans Affairs (VA) has recently learned that an employee took home electronic data from the VA, which he was not authorized to do and was in violation of established policies. The employee's home was burglarized and this data was stolen. The data contained identifying information including names, social security numbers, and dates of birth for up to 26.5 million veterans and some spouses, as well as some disability ratings. As a result of this incident, information identifiable with you was potentially exposed to others. It is important to note that the affected data did not include any of VA's electronic health records or any financial information.

Appropriate law enforcement agencies, including the FBI and the VA inspector General's office, have launched full-scale investigations into this matter. Authorities believe it is unlikely the perpetrators targeted the items because of any knowledge of the data contents.

Out of an abundance of caution, however, VA is taking all possible steps to protect and inform our veterans. While you do not need to take any action unless you are aware of sispicious activity regarding your personal information, there are many steps you may take to protect against possible identity theft and we wanted you to be aware of these. Specific information is included in the enclosed question and answer sheet. For additional information, the VA has teamed up with the Federal Trade Commission and has a Web site (www.firstgov.gov) with information on this matter or you may call 1-800-FED-INFO (1-800-333-4636). The call center will operate from 8 a.m. to 9 p.m. (EDT), Monday-Saturday, as long as it is needed.

Beware of any phone calls, e-mails, and other communications from individuals claiming to be from VA or other official sources, asking for your personal information or verification of it. This is often referred to as information solicitation or "phishing." VA, other government agencies, and other legitimate organizations will not contact you to ask for or to confirm your personal information. If you receive such' communications, they should be reported to VA at 1-800-FED-INFO (1-800-333-4636).

We apologize for any inconvenience or concern this situation may cause, but we at VA believe it is important for you to be fully informed for any potential risk resulting from this incident. Again, we want to reassure you we have no evidence that your protected data has been misused. We will keep you apprised of any further developments. The men and women of the VA take our obligation to honor and serve America's veterans very seriously and we are committed to ensuring that this never happens again.

In accordance with current policy, the Internal Revenue Service has agreed to forward this letter' because we do not have current addresses for all affected individuals. The IRS has not disclosed your address or any other tax information to us.

Sincerely yours,

R. James Nicholson

Enclosure

May 2006

VA Answers to Frequently Asked Questions by Veterans

1. I'm a veteran, how can I tell if my information was compromised?

At this point, there is *no* evidence the missing data has been used illegally. However, the Department of Veterans Affairs is asking all veterans to be extra vigilant and to carefully monitor bank statements, credit card statements, and any statements relating to recent financial transactions. If you notice unusual or suspicious activity, you should report it immediately to the financial institution involved and contact the Federal Trade Commission for further guidance.·

2. What is the earliest date at which suspicious activity might have occurred due to this data breach?

The information was stolen from an employee of the Department of Veterans Affairs during the month of May 2006. If the data has been misused or otherwise used to commit fraud or identity theft crimes, it is likely that veterans may notice suspicious activity during the month of May.

3. I haven't noticed any suspicious activity in my financial statements, but what can I do to protect myself and prevent being victimized by credit card fraud or identity theft?

The Department of Veterans Affairs strongly recommends that veterans closely monitor their financial statements and visit the Department of Veterans Affairs special website on this, http://www.firstgov.gov or call 1-800-FED-INFO (1-800-333-4636).

4. Should I reach out to my financial institutions or will the Department of Veterans Affairs do this for me?

The Department of Veterans Affairs does not believe that it is necessary to contact financial institutions or cancel credit cards and bank accounts, unless you detect suspicious activity.

5. Where should I report suspicious or unusual activity?

The Federal Trade Commission recommends the following four steps if you detect suspicious activity:

Step 1: Contact the fraud department of *one* of the three major credit bureaus:

- Equifax: 1-800-525-6285; http://www.equifax.com; P.O. Box 740241, Atlanta, GA 30374-0241.

- Experian: 1-888-EXPERIAN (397-3742); http://www.experian.com; P.O. Box 9532, Allen, Texas 75013.

- TransUnion: 1-800-680-7289; http://www.transunion.com; Fraud Victim Assistance Division, P.O. Box 6790, Fullerton, CA 92834-6790.

Step 2: Close any accounts that have been tampered with or opened fraudulently.

Step 3: File a police report with your local police or the police in the community where the identity theft took place.

Step 4: File a complaint with the Federal Trade Comission by using the FTC's Identity Theft Hotline by telephone: 1-877-438-4338, online at www.consumer.gov/idtheft, or by mail at Identity Theft Clearinghouse, Federal Trade Commission, 600 Pennsylvania Avenue NW, Washington, DC 20580.

6. I know the Department of Veterans Affairs maintains my health records electronically; was this information also compromised?

No electronic medical records were compromised. The data lost is primarily limited to an individual's name, date of birth, Social Security number, in some cases their spouse's information, as well as some disability ratings. However, this information could still be of potential use to identity thieves and we recommend that all veterans be extra vigilant in monitoring for signs of potential identity theft or misuse of this information.

7. What is the Department of Veterans Affairs doing to ensure that this does not happen again?

The Department of Veterans Affairs is working with the President's Identity Theft Task force, the Department of Justice, and the Federal Trade Commission to investigate this data breach and to develop safeguards against similar incidents. The Department of Veterans Affairs has directed all VA employees complete the "VA Cyber Security Awareness Training Course" and complete the separate "General Employee Privacy Awareness Course" by June 30, 2006. In addition, the Department of Veterans Affairs will immediately be conducting an inventory and review of all current positions requiring access to sensitive VA data. It will also require all employees requiring access to sensitive VA data to undergo an updated National Agency Check and Inquiries (NACI) and/or a Minimum Background Investigation (MBI) depending on the level of access required by the responsibilities associated with their position. Appropriate law enforcement agencies, including the Federal Bureau of Investigation and the Inspector General of the Department of Veterans Affairs, have launched full-scale investigations into this matter.

8. Where can I get further, up-to-date information?

The Department of Veterans Affairs has set up a special website and a toll-free telephone number for veterans which features up-to-date news and information. Please visit http://www.firstgov.gov or call 1-800-FED-INFO (1-800-333-4636) [10].

Identity Theft Victim: The Audra Schmierer Story

You have all heard stories about identity theft on a daily basis. The truth is that it is hard to measure the extent of the damage that it inflicts, not to mention the high costs involved in trying to repair that damage [10].

Our story begins with Dublin, California, stay-at-home mom Audra Schmierer, who tried to apply for a job at a temporary agency in late 2004. When the agency did a background check on her, they were surprised. They asked her why she was applying for a job she already had. To her surprise, an illegal alien (one of 278 across the nation) had stolen her Social Security number to help him or her obtain employment. But this was only the tip of the iceberg [10].

Note: *Instead of 278 illegal aliens using one stolen Social Security number, they now have 27 million newly stolen Social Security numbers from U.S. veterans to choose from.*

The other 277 illegal aliens who used Schmierer's Social Security number to obtain employment in jobs ranging from fast food restaurants to Microsoft were also receiving Social Security and medical benefits from those same jobs. Those benefits were meant for legal U.S. citizens. What is it that they don't understand: Illegal means illegal!!

From here, the story takes a turn for the worse. In May of 2005, Schmierer received a bill from the IRS for $16,000. The bill wasn't hers; but it was attached to her Social Security number. She tracked the illegal alien who filed the phony tax return to Houston, Texas. The illegal alien told her that he purchased her Social Security number at a Texas flea market [10].

She and her husband were also detained at Customs in late 2005 while returning from an overseas business trip. U.S. Customs said that she had a criminal record and was wanted. Another illegal alien that was using her Social Security number was involved in a crime, and she took the blame [10].

By the time January of 2006 rolled around, she had a tax bill from the IRS totaling $1,000,000. What is it that they don't understand: Identity theft is a felony [10].

Anyway, as of this writing, she has a zero balance in her Social Security account. The 35 employers who hired the illegal aliens are refusing to take action to rectify the situation. Those 35 employers (and you know who you are) feel that cheap labor is more important than the inalienable rights of U.S. citizens and legal immigrants [10].

The IRS and the Social Security Administration are also refusing to take action. The Social Security Administration is refusing to give Schmierer a new Social Security number. The FTC, who she has a case with, is also refusing to take action. And, what's even worse, most U.S. Senators support the current Bush administration's immigration policy as well as supporting Social Security benefits for illegal aliens' use of stolen identities [10].

In the end, the only victims here are the real American citizens and legal immigrants [10].

Furthermore, customer service representatives and their managers can usually override verification procedures when they deem it necessary. A caffeine-addled agent working a double shift may be only too eager to use her override privileges to let you (or your would-be doppelgänger) make a purchase [2].

To ensure truly secure credit card transactions, you need to minimize this kind of human intervention in the verification process. Such a major transition will come at a cost that credit card companies have so far declined to pay. They are particularly worried about the cost of transmitting and receiving biometric information between point-of-sale terminals and the credit card payment system. They also fret that some customers, anxious about having their biometric information floating around cyberspace, might not adopt the cards. To address these concerns, let's look at an outline for a self-contained smart card system that could be implemented within the next few years [2].

Here's how it would work. When activating your new card, you would load an image of your fingerprint onto the card. To do this, you would press your finger against a sensor in the card—a silicon chip containing an array of microcapacitor plates [2].

Note: In large quantities, these fingerprint-sensing chips cost only about $5 each.

The surface of the skin serves as a second layer of plates for each microcapacitor, and the air gap acts as the dielectric medium. A small electrical charge is created between the finger surface and the capacitor plates in the chip. The magnitude of the charge depends on the distance between the skin surface and the plates. Because the ridges in the fingerprint pattern are closer to the silicon chip than the valleys, ridges and valleys result in different capacitance values across the matrix of plates. The capacitance values of different plates are measured and converted into pixel intensities to form a digital image of the fingerprint (see Figure 24-7) [2].

Figure 24-7
Fingerprint matching. (Source: Reproduced with permission from IEEE.)

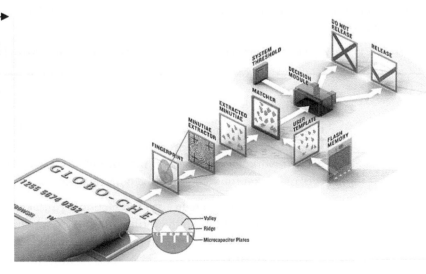

Next, a microprocessor in the smart card extracts a few specific details, called minutiae, from the digital image of the fingerprint. Minutiae include locations where the ridges end abruptly and locations where two or more ridges merge, or a single ridge branches out into two or more ridges. Typically, in a live-scan fingerprint image of good quality, there are 20 to 70 minutiae; the actual number depends on the size of the sensor surface and the placement of the finger on the sensor. The minutiae information is encrypted and stored along with the cardholder's identifying information as a template in the smart card's flash memory [2].

At the start of a credit card transaction, you would present your smart credit card to a point-of-sale terminal. The terminal would establish secure communication channels between itself and your card via communication chips embedded in the card and with the credit card company's central database via Ethernet. The terminal then would verify that your card has not been reported lost or stolen, by exchanging encrypted information with the card in a predetermined sequence and checking its responses against the credit card database [2].

Next, you would touch your credit card's fingerprint sensor pad. The matcher, a software program running on the card's microprocessor, would compare the signals from the sensor to the biometric template stored in the card's memory. The matcher would determine the number of corresponding minutiae and calculate a fingerprint similarity result, known as a matching score. Even in ideal situations, not all minutiae from the input and template prints taken from the same finger will match. So the matcher uses what's called a threshold parameter to decide whether a given pair of feature sets belong to the same finger. If there's a match, the card sends a digital signature and a time stamp to the point-of-sale terminal. The entire matching process could take less than a second, after which the card is accepted or rejected [2].

The point-of-sale terminal sends both the vendor information and your account information to the credit card company's transaction-processing system. Your private biometric information remains safely on the card, which ideally never leaves your possession [2].

But say your card is lost or stolen. First of all, it is unlikely that a thief could recover your fingerprint data, because it is encrypted and stored on a flash memory chip that very, very few thieves would have the resources to access and decrypt. Nevertheless, suppose that an especially industrious, and perhaps unusually attractive, operator does get hold of the fingerprint of your right index finger—say, off a cocktail glass at a hotel bar where you really should not have been drinking. Then this industrious thief manages to fashion a latex glove molded in a slab of gelatin containing a nearly flawless print of your right index finger, painstakingly transferred from the cocktail glass [2].

Even such an effort would fail, thanks to new applications that test the vitality of the biometric signal. One identifies sweat pores, which are just 0.1 mm across, in the ridges using high-resolution fingerprint sensors. You could also detect spoofs by measuring the conduction properties of the finger using electric field sensors from AuthenTec Inc., of Melbourne, Florida. Software-based spoof detectors aren't far behind. Researchers are also differentiating the way a live finger deforms the surface of a sensor from the way a dummy finger does. With software that applies the deformation parameters to live scans, you can automatically distinguish between a real and a dummy finger 85% of the time—enough to make your average identity thief think twice before fashioning a fake finger [2].

No System Is Perfect

No system is perfect, of course, including the one proposed here. Any biometric system is prone to two basic types of errors: a false positive and a false negative. In a false positive, the system incorrectly declares a successful match between the fingerprint of an impostor and that of the legitimate cardholder—in other words, a thief manages to pass himself off as you and gains access to your accounts. In the case of a false negative, on the other hand, the system fails to make a match between your fingerprint and your stored template—the system doesn't recognize you and denies you access to your own account [2].

According to the National Institute of Standards and Technology researchers, a stand-alone fingerprint system might achieve a 1% false-positive rate and a corresponding false-negative rate of 0.1%. If such a system were used in conjunction with the existing means used to secure credit cards (such as PINs and signatures), the system's security could be 100 times as effective, while at the same time incorrectly rejecting just one more transaction per every 1,000 than are rejected today. Credit card users will tolerate this slight additional inconvenience in exchange for far more effective security [2].

How much they would pay for that additional peace of mind is unknown. But certainly, it need not be expensive. Costs are declining for all of the major smart-card components, including flash memory, microprocessors, communications chips, and fingerprint sensors. Indeed, the basic physical card already exists, albeit in the form of a keychain fob from Privaris Inc., in Fairfax, Va. The company's wireless dongle [7] has all the hardware components mentioned here, and it is likely that sufficient sales volume could cut the retail price of the device from $200 to $20 in a couple of years. The dongle uses fingerprint-based user verification to release data, such as an access code, needed to perform a transaction. The fingerprint is sensed, stored, and processed only on the device and is never released, so as to protect the user's privacy. It would be possible to cut costs further by harnessing the mass-market biometric sensors and computing power available in today's cellphones and programming them with data-matching software and digital certificates [2].

A version of the system designed to protect Internet shoppers might be even easier to implement, and less expensive, too. When mulling the costs and benefits of biometric credit cards, card issuers might well decide to first deploy biometric verification systems for Internet transactions, which is where ID thieves cause them the most pain. A number of approaches could work, but here's a simple one that adapts some of the basic concepts from the proposed smart-card system [2].

To begin with, you'd need a PC equipped with a biometric sensing device such as a fingerprint sensor, a camera for iris scans, or a microphone for taking a voice signature. Next, you'd need to enroll in your credit card company's secure e-commerce system. You would first download and install a biometric credit card protocol plug-in for your Web browser. The plug-in, certified by the credit card company, would enable the computer to identify its sensor peripherals so that biometric information registered during the enrollment process could be traced back to specific sensors on a specific PC. After the sensor scanned your fingerprints, you would have to answer some of the old verification questions— such as your Social Security number, mother's maiden name, or PIN. Once the system authenticated you, the biometric information would be officially certified as valid by the credit card company and stored as an encrypted template on your PC's hard drive [2].

During your initial purchase after enrollment, perhaps buying a nice shirt from your favorite online retailer, you would go through a conventional verification procedure that would prompt you to touch your PC's finger scanner. The credit card protocol plug-in would then function as a matcher and would compare the live biometric scan with the encrypted, certified template on the hard drive. If there were a match, your PC would send a certified digital signature to the credit card company, which would release funds to the retailer, and your shirt would be on its way. Accepting the charge for the shirt on the next bill by paying for it would confirm to the card issuer that you are the person who enrolled the fingerprints stored on the PC. From then on, each time you made an online purchase, you would touch the fingerprint sensor, the plug-in would confirm your identity, and your PC would send the digital signature to your credit card company, authorizing it to release funds to the vendor [2].

If someone else tried to use his or her fingerprints on your machine, the plug-in would recognize that the live scan didn't match the stored template and would reject the attempted purchase. If someone stole your credit card number, enrolled his or her own fingerprints on his or her own PC, and went on an online shopping spree, you would dispute the charges on your next bill and the credit card issuer would have to investigate [2].

Summary/Conclusion

As biometric image-based verification becomes an integral part of overall security, biometric systems have to be designed to be more robust to attacks from hackers to prevent break-ins. This chapter highlighted the five weak points in a generic biometric systems model. A challenge/response method to verify a signal from an intelligent sensor has been proposed to alleviate some of the security threats [1].

Biometric verification systems based on available technology would be a major improvement over conventional verification techniques. If widely implemented, such systems could put thousands of ID thieves out of business and spare countless individuals the nightmare of trying to get their good names and credit back. Though the technology to implement these systems already exists, ongoing research efforts aimed at improving the performance of biometric systems in general and sensors in particular will make them even more reliable, robust, and convenient [2].

Remember, no practical biometric system makes perfect match decisions all the time. As a result, thieves occasionally succeed in being positively identified as people they are not, while legitimate users are sometimes incorrectly rejected. That's because two different samples of the same biometric identifier are never identical. There are two main reasons for this.

First, the sensed biometric data might be noisy or distorted—a cut on your finger leaves a fingerprint with a scar, or a cold alters your voice, for example. Noisy data can also result from improperly maintained sensors (say, from dirt on a fingerprint sensor) or from unfavorable sensing conditions, such as poor focus on a user's iris in a recognition system [2].

Second, the biometric data acquired during verification may be very different from the data used to generate the template during enrollment. During verification, a user might touch a sensor incorrectly or blink an eye during iris capture [2].

Some errors might be avoided by using improved sensors. For instance, optical sensors [9] capture fingerprint details better than capacitive fingerprint sensors and are as much as four times as accurate. Even more accurate than conventional optical sensors, the new multispectral sensor from Lumidigm Inc., in Albuquerque, New Mexico, distinguishes structures in living skin according to the light-absorbing and -scattering properties of different layers. By illuminating the finger surface with light of different wavelengths, the Lumidigm sensor captures an especially detailed image of the fingerprint pattern just below the skin surface to do a better job of taking prints from dry, wet, or dirty fingers. As previously mentioned, such sensors are already being used at Walt Disney World to admit paid visitors to the park [2].

Unfortunately, this kind of optical sensor cannot be easily or cheaply manufactured in a form small enough to fit on handheld gadgets or smart cards. Therefore, the system manufacturers will push the makers of capacitive sensor technology and those who develop data-matching algorithms to close the performance gap with these more costly optical sensors while keeping prices low [2].

Systems based on multiple biometric traits could achieve very low error rates, but here, too, costs will be a concern. These multimodal biometric systems make spoofing more difficult, because an impostor must simultaneously fake several biometric traits of a legitimate user. Further, by asking the user to present a random subset of two or more biometric traits (say, right iris and left index finger, in that order), the system can ensure that a live user is indeed present. Of course, that's more burdensome for the legitimate user [2].

Finally, researchers are optimistic that multidisciplinary research teams in both industry and academia can find the right blend of technologies to create practical biometric applications and integrate them into large systems without introducing additional vulnerabilities. The health of the world economy, not to mention our collective peace of mind, may well depend on their efforts [2].

References

1. Nalini K. Ratha, Jonathan H. Connell and Ruud M. Bolle, "A Biometrics-Based Secure Authentication System," IBM Thomas J. Watson Research Center, 30 Saw Mill River Road, Hawthorne, NY 10532 [IBM Corporation, 1133 Westchester Avenue, White Plains, NY 10604] 2003. Copyright © 2001 IBM.

2. Anil K. Jain and Sharathchandra Pankanti, "A Touch of Money," IEEE Corporate Office, 3 Park Avenue, 17th Floor, New York, New York 10016-5997, vol. 43, no. 7, pp. 22–27, July 2006. IEEE Spectrum, Copyright © 2006, IEEE.

3. John R. Vacca, *Electronic Commerce, 4th ed.*, Charles River Media (2003).

4. John R. Vacca, *Net Privacy: A Guide to Developing and Implementing an Ironclad eBusiness Privacy Plan*, McGraw-Hill (2001).

5. John R. Vacca, *Public Key Infrastructure: Building Trusted Applications and Web Services*, CRC Press (2005).

6. John R. Vacca, *Identity Theft*, Prentice Hall, Professional Technical Reference, Pearson Education (2002).

7. John R. Vacca, *Wireless Data Dymistified*, McGraw-Hill (2003).

8. John R. Vacca, *Practical Internet Security*, Springer (2006).

9. John R. Vacca, *Optical Networking Best Practices Handbook*, John Wiley & Sons (2006).

10. Department of Veterans affairs, philadelphia, PA 19255–1498, 2006.

25

How Cancelable Biometrics Work

When you find out that your credit card number has been compromised, you can easily cancel it and obtain a new number. The same can be done with passwords, keys [10], and many other forms of security. So what can you do if your biometric has been compromised? Use a different finger [1]?

Cancelable biometrics refers to a way of designing a biometric system such that the stored templates cannot be converted back into the raw biometric data. The idea is that at some point in the registration process, a transformation is applied to the image, the features extracted from the image, or even to the user's template/model and the recognition process is performed using the transformed data. For good security, the transformation should probably be one-way and the raw biometric image/data should be thrown away [1].

This way, if by nefarious (or plain incompetent) means, someone bad gets access to the templates, they shouldn't be able to re-create the raw biometric data for the purposes of somehow fooling the system (see sidebar, "Fool Me Once—Fool Me Twice") [1].

Fool Me Once—Fool Me Twice

Biometric authentication seems to be, on the face of it, a groovy idea. Do away with hard-to-remember passwords and easy-to-lose keys and cards; authenticate your identity with your voice, or your face, or your fingerprint. What could possibly be wrong with that [2]?

Well, lots of things, actually. High on the list is the fact that if biometric verification is compromised (if someone finds a way to fake your voice or face or finger), you're up a brown and smelly creek without any way to propel your barbed wire canoe [2].

If someone rips off a password of yours, you can change it. If someone steals your credit card, you can cancel it. Lost a key? Change your locks [2].

But, if someone figures out a way to duplicate your fingerprint or voiceprint or retinal or iris ID, there's nothing you can do. Well, OK, you can switch to a different finger or a different eye, but nature puts certain hard limits on how many times you can do that. Once you're out of organs, you're out of luck [2].

The limited number of biometrics each person carries around with them also makes it impossible to have a large number of different biometric keys. It's important to use different passwords, certificates, keys, or what-have-you for different tasks; only things that don't really matter (like your *New York Times* login, for instance) should use the same password. Otherwise the guy that rips off the username and password list from the poorly secured e-store that sold you a T-shirt will also be able to access your bank accounts [2].

All this is only a problem, of course, if biometrics can be duplicated by normal human beings. The marketing departments at biometric ID companies have, historically, insisted that they can't. Sure, maybe the NSA can fake out a finger scanner, but some scroungy little credit card fraudster isn't going to be able to manage it [2].

Of course, there are a few caveats to this. First of all, you need to trust the people who are capturing the raw biometric data (taking your mug shot, scanning your fingerprints, etc.) as they could easily keep the raw data, which could then be compromised through the same nefarious or incompetent means just discussed. Second, you are still well and truly stuffed if someone does get access to you raw biometric data (takes a photo of you) through other means, because someone will find a way to fool the system if they know what it expects to see [1].

So, repeat this: Biometrics are not secret (see sidebar, "Truth Is Stranger than Fiction"). Biometrics cannot be secret. Remember that [1]!

Truth Is Stranger than Fiction

Biometrics are seductive: You are your key. Your voiceprint unlocks the door of your house. Your retinal scan lets you in the corporate offices. Your thumbprint logs you on to your computer. Unfortunately, the reality of biometrics isn't that simple [3].

Biometrics are the oldest form of identification. Dogs have distinctive barks. Cats spray. Humans recognize each other's faces. On the telephone, your voice identifies you as the person on the line. On a paper contract, your signature identifies you as the person who signed it. Your photograph identifies you as the person who owns a particular passport [3].

What makes biometrics useful for many of these applications is that they can be stored in a database. Alice's voice only works as a biometric identification on the telephone if you already know who she is; if she is a stranger, it doesn't help. It's the same with Alice's handwriting; you can recognize it only if you already know it. To solve this problem, banks keep signature cards on file. Alice signs her name on a card, and it is stored in the bank (the bank needs to maintain its secure perimeter in order for this to work right). When Alice signs a check, the bank verifies Alice's signature against the stored signature to ensure that the check is valid [3].

There are a bunch of different biometrics. This chapter has mentioned handwriting, voiceprints, and face recognition. There are also hand geometry, fingerprints, retinal scans, DNA, typing patterns, signature geometry (not just the look of the signature, but the pen pressure, signature speed, etc.), and others. The technologies behind some of them are more reliable than others, and they'll all improve [3].

"Improve" means two different things. First, it means that the system will not incorrectly identify an impostor as Alice. The whole point of the biometric is to prove that Alice is Alice, so if an impostor can successfully fool the system, it isn't working very well. This is called a false positive. Second, "improve" means that the system will not incorrectly identify Alice as an impostor. Again, the point of the biometric is to prove that Alice is Alice, and if Alice can't convince the system that she is her, then it's not working very well, either. This is called a false negative. In general, you can tune a biometric system to err on the side of a false positive or a false negative [3].

Biometrics are great because they are really hard to forge: It's hard to put a false fingerprint on your finger, or make your retina look like someone else's. Some people can mimic others' voices, and Hollywood can make people's faces look like someone else, but these are specialized or expensive skills. When you see someone sign his or her name, you generally know it is that person and not someone else [3].

Biometrics are lousy because they are so easy to forge: It's easy to steal a biometric after the measurement is taken. In all of the applications discussed previously, the verifier needs to verify not only that the biometric is accurate but also that it has been input correctly. Imagine a remote system that uses face recognition as a biometric. In order to gain authorization, you take a Polaroid picture of yourself and mail it in. They'll compare the picture with the one they have on file. What are the attacks here [3]?

Easy. To masquerade as Alice, take a Polaroid picture of her when she's not looking. Then, at some later date, use it to fool the system. This attack works because while it is hard to make your face look like Alice's, it's easy to get a picture of Alice's face. And since the system does not verify that the picture is of your face, only that it matches the picture of Alice's face on file, you can fool it [3].

Similarly, you can fool a signature biometric using a photocopier or a fax machine. It's hard to forge the vice-president's signature on a letter giving you a promotion, but it's easy to cut his signature out of another letter, paste it on the letter giving you a promotion, and then photocopy the whole thing and send it to the human resources department . . . or just send them a fax. They won't be able to tell that the signature was cut from another document [3].

The moral is that biometrics work great only if the verifier can verify two things: One, that the biometric came from the person at the time of verification; and two, that the biometric matches the master biometric on file. If the system can't do that, it can't work. Biometrics are unique identifiers, but they are not secrets [3].

Tip: *Please repeat the preceding sentence until it sinks in.*

Here's another possible biometric system: thumbprints for remote login authorizations. Alice puts her thumbprint on a reader embedded in the keyboard (don't laugh, there are a lot of companies who want to make this happen). The computer sends the digital thumbprint to the host. The host verifies the thumbprint and lets Alice in if it matches the thumbprint on file. This won't work because it's so easy to steal Alice's digital thumbprint, and once you have it it's easy to fool the host again and again. Biometrics are unique identifiers, but they are not secrets [3].

Which brings this discussion to the second major problem with biometrics: It doesn't handle failure very well. Imagine that Alice is using her thumbprint as a biometric, and someone steals it. Now what? This isn't a digital certificate, where some trusted third party can issue her another one. This is her thumb. She only has two. Once someone steals your biometric, it remains stolen for life; there's no getting back to a secure situation [3].

Note: *Other problems can arise: It's too cold for Alice's fingerprint to register on the reader, or her finger is too dry, or she loses it in a spectacular power-tool accident. Keys just don't have as dramatic a failure mode.*

A third, more minor, problem is that biometrics have to be common across different functions. Just as you should never use the same password on two different systems, the same encryption key should not be used for two different applications. If your fingerprint is used to start your car, unlock your medical records, and read your e-mail, then it's not hard to imagine some very bad situations arising [3].

Biometrics are powerful and useful, but they are not keys. They are useful in situations where there is a trusted path from the reader to the verifier; in those cases all you need is a unique identifier. They are not useful when you need the characteristics of a key: secrecy, randomness, the ability to update or destroy. Biometrics are unique identifiers, but they are not secrets [3].

It's possible that cancelable biometrics (see sidebar, "Distorted Mirrors") are actually not the same thing. But, anyway: To compromise the system in the preceding sidebar (assuming you have the altered biometric data), you still need to reconstruct the original biometric to present to the scanner—yes? Are you missing something? Just having access to the altered biometric is akin to having access to a user's password hash, but not their password [1].

Distorted Mirrors

A trick reminiscent of a fun-house mirror might improve the security and privacy [9] of the access-control technology that examines fingerprints, facial features, or other personal characteristics. In such systems, known as biometrics, a computer generally reduces an image to a template of "minutia points"—notable features such as a loop in a fingerprint or the position of an eye. Those points are converted to a numeric string by a mathematical algorithm, then stored for later analysis [4].

But those mathematical templates, if stolen, can be dangerous. So researchers have developed ways to alter images in a defined, repeatable way, so that hackers who managed to crack a biometric database would be able to steal only the distortion (see Figure 25-1), not the true, original face or fingerprint [4].

It is widely believed that biometric fraud will become more sophisticated (and problematic) as border crossings, passports, financial networks, personal computers and even checkout counters increasingly use the technology. According to industry analysts, worldwide biometric industry revenue is expected to soar from $2.6 billion in 2006 to $6.4 billion in 2011, with government and law enforcement accounting for almost half of the total [4].

Figure 25-1 *To protect biometric information, researchers have developed ways to alter images so that hackers who crack a biometric database would be able to steal only the distortion. (Source: Reproduced with permission from USA Today.)*

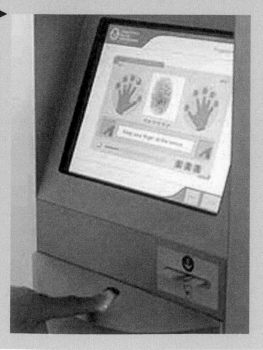

Let's face it: When it becomes worth hacking, it will be done. The threat right now might not be massive, but the threat will be large very soon [4].

Although it is considered impossible to take an image's minutia points and re-create the original, it is possible to concoct an image that shares those points and use it to trick a biometric system. This chicanery requires either hacking into a biometric-equipped network or using a low-tech scam such as making a fake finger out of something like latex or gummy bears [4].

IBM's solution is to make biometric readers distort the image before it is scanned. For example, a face might be made to appear lumpy, or squished up around the eyes. Then a template of the distorted image would be stored [4].

When someone returned to the scanner, the real-life image would be transformed according to the same patterns, creating a match with the tweaked image in the database. The original image isn't stored anywhere. And even if hackers could obtain the altered biometric, it would be of limited use as long as individual organizations maintained their own formulas for transforming images before scanning [4].

Therein lies the real advantage of the method. While a standard biometric can't be torn up and reissued like a credit card or password (since it's based on unchanging aspects of a person's physical appearance), distortion makes that possible. A bank or an office building that had its biometrics compromised could register new ones simply by changing the way it transforms images [4].

That's why this is called "cancelable biometrics." The method has been discussed in research circles for several years, and at least one biometric vendor, iris-scanner Iridian Technologies Inc., offers a cancelable system. Iridian alters the computer-generated template rather than the original image, but the effect is the same. You can't take a biometric out of one application and replay it in another [4].

Perhaps the biggest benefit could be to improve public perception about what happens to biometric data behind the scenes as the technology becomes more widespread. If an organization can check only its version of distorted biometrics, that could reduce fears (some realistic, some paranoid) that government or big companies might maintain a vast database of biometric data for intrusive tracking or marketing purposes [4].

The system could be understood as being more privacy-protected by the normal, everyday consumer. Even so, the distortion approach might not necessarily offer significantly better privacy than systems in which biometric data are not stored in vulnerable, centralized databases, but rather on chip-embedded "smart" cards that people carry with them. In that scenario, the biometric reader simply determines that the person with the card is the person originally granted the card [4].

The cancelable method is a smart way to add a layer of protection to a technology that has some security holes despite being hailed as a huge improvement over more commonly used security measures. This is probably a nice thing to have, but it doesn't resolve all the issues [4].

After all, biometrics are not secret—they're based on physical characteristics that you carry around in plain sight. There's no guarantee someone couldn't lift your real-life fingerprint or take a picture of your face, then figure out a way to present those images to a biometric system [4].

But, in the end, there's no reason to pick on biometrics. Your Social Security number is not secret. Your mother's maiden name isn't secret. What's worse, passwords aren't secret [4].

Now, if you can easily reverse the alteration (which it seems you would have to do to compromise the system), then you have a non altered biometric, and assuming you could fool the scanners with it, you could get into the system no matter what they changed the new alteration to. You could also get into any other system the victim uses the biometric to access (once again assuming you could fool the scanners) [1].

But, maybe all of this might be wrong here. Maybe they can cut off the compromised user by changing the way they transform images. Perhaps they only have a few compromised users, because getting everyone to register their biometrics again on Monday morning will be a giant pain (remember, the original image isn't stored anywhere) [1].

Again, biometrics are not secret. Biometrics cannot be secret [1].

Now, let's look at the research that's going on in the area of replaceable/cancelable biometrics. IBM or "Big Blue" is working toward replaceable biometrics.

Replaceable Biometrics

Biometric security systems have one particularly critical vulnerability: How do you replace your finger if a hacker figures out how to duplicate it? An IBM research team working on that problem indicates it's recently cracked a major problem in the area of "cancelable biometrics" [5].

Biometrics is more personal than a number that somebody assigned to you. You cannot cancel your face. If it is compromised, it is compromised forever [5].

IBM's idea for navigating that obstacle is to construct a kind of technological screen separating a user's actual biological identification information from the records stored in profile databases. The company is developing software to transform biometric data such as fingerprints into distorted models that preserve enough actual identification markers to make the distortion repeatable (see sidebar, "Biometrics of Hazard") [5].

Biometrics of Hazard

When thieves in Malaysia made off with a car recently, they didn't stop with the machine itself; its owner also lost the tip of his index finger, which the car's fingerprint recognition system needed in order to engage the ignition. The incident has sparked calls for liveness detection in biometric security systems—for example, fingerprint recognition systems that check for perspiration to ensure that the finger in question is still attached to a (living) body [6].

So what? Biometrics is a very fine security technology in principle, but it does pose some subtle hazards—not least of which is the potential loss of key body parts [6].

Note: Dan "Da Vinci Code" Brown's Angels and Demons, for example, has a very ... intense ... scene involving a retina scanner.

And of course, many people have physical deficits (lost limbs, worn finger pads) that render them "invisible" to some biometric security systems. Face recognition (pace Hannibal Lecter) seems like one good solution to the problem, although the technology isn't (yet) as robust as one might wish [6].

Okay, so what's my solution? You may not like it: X-ray skeleton recognition. That's right: Your body should be turned into one giant biometric key and irradiate it each time you need to establish your identity. There's even a motto for the system's manufacturer: "It can be kidnapped, but it's really hard to remove." Ah, once again, there's that sickly sweet smell of an indefensible patent application [6].

Organizations that store profiles can then retain just the distorted model, so that if their databases are hacked, the hacker only has access to that organization's profile, rather than to a user's actual fingerprint. The key is that it needs to be irreversible. Otherwise, a hacker can simply reverse-engineer the distorted models to re-create a user's biometric data [5].

IBM researchers have been working for years on the cancelable biometric problem, but a big breakthrough came after they began collaborating with other researchers. They got them together with the cryptographers and applied cryptographic thinking. You think that's irreversible? Ha! Here's how you reverse it [5].

Very recently, the partnership paid off in algorithms that IBM is reasonably confident are genuinely irreversible. A software demo the company released information about is functionally ready for trials. The big technical obstacle was beaten down. Now it's just getting it into the right product or service. IBM Global Services and the company's Tivoli security and systems management software are two likely areas [5].

IBM's system wouldn't entirely solve the replaceability problem of biometrics: If a hacker got hold of a user's fingerprint and made a passable model, he or she could still wreak havoc with it. What IBM's technology could do, however, is significantly narrow hackers' opportunities to gain access to such data. If a user's fingerprints (or facial photographs, iris scans, or any other biological marker) aren't stored in any of the systems she uses them to access, cracking those systems won't give the hacker keys to the victim's biometric kingdom. If a hacker did get in (and the frequency with which companies sheepishly confess to database hacks and inadvertently exposed personal information illustrates the reality of that risk), IBM's system would let a user quickly cancel the compromised biometric profile and generate a new one, akin to replacing a lost or stolen credit card [5].

This technology is being adopted by businesses such as retailers that would benefit from access to customers' biometrics (several stores have run trials of fingerprint-based payment systems), but need to convince those customers their data will be safe. Right now, biometric hacking is only a theoretical problem. Once biometric security gains critical mass, however, attacks will follow [5].

In general, no one is stealing fingerprints. Well, hackers go where the money is. Who would have foreseen phishing? Once there's value, and once people show that something can be done, it will be [5].

Let's take a brief look at a cancelable biometric algorithm. Basically, this is an algorithm of hashing fingerprint data and performing fingerprint matching using hashed values.

Cancelable Biometric Algorithm

Privacy protection is one of the main concerns for biometric verification and identification systems. Biometric templates typically are stored unprotected in a central database. If the database is compromised and the intruder obtains a person's biometric template, it will be impossible to change it for the rest of that person's life. Even if stored templates are encrypted, matching is still performed using decrypted templates, and the decryption process can be compromised as well [8].

Figure 25-2   *An algorithm of hashing fingerprint data and performing fingerprint matching using hashed values. (Source: Adapted with permission from the University at Buffalo.)*

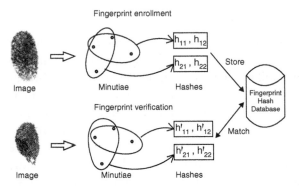

Recently, researchers have developed an algorithm of hashing fingerprint data and performing fingerprint matching using hashed values (see Figure 25-2) [8]. Hashing can be performed on the scanner device, and only hashed values are transmitted and stored in the database. The hashing function is a one-way function; and, given hash values, it is impossible to reconstruct the original template. In case hash values are compromised, a person will be re-enrolled using a new hash function. Since different hash functions are used for old or compromised hash values and new hash values, no match between them is possible [8].

Note: The hash function is a public function. Thus it is possible to simply store information about a specific hash function used together with a fingerprint template in the database server. During verification, the server might instruct the client scanner on which the hash function should be used. Different hash functions can be used for different enrolled persons.

The main idea of the algorithm is that hash values are constructed for the localized minutia subsets, with the condition that the number of hash values is less than the number of minutia. This makes it impossible to algebraically reconstruct localized minutia information. Matching hashes of localized subsets produce a set of match confidences and corresponding transformation parameters: rotation and translation. The presence of many localized matches with similar transformation parameters indicates a global fingerprint match. Global reconstruction of the original minutiae is extremely difficult because of the uncertainty about which minutia points belong to which localized subset. Currently, an achieved equal error rate of the proposed algorithm is 3% on FVC2002's DB1 database (2,800 genuine and 4,950 tests), which is slightly worse than the performance of a similar algorithm using the same minutia extraction algorithm and full information for minutia matching [8].

Summary/Conclusion

Large-scale deployment of biometric systems raises concerns that go beyond ensuring the security of the transaction; they involve privacy of the original biometric data collected from the users. People have legitimate concerns about the use of their biometric data without their permission. This is exacerbated by the fact that biometric data (fingerprint or face), unlike passwords, cannot be changed if compromised. This chapter proposed combining the fingerprint and signature data to construct a new cancelable biometric in a unique way to mitigate these issues.

In other words, your fingerprints are yours and yours alone, and that makes them a useful tool for confirming the identity of people doing things like conducting secure banking transactions or passing through corporate security checkpoints. The trouble is, it's theoretically possible for a hacker to break into the software of, say, your employer, steal a copy of your stored fingerprint, and later use it to gain entrance [7].

So, researchers at IBM have come up with "cancelable biometrics": If someone steals your fingerprint, you're just issued a new one, like a replacement credit card number. The IBM algorithm takes biometric data and runs it through one of an infinite number of "transform" programs. The features of a fingerprint, for example, might get squeezed or twisted. A bank could take a fingerprint scan when it enrolls a customer, run the print through the algorithm, and then use only the transformed biometric data for future verification [7].

If that data is stolen, the bank simply cancels the transformed biometric and issues a new transformation. And, since different transformations can be used in different contexts (one at a bank, one at an employer), cross-matching becomes nearly impossible, protecting the privacy of the user [7].

Finally, the software makes sure that the original image can't be reconstituted from the transformed versions. IBM hopes to offer the software package as a commercial product in 2009 [7].

References

1. David Dean, "Cancelable Biometrics," Speech, Audio, Image and Video Technologies Research Laboratory, Queensland University of Technology, Brisbane, Queensland, Australia, February 25, 2006.

2. Daniel Rutter, "DigitalPersona U.are.U Personal Fingerprint Scanner," March 12, 2005. Copyright © Daniel Rutter.

3. Bruce Schneier, "Biometrics: Truths and Fictions," Crypto-Gram Newsletter, 1090 La Avenida Avenue, Mountain View, CA 94043, August 15, 1998. Copyright © Bruce Schneier.

4. Brian Bergstein, "Distorting Biometric Images Enhances Security, Privacy," The Associated Press, USA Today, 7950 Jones Branch Drive, McLean, VA 22108-0605, August 28, 2005. Copyright © 2006 USA TODAY, a division of Gannett Co. Inc.

5. Stacy Cowley, "IBM Works Toward Replaceable Biometrics," InfoWorld, InfoWorld Media Group, 501 Second Street, San Francisco, CA 94107, August 17, 2005. IDG News Service, Copyright © 2006. All Rights Reserved.

6. Ed Gottsman, "On Keeping It All Together," ZDNet, CNET Networks, Inc., 235 Second Street, San Francisco, CA 94105, January 18, 2006. Copyright © 2006 CNET Networks, Inc. All Rights Reserved.

7. David Talbot, "Changeable Fingerprint," *Technology Review*, One Main Street, 7th Floor, Cambridge, MA 02142, January 2006.

8. Sergey Tulyakov, "Cancelable Biometrics," University at Buffalo, SUNY, 408 Capen Hall, Buffalo, NY 14260-1608, 2006. Copyright © 2004, University at Buffalo. All Rights Reserved.

9. John R. Vacca, *Net Privacy: A Guide to Developing and Implementing an Ironclad eBusiness Privacy Plan*, McGraw-Hill (2001).

10. John R. Vacca, *Public Key Infrastructure: Building Trusted Applications and Web Services*, CRC Press (2005).

Part 8: Large-Scale Implementation/Deployment of Biometric Technologies and Verification Systems

26

Specialized Biometric Enterprise Deployment

There is a narrow view of the enterprise that exists today that only includes the network and computer realms within an organization. In order to have an end-to-end sense of control over your users and assets within a particular organization, the view must be broadened a bit. The enterprise should, at a minimum, include the following realms, as shown in Figure 26-1 [1]:

1.   Physical

2.   Network

3.   Computer [1]

Securing the physical realm consists of protecting a facility or a tangible asset from entry or use. For example, a building typically is secured from entry with a lock that requires a key for ingress. Usually, when securing network and computer assets, they are grouped together. They are related, but very different when concerned with the task of securing them. The network provides a gateway of use for a computer asset. In other words, the network is the first line of defense against unauthorized use of a computer asset. Securing it is an all-or-nothing proposition. An individual can be given—or denied—rights to access it. The computer asset is not that simple. The computer realm can be further broken down into the following sub-realms:

1.   Login

2.   Application

3.   Data [1]

The login security of the computer asset protects it from any unauthorized access. Again, like the network asset, it is all or none. Application security

Figure 26-1
*Realms. (Source:
Adapted with
permission from
N-Cycles, Inc.)*

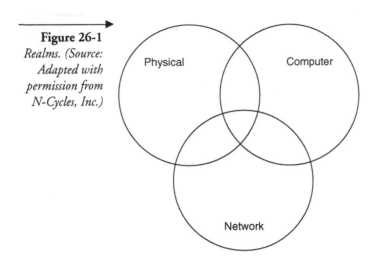

is applied to each individual application that is being run on a computer asset. Each application should have security mechanisms that grant or deny access to the use of the application. This sub-realm is a bit different in the sense that it does not exhibit the all-or-nothing access characteristics of the previously discussed realm. Security for the application asset is going to depend on who you are and where you are (internal or external). For example, a manager (who you are) will have access to the payroll function of a Web-based corporate enterprise application that an employee would not have. Also, it might be a requirement for some types of sensitive functions on the same application to be "hidden" if the manager is accessing the application from his or her home computer versus his or her office computer. Finally, data is the last realm. The data will typically be the most sensitive part of the entire system. This sub-realm exhibits the characteristic of access being based on who you are and where you are. Finally, consider the following broad grouping characteristics of security challenges that are employed in the enterprise:

- Something you know (password, PIN)

- Something you have (token, smart card, certificate, etc.)

- Something you are (fingerprint, face, voice, etc.) [1]

In a modern enterprise security system, all three of these broad groupings will be employed. The user's actions and level of desired access will dictate the security policy employed and determine the mix (one, some, or all) of security challenges [1].

Any single layer can be made more secure by increasing the complexity of the challenge. However, stronger security can also be less convenient for users. Therefore, the right mix of challenges and complexity of the challenge will depend on the particular need [1].

Biometrics: Life Measures

As the name biometrics suggests, the idea behind this emerging technology is to map measurements of human physical characteristics to human uniqueness. If this can be accomplished in a reliable, repeatable fashion, the verification and identification of human individuals by machine becomes a reality. To that end, biometrics is a fusion of human physiology, pure mathematics, and engineering. The idea of specialized biometrics is very simple to grasp, but the implementation/deployment can be daunting and very difficult to realize. The difficulty does not come from the gathering of the actual measurements, but from the analysis of these measures. As with most pattern recognition problems, more data can always be gathered. The problem is what to do with it after it has been gathered. Enough measurements have to be taken to assure uniqueness of each individual. There is a fine line between not having enough data and having too much. Too much data can cause an "aliasing" effect, where individual uniqueness is lost [1].

The application of the specialized biometric device has important ramifications on how much data should be collected. Simply verifying someone's identity is much less complex than identifying a person. Verification versus identification might seem like semantics, but think about the difference between checking someone's driver's license photo and recognizing someone in a packed room who you've never met. Verification involves having someone tell a biometric system she is Jane Doe and then using one or more set of specialized biometric information to verify that she is in fact Jane Doe. Identification is Jane Doe walking up to a set of specialized biometric sensors and being recognized as Jane Doe. If you have seen the movie *Minority Report*, you remember how the mall stores biometrically identify Tom Cruise as he moves from store to store. The increased complexity of identification in our world means that reliable identification is still some years away from being perfected, which is why the Super Bowl organizers elected not to try to use specialized biometric identification techniques. On the other hand, verification is extremely reliable today, and for some security applications is quite appropriate for either increased security or ease of use [1].

Depending on the application and objective, different "form factors" are more appropriate. The predominant specialized biometric form factors today are:

1.   Handprint

2.   Fingerprint

3.   Retina

4.   Iris

5.   Voice/Speech

6.   Handwriting/Signature

7.   Face

8.   Movement patterns (typing, walking, etc.) [1]

The Case for Biometrics in Enterprise Security

As previously mentioned, the use of specialized biometric verification and identification have so far been limited to a few movies and TV shows. The public seems to view specialized biometrics as futuristic, if not scary, even though these technologies have been around for several years. The reality is this: The form factors discussed in this chapter are real, and they work. The issue becomes how to apply them appropriately. Companies want to rush to the market with a really cool technology without a real problem to solve. This was the case with early biometric solutions. The early implementations were burdened with the fact that they were not as reliable as hoped for, and prospective customers did not see the need to deploy yet another security technology that at the time only looked like a fancy replacement for passwords during computer or network login. This is all beginning to change. The important thing to focus on is the application of the technology, which is much more robust than it was in years past. Some of the technologies, fingerprint, for example, are beginning to put up 99% correct verification rates [1]. Other technologies, such as facial and speech recognition, will always be plagued by ambient environmental factors such as lighting and limited bandwidth communications channels (cellphones). These technologies will not be able to function alone in an enterprise security system, but can be an important part of a group of technologies that are used in conjunction with one another to strengthen the whole. With all of this said, let's look at a few factors to consider when applying specialized biometrics [1].

Convenience

Convenience is the number one factor for most customers concerning the deployment of a specialized biometric system. Security systems will be forced to become more stringent in the days to come. This means that, at the low end, password policies will be modified to increase the frequency of changes to a user's password. Most organizations today are forcing users to change passwords at least every 30 days. Most of these policies also do not allow the user to reuse the past five passwords. This gives the user one extra headache and a reason to circumvent the security practices. This also results in more calls being made to the help center to reset forgotten passwords. Frustration rises, and time is wasted to achieve the higher level of security. Passwords are also going to become longer and be machine generated. This will really cause user frustration and loss of efficient use of protected physical/computer/network assets. The password is the predominantly deployed verification mechanism in all enterprise security today. A password is considered "something that you know." The convenience provided by a specialized biometric device that represents the user's identity ("something you are") is mapped to the user's password or PIN, and has a very high rating. If the user obtains use of a physical/computer/network asset by only using a specialized biometric, the convenience of the security system is greatly increased. Further, the security of the system can be bolstered without affecting the user. The password could be changed every 10 minutes and made to be 32 characters long, if the security policy so demanded. The user would only have to bring their specialized biometric along [1].

Security

Incorporating specialized biometric devices into the enterprise security architecture increases security by eliminating the ability to share passwords and making it much more difficult to counterfeit or steal the security key (see sidebar, "Deploying Specialized Biometric Security to the Enterprise"). The level of security provided by a device also depends on the number of "reference points," which are the individual metrics taken in each scan. For instance, iris scanners capture 200+ while fingerprint readers typically only capture around 80. However, the effectiveness of the reference points also depends on the algorithms used. More reference points can mean more false negative identifications. In other words, better accuracy can actually result in rejecting the right person. While more reference points theoretically means a better "signature," it can also mean that there are more chances for failure in the secondary scan. This problem is more pronounced in difficult environments or where ambient noise or light can impact the scanning environment. The original iris scan could store all 200 points correctly, but if the person using the device is not positioned correctly, then the scanner could not pick up each reference point properly. The individual sensitivity settings on a device controls whether it will err on the side of caution (rejection) or convenience (acceptance). Setting the sensitivity too high can result in too high a false-rejection rate (FRR), and setting it too low can increase the false-acceptance rate (FAR). When used with multiple form factors, lower individual levels can provide fewer false acceptances or rejections because of the multiple points of reference [1].

Deploying Specialized Biometric Security to the Enterprise

Warriors have long used emblems, uniforms, and tattoos to physically identify themselves to their compatriots. Secret passwords were in use long before the first person logged in at a keyboard [2].

Today, the world of enterprise security is increasingly incorporating specialized biometric identifiers as an additional weapon within the security arsenal. According to industry analysts, the worldwide market for specialized biometric devices grew 89% in 2006 to reach $3.4 billion. And analysts estimate a further expansion to $6.8 billion by 2010 [2].

The largest share of that money (50%) goes for fingerprint recognition systems, followed by facial recognition (14%). While these two are the most popular, there are other methods that analyze a person's physical or dynamic characteristics. Physical specialized biometric methodologies also look at:

- **Eyes:** Examining the lines of the iris or the blood vessels in the retina;
- **Hands:** Taking a 3D image and measuring the height and width of bones and joints;
- **Skin:** Analyzing surface texture and thickness of skin layers [2].

When looking at strong verification, you want two out of three factors—something you have, something you are, and something you know. While eyes, hands, and skin are commonly used as specialized biometric identifiers, more dynamic methodologies also are being introduced, such as:

- **Voice:** Detects vocal pitch and rhythm;
- **Keystroke dynamics:** Analyzes the typing speed and rhythm when the user ID and password are entered;
- **Signature:** Matches the signature to one on record, as well as analyzing the speed and pressure used while writing;
- **Gait:** Measures length of stride and its rhythm [2].

To keep performance high and storage requirements manageable, today's specialized biometric technologies do not have to store or analyze a complete picture of the body part or the physical feature being used. Imagine the processing power that would be needed to store a high-resolution picture of someone's face and then compare it with a live image, pixel by pixel [2].

Instead, each method reduces the body part or activity to a few essential parameters and then codes the data, typically as a series of hash marks. For example, a facial recognition system may record only the shape of the nose and the distance between the eyes. That's all the data that needs to be recorded for an individual's passport, for example [2].

When that person comes through Customs, the passport doesn't have to include all the data required to reproduce a full-color picture of the person. Yet, armed with a tiny dose of key specialized biometric information, video equipment at the airport can tell whether the person's eyes are closer together or if his nose is slightly wider than the passport says they should be [2].

None of these specialized biometric systems are infallible, of course, though the rates of false negatives and false positives have markedly improved. One of the problems with fingerprint readers, for instance, is that they couldn't distinguish between an actual fingerprint and the image of one. In the recent movie *National Treasure*, Nicholas Cage's character lifted someone's fingerprint off a champagne glass and used it to gain access to a vault. That is not pure fiction [2].

Japanese cryptographer Tsutomu Matsumoto lifted a fingerprint off a sheet of glass and, following a series of steps, created gelatin copies. He then tested these on 11 fingerprint readers, and each accepted the gelatin prints [2].

Outside the lab, Malaysian thieves chopped the fingertip off a businessman and used it with the fingerprint reader on his Mercedes. But neither of those methods would work with higher-end fingerprint readers [2].

The latest fingerprint readers are incorporating more advanced features, such as making sure the finger is a certain temperature. Everyone's hand is different, as some are consistently warm or cold. In addition, the readers can check for a pulse and tell how much pressure is being applied [2].

Such sophistication, however, has its drawbacks. Authorized users may find themselves locked out even when the devices are working properly. Why? Tiny changes, due to accidents or injuries, can change a specialized biometrics profile, rendering it effectively obsolete [2].

The thing to keep in mind with any specialized biometrics is that your ID does change over time. If you cut your finger, your specialized biometric may not be the same. Or your early-morning voice may be different than after talking for eight hours [2].

Specialized Biometrics in the Enterprise

While specialized biometric verification certainly adds an extra layer of security, it would be a mistake to implement a high-end system and then feel that break-ins instantly would be consigned to the history books. It takes back-end integration, constant vigilance, and consistent user involvement to keep an enterprise secure [2].

Security is a user issue and must go all the way to the desktop. The philosophy here is to do defense in depth. Therefore, you must have a very layered architecture and assume that any layer will fail some day [2].

The most popular specialized biometric tool at the moment is the fingerprint reader. Some even use USB drives, and some keyboards and laptops come with them built in. These devices have come way down in price. As a stand-alone device, the unit price has dropped below $100. But, in an enterprise setting, that is just the start of the costs [2].

Often, companies look at specialized biometrics as being ultra sexy, cool technology, but forget that there are integration issues. IT departments have to ensure, for example, that back-end security systems can accommodate specialized biometric verification, and scale to the required number of users. Plus, if fingerprint readers are not incorporated into the laptop or desktop, it adds to the number of devices that need to be supported by IT [2].

There is little point, then, in adopting a stand-alone specialized biometric system that cannot easily be assimilated into the organization's existing security fabric. Security is no longer something you can address as an afterthought. It needs to be built into the infrastructure to deal with pervasive threats [2].

The good news is that the specialized biometric verification techniques are no longer so leading-edge that they are difficult to marry with traditional security safeguards. Today's systems are well enough developed that they can be incorporated into enterprise systems without too much effort [2].

A strong verification system is what you want to focus on, and biometrics can be part of that. But the user should still have to memorize something or have a token, and you need to make sure that policies and the management structure relating to it are firmly in place [2].

Since specialized biometrics rely on "something you are," obviously, other people will not be able to use them. Some movies have included examples of killing a person and then using the deceased's finger, hand, and so on. In one of the *Mission Impossible* movies someone used a molded rubber copy of a person's hand. Both of these approaches would be extremely difficult, if not impossible, to use to defeat today's specialized biometric sensors. In the future, multiple form factors will be used at the same time to reduce the likelihood even further. Multiple form factors refers to using, say, fingerprint, voice, and movement pattern recognition at the same time to make it that much harder to effectively counterfeit someone's identity [1].

Usability

Various specialized biometric sensors require more or less involvement on the user's part. Even more importantly, the nature of the signatures collected impact the ease of enrollment and implementation of the equipment. Another important aspect of usability is whether a device is intrusive or nonintrusive. Intrusive devices such as retina, fingerprint, and handprint require users to touch or be very close to the sensor. Nonintrusive devices such as iris, facial recognition, and voice generally can operate at less intrusive distances. The level of the intrusiveness varies by device (fingerprint is much less intrusive than a retina scanner). From here on, as various form factors are being discussed, comments will be made on the particular usability challenges inherent in each choice [1].

Present State of Enterprise Specialized Biometric Security

The next part of the chapter provides an overview of the main types of device "form factors" available for practical use today. Each one is in various stages of refinement, although all have some useful applications. In addition to discussions of each one, they will also be rated along the same dimensions defined in the preceding: convenience, security, and usability. Of course, your particular circumstances may modify these ratings, but by providing an overall sense of the relative characteristics, this part of the chapter will hopefully provide a better idea of the general tradeoffs among the devices [1].

Handprint

Handprint is probably most familiar from spy movies, where top-secret rooms have a pad for handprint use. While the actual details are different in reality, the basic idea is the same. Handprint is usually most appropriate for fixed physical locations requiring very high assurance of identity, since it combines the specialized hand biometric with essentially five different specialized fingerprint biometrics. Fairly large physical assets such as buildings are necessary, simply because of the size of the sensor. Imagine how awkward a full handprint sensor would be on a desktop, let alone a notebook computer. The security and reliability can be even further enhanced by combining a handprint with any of the other form factors (see Table 26-1) [1]. Cost is another factor that limits use of handprint readers to mostly larger physical assets. The system vendors typically specialize in door lock systems.

Table 26-1   *Handprint Form Factor Ratings*

Form Factor	Rating
Convenience	Moderate
Security	Moderate
Usability	Moderate

While there are many vendors incorporating handprint readers into attendance or security products, the main provider of the underlying technology is Recognition Systems. Handprints continue to be used primarily for traditional applications in data rooms, sensitive office zones/buildings, national security/intelligence facilities, and vaults. However, handprint reader use for normal commercial and light industrial building access is waiting for identification algorithms to become reliable, so that building managers can stop issuing access cards [1].

Fingerprint

Specialized fingerprint biometrics involve a finger-sized identification sensor with a low-cost specialized biometric chip. Fingerprint provides the best option for most uses of specialized biometric verification, especially attached to specific computer and network assets. The relatively small size and low cost allow them to be easily incorporated into devices. They are fairly reliable. Many PC manufacturers are experimenting with integrating the devices either on keyboards, mice, or the actual computer case. Dell seems to be the furthest along, although all are working on it. For now, actual implementations have used third-party devices such as Identix, Authentec, Veridicom, Secugen, Sony, or Infinieon. Most stand-alone fingerprint readers sell for $75–$150 each at retail, making it one of the most affordable form factors. In addition, many integrated devices such as time clocks are incorporating specialized biometric fingerprint readers [1].

Specialized fingerprint biometric devices must be distinguished from simple fingerprint recorders. Some banks are starting to implement simple fingerprint recorders to provide more security in check-cashing operations. This is simply taking a snapshot of the fingerprint to aid in tracking and prosecuting check fraud. Most users choose to run pilot programs and/or implementations with the specialized biometric verification devices because of the greater security and convenience they provide (see Table 26-2) [1].

Table 26-2   *Fingerprint Form Factor Ratings*

Form Factor	Rating
Convenience	High
Security	Moderate
Usability	High

Retina

Retina scanning involves examining the unique patterns on the back of a person's eye. The retina is the part of the eye that translates light into the electrical impulses sent to the brain. Because of the complexity of current scanners, most retina-specialized biometric devices require a relatively large footprint. Some manufacturers are working on ways to install or simply place retina scanners on top of computer monitors. However, most are still used to protect fixed physical assets. Using a retina scanner is less convenient because the user must position himself a certain distance away from the scanner and then rest his head on a support or look into a hood (see Table 26-3) [1]. This is necessary in order to effectively read the back of the eye [1].

According to industry analysts, the leading provider of retina scanners is EyeDentify. EyeDentify pulled its retina product off the market in order to try to reduce the cost from around $2,000 per unit to the $400–$500 range [1].

Iris

Iris scanning is similar to retina scanning, but the scanner is looking at the unique patterns on a person's iris. This is the "colored" part of the eye that is visible. Retinas are on the inside back of the eyeball. A key benefit for iris over

Table 26-3   *Retina Form Factor Ratings*

Form Factor	Rating
Convenience	Low
Security	High
Usability	Moderate

Table 26-4   *Iris Form Factor Ratings*

Form Factor	Rating
Convenience	Low
Security	High
Usability	Moderate

retina is that iris scanners do not need to be nearly as close to the eye and do not need the eye to be as precisely positioned (see Table 26-4) [1].

According to industry analysts, the leading providers of iris scanning hardware are Panasonic and Diebold. The company has experimented with adding iris scanners as an integrated option as part of its line of ATMs [1].

Voice/Speech

Specialized biometric verification using speech is uniquely appealing simply because no specialized recording device needs to be used. Specialized biometric verification using a voiceprint is completely a matter of the algorithms and analysis software. This opens up the possibilities of being able to use it for phone-based applications such as voice response systems and time card entry [1].

The possibility of using voice verification to make secure remote data reporting applications more convenient in the criminal justice and healthcare industries is extremely promising as well (see Table 26-5) [1]. Sexual offender databases could be made much more reliable if each offender had to call in periodically to provide updated contact information. The entry could be verified via his or her unique voice pattern. Any offenders who missed their deadline to call in would be flagged for further investigation. In healthcare, people on home care or hospice could use voiceprint-secured telephone systems to report

Table 26-5   *Voice/Speech Form Factor Ratings*

Form Factor	Rating
Convenience	High
Security	Moderate
Usability	Moderate

progress or request prescription refills. In a similar way, home care nurses could use voiceprint-verified systems to report after each patient stop [1].

Specialized voiceprint biometric identification has been developed most extensively by the NSA in order to assist in electronic espionage, but as commercially available software continues to evolve, an increasingly wide range of applications will become feasible. This will enhance the convenience of voice-based verification systems as well as enable new applications. For instance, marketers would love to enhance telesales centers with the ability to identify the caller by his or her voiceprint in the first few seconds of a call in order to supply the telesales agent with all available information about that individual. This would reduce the inaccuracy of relying on caller ID to guess who is calling [1].

Handwriting/Signature

Specialized biometric verification via handwriting or signature must be distinguished from simple signature capture pads. Unlike a signature capture pad, which simply records an image of what the person wrote, specialized biometric enabled capture pads actually record the pressure, distance of strokes, and speed of writing. These data points enable specialized biometrics by verifying whether the person writing the signature is indeed the same person who supplied the original enrollment sample. Depending on the threshold settings used, the specialized biometric device could flag potential forgers. Even if a forger duplicated the exact image of a signature, the pressure and speed would be different from the genuine signature. However, the tradeoff between false positives and false negatives is particularly fuzzy here because people vary the way that they sign their names, particularly at younger ages. Setting the threshold too tight will cause genuine signatures to be rejected. Setting the threshold too loose will let forged signatures pass [1].

Specialized biometric signature verification is particularly interesting to the financial and legal communities because it is substantially less obtrusive and requires less behavior modification. It still feels like a signature—just digitally captured. However, for frequent verifications such as computer and network logins or physical asset access, signatures are less ideal because they take longer than simply using a thumbprint reader (see Table 26-6) [1].

Tip: Vendors of signature verification solutions include Cyber-Sign and Communication Intelligence Corporation.

Table 26-6   *Handwriting/Signature Form Factor Ratings*

Form Factor	Rating
Convenience	Low
Security	High
Usability	Moderate

Face

Face recognition involves scanning the unique features of a person's face. Because some aspects change over time, this is a less reliable form factor. Face recognition is less attractive for up-close verification than for long-distance identification (see Table 26-7) [1]. Once a person is close enough to a physical asset in order to get a high-quality specialized biometric scan, other form factors are viable and are currently much more reliable. However, eye, hand, and finger are practically worthless at a distance, and the quality of voiceprint identification degrades rapidly with distance. Therefore, face recognition promises to be the best bet for remote identification. Security teams for Super Bowl XXXVI considered using rough forms of facial identification to help spot terrorists, but shelved the idea because of current limitations. It will be several years before remote facial recognition can be cost-effectively used to monitor workplaces or remote physical assets [1].

Similarly, nonsecurity applications for facial recognition are probably even further off, mainly for privacy reasons [4]. CRM specialists drool at the thought of being able to record and then use facial prints to identify customers when they enter a store or restaurant [1].

Table 26-7   *Face Form Factor Ratings*

Form Factor	Rating
Convenience	High
Security	Low
Usability	Low

Table 26-8   *Movement Form Factor Ratings*

Form Factor	Rating
Convenience	High
Security	Moderate
Usability	Low

Movement Patterns

The movement pattern specialized biometric form factor is a little harder to grasp (see Table 26-8) [1]. It involves monitoring the way that a person moves (types, walks, etc.) and guessing their identity. The measurements involved are more complex because they must combine spatial and time series data. Also, the scanning required to accurately read the movements still depends on ensuring a consistent angle of observation. The subject must walk by the sensor at the same angle as measurements are taken. For this reason, typing is probably the most promising current form factor, since the observation area is relatively fixed [1].

Example Enterprise Scenarios

Now, let's look at some specialized biometric enterprise scenarios. The following examples are covered in this part of the chapter:

- Web portals
- Single sign-on (SSO)
- Inter-enterprise

Web Portals

The most obvious scenario for biometrically securing a Web portal is for online banking or online financial aggregation. As larger banks such as Citibank and Bank of America continue to push account aggregation, these two areas will gradually merge. Consumers may be more willing to aggregate their online accounts with an institution that protects the aggregation point with specialized biometric verification. This could potentially increase the switching costs for users even further by making it harder for consumers to change account aggregation points, if not entire banks. Currently, customer loyalty is highest for consumers using online bill payment; and, this could increase by securing it through specialized biometric verification.

However, the traditional mega portals are also strong contenders to benefit from specialized biometric security. Yahoo, AOL, and MSN are all trying to push different single sign-on strategies to make it harder to use competing services. Yahoo has the Yahoo ID and Microsoft has Passport. However, adoption has been slower than each company would prefer, perhaps because users are hesitant to trust one particular entity as their online gatekeeper. At least part of this fear is that it makes their personal information more susceptible to compromise since it is consolidated in one single location. There are other privacy reasons, but the security reason is a major factor [1].

The main factor affecting all of these scenarios (and severely limiting the adoption by other types of public Web portals) is the simple question of enrollment and access. Before a specialized biometric can be used, it must be enrolled (recorded). Enrollment can be either handled in a secured location or remotely via "self-enrollment." Both have limitations. Physical enrollment provides the best assurance of security, because a person must travel to a physical location and present some other type of identity verification to the person handling the enrollment. This provides the highest probability that John Doe's specialized biometric is indeed John Doe's. Self-enrollment involves generating a temporary password that John Doe would use to log in for the first time and then use his or her local specialized biometric reader to enroll. This is more convenient, but introduces a greater probability of fraud. Identity thieves could steal personal information and open an account using the self-enrollment approach. That is much more difficult with physical enrollment. Physical enrollment would be very costly for the true Web portals unless they contracted with some other type of business such as Kinkos that already has a communications infrastructure and a national presence. However, this would not be as secure as true physical enrollment. Banks would probably have the easiest time implementing physical enrollment because most already have branch networks near or within the areas where customers live and work [1].

Access might be a harder problem to overcome, at least at current specialized biometric scanner prices. One of the benefits of Web portals is that they provide almost universal access from any Internet-connected computer. Current specialized biometric devices are not compatible with each other. This means that if a bank wanted to offer specialized biometric security for its Internet banking, then it would have to physically ship a compatible, specialized biometric scanner to the consumer, who would then have to install it before using it. Even if the bank wanted to invest in the ability to handle a variety of form factors and manufacturer's devices, the user would still need to access his or her account from a computer, using a device that was compatible with the one that he or she used to enroll on. This means that if John Doe enrolled on his home computer, he could most likely not check his account balance on his office computer, let alone a friend's computer. The access problem will not be solved until a global standard for specialized biometric signatures is agreed upon or a single device manufacturer obtains massive market dominance [1].

Mainly because of the current incompatibility of devices, most industry analysts do not see widespread use of specialized biometric security for public Web portals for quite some time. More limited use within enterprises are certainly more feasible, but they still present problems any time that a user would need access to the portal from a computer outside of the enterprise [1].

Single Sign-On (SSO)

Traditional single sign-on initiatives are concerned with consolidating every computer-based verification into a single set of credentials, so that a person only has to remember one password or token. However, specialized biometric security devices allow the concept of single sign-on to extend to the physical layer as well. A person would only have to enroll once to let his or her specialized biometric characteristics give access to every door, computer, or application that he or she needs access to. Fingerprint readers make sense in this environment, because they can be deployed relatively cheaply and in a variety of different type of locations. More importantly, they can all be integrated into the underlying network, computer, and physical security systems. It is feasible for an organization to have a central enrollment point for biometric verification of both network passwords and physical security doorlocks. In order to be effective, the company must be able to control the access privileges; and, in order to be administratively cost-effective, they need to be integrated with network, application, and physical security systems. So, as other organizations attempt to take advantage of the security advantages provided by specialized biometric verification, the flexibility to integrate directly down to the application level will be critical to avoid impacting workflow speed and operator frustration. The necessity of making verification dependent on an almost infinite different combination of business rules increases the need to have complete control over the device and identity storage methodology selected within the enterprise [1].

Inter-Enterprise

At the present time, no financial institutions are using real-time specialized biometric identity verification at the time of transaction, mainly because of the enrollment problems. In order for it to be feasibly convenient, a person would need to be able to enroll once and then verify a transaction at many (if not all) potential transaction locations or websites. A couple of start-up companies are offering biometric verification-based check-cashing systems to localized retail stores (primarily liquor stores and grocery stores). Most of these systems require separate enrollment at each location and are not integrated with the rest of the enterprise. However, BioPay provides an inter-enterprise solution to biometrically verify check transactions. The company has inked a deal with Kroger to run tests with a centralized database, so that customers can use a variety of Kroger locations after enrolling only once. The system is sufficiently integrated to allow customers to both cash checks and verify check purchases with a specialized biometric signature. BioPay is attractive because it offers a centralized database of bad check writers that depends on the specialized biometric signature and not the particular account number or name [1].

Finally, two different primary uses for fingerprint verification are evolving: centralized third-party inter-enterprise transaction facilitators, and enterprise-specific workflow security. The first case will be for verification of identity, primarily in consumer-focused transactions. Consumers would rather enroll once with a trusted provider and then be able to use verification devices at a variety of transaction origination points. Consumers will be reluctant to enroll at all unless they trust the institution enough to safeguard their most private financial and/or medical information. Also, consumers will quickly become annoyed if each institution requires separate fingerprint enrollment because of the relatively larger hassle of having to be physically present to enroll. Besides the check cashing/payment security that BioPay provides, ATM transaction security is another prime example of a situation where a centralized provider will make sense, unless a specialized biometric signature interchange standard emerges. For specialized biometric verification to make sense to consumers, they must first be able to enroll at their home institution in order to use any ATM location. Prescription identity verification is another area where a centralized third-party verification system makes economic sense, because a patient needs to pick up prescriptions from a variety of locations and providers. They would rather not enroll with each chain separately, although with the consolidation in the pharmacy business, this may become less of an issue. For intra-enterprise security applications, the details of the implementation can be controlled within an organization, while the scope of integration and application replacement will depend on the scope of the organization and the desired level of deployment.

Summary/Conclusion

As mentioned in previous chapters, throughout the last 30 years, James Bond, *Star Trek*, *La Femme Nikita*, *Stargate SG1*, *Stargate Atlantis*, *Alias*, *24*, and countless other stories of intrigue and science fiction have heralded the use of biometrics. Today, you're finally getting a glimpse of how easy advanced security can be when an individual's unique physical characteristics are electronically stored and scanned. Recent advances have made specialized biometrics more reliable, accurate, scalable, and cost-effective for the enterprise. Nevertheless, the technologies remain too expensive for most organizations to deploy widely, so biometrics are only ideal for environments with the highest of security needs [3].

Fingerprint identification, hand geometry, voice verification, retina or iris scanning, and facial recognition are the specialized biometric techniques most likely to be used in an enterprise, but not the only ones. Extreme methods, including DNA, ear lobe, and typing-pattern recognition, can now be used in circumstances that require extremely high security measures, such as monitoring access to a missile launching system [3].

Deploying any method of biometrics can offer a greater sense of security, but the high cost of purchasing the hardware, installing and integrating it into enterprise systems, and training end-users is still prohibitive—and often is not offset by the increased security levels. General day-to-day issues, such as how to enroll a new finger in the system if a user injures his verification finger, are easy to overlook [3].

Mapping the Body

Although many parts of the human body can provide data for electronic identification, users remain most comfortable offering their fingertips. Fingertip scanners are the most commonly used form of biometrics (and the least expensive and easiest to deploy), but not all scanners are the same. Some match the ridges in a thumbprint, others are straight pattern-matching devices, and still others take unique approaches such as ultrasonics [3].

A more accurate system is hand geometry. Because this area has not seen the dramatic price decreases of fingerprint scanners, however, hand geometry is usually deployed only in the most sensitive areas of the enterprise, such as vaults or data centers [3].

Banks are great candidates for voice verification. Simply allowing customers to change their personal identification numbers (PINs) by voice could save banks thousands of dollars. But despite its potential for improving customer service, voice verification techniques are largely limited to use on internal networks. The variability of telephone handsets and line quality creates significant challenges for deploying it over public networks [3].

One of the most advanced but most intrusive specialized biometrics is retina scanning, which scans the unique patterns of the retina with a low-intensity light source. By contrast, iris scanning uses a camera and requires no intimate contact with the reader; its ease of use and system integration have traditionally been poor but are improving dramatically with recent developments. Nevertheless, its high cost will continue to limit iris scanning to extremely sensitive areas [3].

Facial recognition, which compares a user's facial characteristics with the stored results of an algorithm calculation (similar to a data hash), offers the ultimate security. Some systems match two static images, and others claim to be able to unobtrusively detect the identity of an individual within a group. But facial recognition has had only very limited success in enterprise applications, such as access to nuclear facilities, because of its cost and complexity [3].

Getting Under Its Skin

When deciding to add specialized biometrics to your mix of enterprise security, selecting the type of technology to use is only half the battle. You must also consider identification versus verification, template storage, and network impact [3].

Cost, processing speed, and fewer false positives make verification systems the popular choice. Most specialized biometric products use verification: The user's specialized biometric template is retrieved from storage by a PIN, token, or smart card and quickly compared to a live sample. Identification solutions, on the other hand, compare the live sample to the entire database. If the comparison parameters are loosely defined, the search may match more than one live user to the same data. Identification works well with a small group of users, but when you get into the thousands, the time, processing power, and cost needed to scan the entire database can be excessive [3].

If you choose a specialized biometric verification solution, you'll next have to figure out how to store the templates, which contain data defining the users' characteristics. If templates or the database containing the templates are compromised, attackers could easily inject unauthorized templates to gain access to the protected network, system, or application. Templates are frequently stored on a centralized database, but they can also be kept on the specialized biometric reader or in a portable token [3].

Centralized databases create many security risks and can add a substantial amount of traffic to your network. If the database is compromised, your entire specialized biometric solution may also be compromised. Storing the template locally decreases processing time but may introduce difficulties if users move from machine to machine. Using smart cards may be the best way to go because they give the user portability and do not require centralized storage. But if the token goes missing the user must re-enroll, which can be a costly, time-consuming process [3].

If your company has decided to use specialized biometrics to defray costs, you may require a different specialized biometric technology than a company that primarily wants to boost security. Because the average corporation spends $150–$200 per user per year resetting passwords, according to industry analysts, the initial setup costs for biometrics can quickly be recovered by the annual savings of not resetting passwords. If this is your reasoning, your specialized biometric solution should be easy to use and not be very intrusive. But if you are implementing specialized biometrics for security reasons, your comparison parameters should be biased to deny access [3].

Specialized biometrics are still in the early adopter stage, but significant technological advances are beginning to make them a viable, if costly, solution in particular areas of the enterprise. Standards that allow interoperability among readers, increased accuracy and reliability, and lower costs will make specialized biometrics a more practical alternative—someday [3].

Finding Specialized Success

Travel and immigration, healthcare, and financial services sectors (industries that require übersecurity to protect sensitive data and confirm individuals' identities) will get the most benefit from specialized biometric technologies. The travel and immigration industries have started to apply biometrics in some of the most interesting ways. The INS Passenger Accelerated Service System (INSPass) allows travelers to bypass immigration lines by using biometric terminals. Airlines are also considering using biometrics to help identify passengers before boarding [3].

Working to become compliant with Health Insurance Portability and Accountability Act (HIPAA) regulations, healthcare companies are seeking more secure solutions. Although the technology is not specifically required by HIPAA, many organizations are looking at biometric techniques to secure access to confidential patient data. Some are also using biometrics to authenticate drug prescriptions. Doctors who enter prescriptions online use biometrics to identify themselves to the system and authorize the transaction [3].

Finally, financial services companies are turning to biometric technologies. Some are considering using biometrics at ATM machines, but others use biometrics in only the most sensitive situations, such as when allowing access to vaults or approving the execution of huge transactions [3].

References

1. Hunter Purnell and Dan Marks, "Enterprise Biometric Security," N-Cycles, Inc., 5350 Poplar Ave, Suite 800, Memphis, TN 38119, January 2003. Copyright © 2001, 2006 N-Cycles, Inc. All Rights Reserved.

2. Drew Robb, "Adding Biometrics to Enterprise Security Arsenal," Jupitermedia, 23 Old Kings Highway South, Darien, CT 06820, October 18, 2005. Copyright © 2006 Jupitermedia Corporation. All Rights Reserved.

3. Mandy Andress, "Biometrics at Work?" *Crypto-Gram Newsletter*, InfoWorld Media Group, 501 Second Street, San Francisco, CA 94107, May 25, 2001. Copyright © 2006, IDG Network. All Rights Reserved.

4. John R. Vacca, *Net Privacy: A Guide to Developing and Implementing an Ironclad eBusiness Privacy Plan*, McGraw-Hill (2001).

27

How to Implement Biometric Technology and Verification Systems

Clearly a need exists to accelerate the development and implementation of biometric technologies and verification systems to a point where they can be used by the masses reliably. Very few will argue that while biometric technologies and verification systems have been in development for years at the algorithm level, very little has been done in terms of actual implementable/deployable applications.

In reality, the biometric industry is still years away from the point where individual technologies can be used reliably, in any environment, on a standardized base structure. Up until 9/11, most of the 200-odd algorithm developers hadn't even produced a prototype system, let alone an implementable/deployable one.

In analyzing the benefits of a multiple biometric verification system, one needs to get back to the basics of biometric verification system performance. The base measurements for verification systems still apply on multiple biometric verification systems.

While both FAR and FRR are commonly used in the evaluation process of biometric technologies and verification systems, the real-world evaluation is somewhat different. Lab results and theoretical figures pertaining to the varied technologies are completely different than those figures that can be expected in the real world.

Biometric technology and verification systems never perform as well as the vendors claim, since the real world does not provide for perfect conditions. There are now two generally accepted principles surrounding the use of biometrics in the mainstream: There is no single biometric technology and verification system that works perfectly; and there is no single biometric technology and verification system that works, even imperfectly, in all environments.

So, whether the end application uses a single, dual, or multiple approach to biometric implementation/deployment, it is essential to choose the correct biometric or combination of biometrics for the application. With nine or 10 implementable/deployable disciplines and over 700 vendors of algorithms and/or devices, the end-user has a plethora of choices, and subsequent decisions to make.

Clearly, the solution to the dilemma posed to the end-user pertaining to biometric technology, verification systems, and vendor choice is the biometric platform. Unlike traditional biometric middleware, the biometric platform is able to combine multiple biometric technologies and verification systems together on a software, firmware, and hardware basis, as opposed to purely on a software level. The advantages of the biometric platform over a separate disparate approach is as follows:

- Multiple biometric disciplines can be supported;
- Biometric fusion is achievable (see sidebar, "Biometric Fusion");
- Devices and algorithms are easily interchanged;
- Full-system redundancy is achievable.

Biometric Fusion

All biometric systems have some weaknesses [1], so it is difficult to obtain a biometric system that accomplishes the four most desirable points for a biometric-based security system:

- **Universality:** All the persons should have the selected biometric identifier.
- **Distinctiveness:** A biometric characteristic too close to two persons to be confused should not exist.
- **Permanence:** The biometric identifier should remain the same for long periods of time, enabling user verification years after the registration of the user in the database.
- **Collectability:** The biometric should be measurable quantitatively [1].

There are several scenarios that show users having a difficult time. Table 27-1 summarizes some drawbacks of the well-known biometric systems [1]. This list skips those situations where the user is not collaborative enough or some unavoidable environment changes take place (different illumination, ambient noise, etc.). Obviously in these situations, data fusion can also facilitate the recognition process [1].

Another problem is a hacker trying to illegally access a biometric system that relies on a single biometric characteristic. A single biometric system can be fooled in several ways. The combination of different systems can improve the security level of only one system. For example, in a biometric system consisting of a fingerprint and voice analysis, it is more difficult to imitate the fingerprint and voice of a given user

Table 27-1   *Drawbacks of the Main Biometric System*

Biometric Technology	Weaknesses
Fingerprint	Certain users do not have suitable fingerprints (elderly people, some Asian populations, manual workers with acid, cement, etc.).
	Some fingerprint scanners cannot acquire fingerprints that are too oily, dry, wet, warm, etc.
	Temporary or permanent damage can make fingerprint recognition impossible.
Face	Changes in hairstyle, makeup, facial hair, etc.
	Addition or removal of glasses, hats, scarves, etc.
	Dramatic variations of weight, skin color change due to sun exposure, etc.
Iris	Eye trauma is rarely present, but still possible. Although this system is quite robust, it is not popular—nor are the sensors widely introduced.
Voice	Illness can modify the voice (cold, flu, aphonia, etc.).
	Acquisition devices and environments can vary significantly, for instance, in mobile phone access. This degrades the recognition rates.
Hand Geometry	Weight increase or decrease, injuries, swelling, water retention, etc., can make recognition impossible.
	Some users can be unable to locate the hand geometry due to paralysis, arthritis, etc.

than if just using one biometric characteristic. Or, if a person presents low-quality fingerprints, he or she can be recognized by means of his or her voice [1].

The key point to overcome these drawbacks, or at least to mitigate them, is to use a combination of different information. This is done by live beings in order to improve our knowledge of the surrounding world. Some examples are:

- The combination of information sensed by two ears lets us identify the arrival direction of the sound; two eyes let us identify the depth of a scene and obtain a three-dimensional image.

- Simultaneously touching and looking at an object yields more information than just using only one sense.

- In a democracy, the final decision of who the governor is consists of the combination of millions of people's decisions [1].

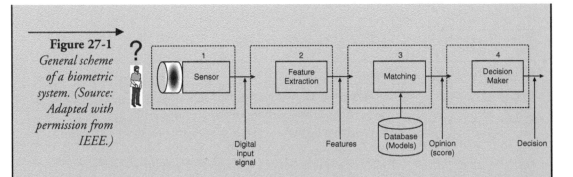

Figure 27-1
General scheme of a biometric system. (Source: Adapted with permission from IEEE.)

A similar strategy can be adopted to improve a biometric system. Figure 27-1 shows the scheme of a general biometric system [1]. Four main parts corresponding to different data fusion levels can be identified. In all cases, the system can be classified as either [1]:

- **Unimodal biometric system:** It relies on a single biometric characteristic.

- **Multimodal biometric system:** It uses multiple biometric characteristics, like voice plus fingerprint; or face plus iris [1].

Usually the unimodal systems are easier to install, the computational burden is typically smaller, they are easier to use, and they are cheaper because just one sensor (or several sensors of the same kind) are needed. On the other hand, a multimodal system can overcome the limitations of a single biometric characteristic [1].

Data Fusion Levels

Considering the main blocks plotted in Figure 27-1 [1], the following levels can be defined:

1.   Sensor level

2.   Feature level

3.   Opinion level

4.   Decision level

Sensor Level

In this level, the digital input signal is the result of sensing the same biometric characteristic with two or more sensors. Thus, it is related to unimodal biometrics. Figure 27-2 shows an example of sensor fusion that consists of sensing a speech signal simultaneously with two different microphones [1]. The combination of the input signals can provide noise cancellation, blind source separation [1], etc.

Another example is face recognition using multiple cameras that are used to acquire frontal and profile images in order to obtain a three-dimensional face model, which is used for feature extraction.

Although this fusion level is useful in several scenarios, it is not the most usual one.

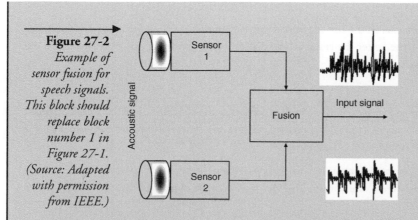

Figure 27-2
Example of sensor fusion for speech signals. This block should replace block number 1 in Figure 27-1. (Source: Adapted with permission from IEEE.)

Feature Level

This level can apply to the extraction of different features over a single biometric signal (unimodal system) and the combination of feature levels extracted from different biometric characteristics (multimodal system). An example of a unimodal system is the combination of instantaneous and transitional information for speaker recognition [1].

Figure 27-3 shows an example that consists of a combination of face and fingerprint at the feature level [1]. This combination strategy is usually done by a concatenation of the feature vectors extracted by each feature extractor. This yields an extended size vector set. One drawback of this fusion approach is that there is little control over the contribution of each vector component on the final result, and the augmented feature space can imply a more difficult classifier design, the need for more training and testing data, etc. Second, both feature extractors should provide identical vector rates. This is not a problem for the combination of speech and fingerprint, because one vector per acquisition is obtained. However, it can be a problem for combining voice with another biometric characteristic, due to the high

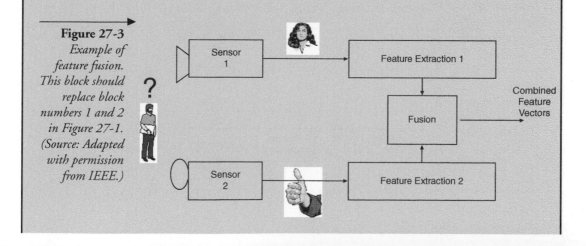

Figure 27-3
Example of feature fusion. This block should replace block numbers 1 and 2 in Figure 27-1. (Source: Adapted with permission from IEEE.)

number of vectors that depend on the test sentence length. Although it is a common belief that the earlier the combination is done, the better the result achieved, state-of-the-art data fusion relies mainly on the opinion and decision levels [1].

Opinion Level

This kind of fusion is also known as confidence level. It consists of the combination of the scores provided by each matcher. The matcher just provides a distance measure or a similarity measure between the input features and the models stored on the database [1].

It is possible to combine several classifiers working with the same biometric characteristic (unimodal systems) or to combine different ones. Figure 27-4 shows an example of multimodal combination of face and iris [1].

Before opinion fusion, normalization must be done. For instance, if the measures of the first classifier are similarity measures that lie on the [0, 1] range, and the measures of the second classifier are distance measures that range on [0, 100], two normalizations must be done. The similarity measures must be converted into distance measures (or vice versa); and the location and scale parameters of the similarity scores from the individual classifiers must be shifted to a common range [1].

After the normalization procedure, several combination schemes can be applied [1]. The combination strategies can be classified into three main groups:

- Fixed rules

- Trained rules

- Adaptive rules

Figure 27-4
Example of opinion fusion. This block should replace block numbers 1, 2, and 3 in Figure 27-1. (Source: Adapted with permission from IEEE.)

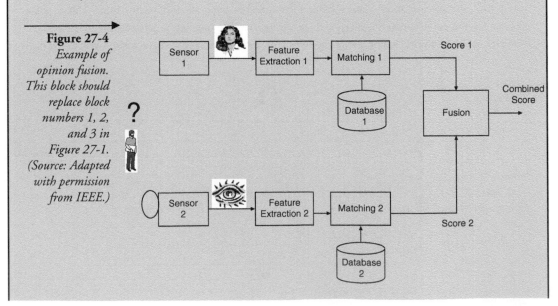

Fixed Rules

All the classifiers have the same relevance. An example is the sum of the outputs of the classifiers [1].

Trained Rules

Some classifiers should have more relevance on the final result. This is achieved by means of some weighting factors computed using a training sequence [1].

Adaptive Rules

The relevance of each classifier depends on the instant time. This is interesting for variable environments. For instance, a system that combines speech and face can detect those situations where the background noise increases and then reduce the speech classifier weight. Similarly, the face classifier weight is decreased when the illumination degrades or there is no evidence that a frontal face is present [1].

The most popular combination schemes are weighted sum, weighted product, and decision trees (based on if-then-else sentences). Figure 27-5 shows an example of data fusion using a decision tree [1].

Decision Level

At this level, each classifier provides a decision. On verification applications, it is an accepted/rejected decision. On identification systems, it is the identified person or a ranked list with the most probable person on its top. In this last case, the Borda count method [1] can be used for combining the classifiers' outputs. This approach overcomes the scores normalization that was mandatory for the opinion fusion level. Figure 27-6 shows an example of the Borda count [1]. The Borda count assigns a score that is equal to the number of classes ranked below the given class.

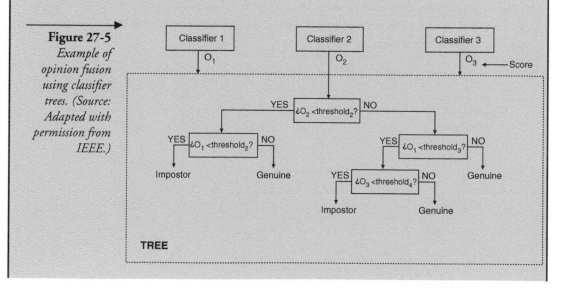

Figure 27-5
Example of opinion fusion using classifier trees. (Source: Adapted with permission from IEEE.)

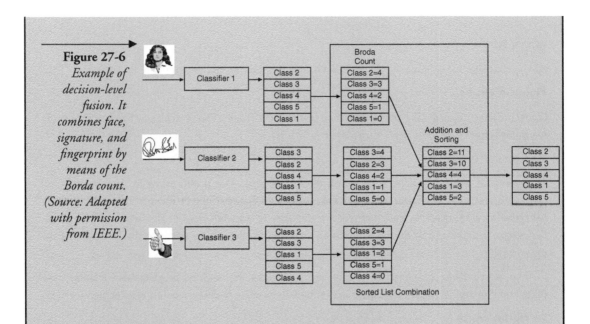

Figure 27-6 *Example of decision-level fusion. It combines face, signature, and fingerprint by means of the Borda count. (Source: Adapted with permission from IEEE.)*

One problem that appears with decision-level fusion is the possibility of ties. For verification applications, at least three classifiers are needed (at least two will agree and there is no tie), but for identification scenarios, the number of classifiers should be higher than the number of classes. This is not a realistic situation, so this combination level is usually applied to verification scenarios [1].

An important combination scheme at the decision level is the serial and parallel combination, also known as "AND" and "OR" combinations. Figure 27-7 shows the block diagram [1]. In the first case, a positive verification must be achieved in both systems, while access is achieved in the second one if the user is accepted by one of the systems [1].

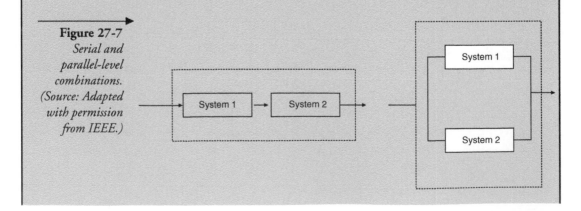

Figure 27-7 *Serial and parallel-level combinations. (Source: Adapted with permission from IEEE.)*

> The AND combination improves the false-acceptance ratio (FAR) while the OR combination improves the false-rejection ratio (FRR). Simultaneously combining serial and parallel systems, it is possible to improve both rates. In this case, if each system on its own yields a 1% FAR and 1% FRR, the combined system yields FAR = 0.0882% and FRR = 0.0002% [1].

The biometric platform also enables the interchangeability of biometric devices and components, much like a computer user is able to unplug a mouse and plug in a trackball device without fear of affecting the operation of the system. When the end-user would like to install a stand-alone device such as a hand-geometry reader within the platform, the same device that is sold by the vendor is simply connected to the platform through an interconnection module that converts the signals sent to and from the device and activates the device on the platform.

So, while biometric standards are being developed at a greater pace than ever before, implementation/deployments of solutions by 2010 will almost certainly lack the benefit of these standards. In any event, this chapter deals with the implementation of social, economic, legal, and technological aspects of biometric and verification systems.

Social Aspects of Biometric Technologies and Verification Systems

Out of the many different social issues to be discussed when reflecting upon the implementation of biometrics, the following main themes will be touched upon:

- Clarity of purpose in relation to technology implementations;
- Interoperability and equivalence of performance and process;
- Biometrics as an enabler for other aspirations;
- Human factors, social inclusion, and exclusion;
- Impact upon the trust model between citizen and state [2].

It will be argued that there are many factors outside of the technical design or provision of systems that must be considered if current aspirations are to be realized in an ethical, responsible, and sustainable manner. In the current rush to introduce biometrics and related technology to a number of processes in the public sector, there is a danger that such matters will not be fully understood or catered for. There is an additional danger that incorrect assumptions are made as to the real value of a biometric identity verification check and what this actually means. Therefore, Europe faces a challenge to understand better the longer-term importance of the implementation of biometrics in order to ensure its benevolent deployment. Such matters need to be taken fully into account [2].

Clarity of Purpose in Relation to Biometric Implementations

One of the concerns often expressed in relation to public sector implementations of strong identity verification technology is that of function creep (that technology and processes introduced for one purpose will quickly be extended to other purposes that were never discussed or agreed to at the time). For example, let's consider the new generation of travel documents, which will incorporate a chip and up to three biometrics. What precisely is the purpose of introducing these technologies to the travel document? If it is to verify that the individual presenting the document is the same individual to whom it was originally issued, then let's be clear about that purpose and develop the technology infrastructure accordingly. This would be a distinct purpose that may be easily articulated and that most likely would be accepted by the majority of law-abiding citizens. Similarly, if a biometrically equipped national identity card is primarily used for the purpose of verifying that the individual presenting it is indeed the authorized holder, then let's be clear about that purpose. Identity verification via the use of a token, be it a passport, national identity card, or a commercially issued token should be contained as a specific function [2].

In many instances, an important distinction needs to be made between identity verification and entitlement. The entitlement or benefit associated with the transaction in question should not be confused with the identity verification function [2].

Similarly, the identity verification function should not be extended into areas that are not directly concerned with or expressly necessary for the transaction. In the case of a travel document being presented at a border crossing point, for example, the identity verification function might be a self-contained transaction, the result of which enables a trained officer to reach a decision about entitlement. Many would be of the opinion that the same transaction should not be extrapolated into areas of general law enforcement or other public and private service areas that have nothing whatsoever to do with the distinct immigration process. This could give rise to general public confusion around such matters and will reflect poorly upon government departments seeking to introduce such technologies. Clarity of purpose should be a key factor in deliberations and, furthermore, clarity of purpose should be properly articulated and communicated in relation to every single program under consideration. Broad and emotive statements around "fighting terrorism" or "making the world a safer place" are not the most adequate labels with which to introduce these programs [2].

Interoperability and Equivalence of Performance and Process

This is an area that, even at this relatively late stage in related developments, is seldom understood. Many consider the use of the word "interoperability" to refer to purely technical matters. The greater interoperability however, lies in the interoperability of process and, where applicable, supporting legislation. This is especially relevant to international situations such as border control and the use of nationally issued documents in other countries. Let's consider, for example, a biometric identity verification check that returns a negative result. Is this result understood and interpreted in the same way throughout Europe? Or between Europe and the Americas? Or, in the Asia-Pacific region? If not, what are the consequences of such regional interpretation [2]?

Bear in mind that a failed verification transaction does not necessarily mean you are dealing with the wrong person—there are many types of potential errors and many reasons for them. A great deal of confusion could ensue in this respect when usage starts to scale upward. From a travel perspective, it raises interesting questions with respect to multisegment journeys that cross several geographic boundaries, and where the same individuals might be treated quite differently at different points along the way, irrespective of their legitimate entitlement to cross the borders in question. From a social services and entitlement perspective, it also raises interesting questions, both within a single member state and between member states. However, even this scenario assumes a common level of performance (of the biometric identity verification transaction), which will certainly not be the case in practice [2].

Equivalence of performance across multiple nodes is a factor that has not been properly understood, nor addressed. How is the biometric technology at individual points of presence calibrated? To what specification? Who has control over this? How is realized performance measured? How is this coordinated between nodes? You must also take into consideration nontechnical factors such as the physical and technical environment, user psychology, and human factors such as age, ethnicity, gender, disabilities, and so forth, all of which will be proportionally different at different points of presence. This will lead to possibly significant differences in realized performance across nodes [2].

This in turn will lead to differences in the user experience and therefore user perception. Habituated users of related systems within the public sector will quickly notice differences in both realized performance and local administration response between points of presence. If the broader situation appears uncoordinated, with little equivalence of process in the way the individual is treated by the local administration, this will itself have a societal impact as citizens begin to question the effectiveness of such systems. There are ways of assuring equivalence of realized performance across nodes that take environmental and human factors into consideration. However, equivalence of process and response are matters that must be addressed by the agencies concerned [2].

Biometrics as an Enabler for Other Aspirations

Some initiatives that publicly focus upon biometrics and tokens (such as identity cards) seem to be less focused on identity verification in relation to specific transactions than on collecting citizen information for inclusion in various databases. This is currently an area of concern to many, especially where there are aspirations to share this data not only between government agencies but also between countries. Furthermore, the distinction between official and commercial databases and data management is by no means clear, with many suggestions of private sector involvement [2].

When data is shared between databases and between countries (whether specifically "pushed" or simply made available via the granting of third-party access), this calls into question many aspects of data protection and privacy [5]. In such cases, individuals have no control over their personal data, for what purpose it is being used, or who has access to it. The provisions of national data protection acts become meaningless when data crosses national borders. Furthermore, the ability of the individual to challenge incorrect assumptions with respect to their own data is highly questionable—assuming that they even have knowledge of such a situation [2].

There may be legitimate reasons for establishing databases of citizen information, but these should be clearly articulated, as should the detail of how such databases will be used and for what specific purpose. You should not confuse this broader data issue with the provision of biometric technology. Furthermore, aspirations to include biometric data in such databases should be considered very carefully, especially with regard to the specific purpose and use of this data. In some instances, this may be very clear. For example, if biometric data were included in a passport agency database in order to guard against multiple applications, then the majority of citizens would understand and support such usage, provided they were confident that this same data were not automatically shared with other agencies without their knowledge. If the precise purpose of holding such data is not clear, or considered ethical and responsible, then this may create a negative impression among citizens. Similarly, the blurring of government agency functionality, for example, between immigration and law enforcement, may well be considered negatively by citizens. It is therefore important to be very clear about the purpose of introducing a biometric and exactly how this relates to existing and proposed databases, including any proposed sharing of data [2].

Human Factors, Social Inclusion, and Exclusion

The importance of human factors such as age, ethnicity, gender, and disabilities also raises the possibility of inclusion or exclusion from widespread applications and, crucially, assumptions and processes that might ensue as a result. There are many reasons why, for a given individual, it may be extremely difficult to consistently give a live biometric sample or to otherwise participate in an automated biometric identity verification process. Resulting errors from such difficulties will not necessarily mean that you are dealing with the wrong person or that any attempt at fraud is being pursued. An individual who managed to enroll into a given system may repeatedly fail biometric identity verification checks, or simply fail to interface with the technology involved (such as a kiosk or automated barrier) for a variety of reasons [2].

Some of these reasons may be immediately obvious, such as physical disabilities and, if exception handling processes have been properly conceived, these might be dealt with appropriately. Other disabilities may be less obvious, such as memory retention or learning difficulties, degrees of autism, personality disorders, and other psychological effects. There are also physiological issues such as degenerative illnesses, which may gradually reduce an individual's ability to consistently interface with the technology and associated process. The proportion of individuals so affected will no doubt vary according to region and the nature of the system under consideration but in some cases may be materially significant, perhaps leading to incorrect assumptions [2].

In addition, you shall most likely discover a number of individuals whose biometric trait is sufficiently indistinct, or otherwise unusual, to cause problems in enrollment and, or subsequent identity verification. Fingerprints might be weak or the skin texture not ideally suited to the sensors being used. Facial features may be obscured or skin tone may cause problems with specific cameras and local lighting or other environmental conditions. Individual eyes may prove difficult to enroll into iris recognition systems. Medical conditions such as arthritis may make it difficult for individuals to use hand geometry devices. Also, there may be behavioral issues that make it difficult for individuals to consistently provide a biometric. Many such conditions may be discovered at the time of enrollment if your registration processes are properly considered and implemented [2].

Moreover, you shall have to consider exception handling processes for individuals who have difficulty with automated processes. The proportionality of this factor will become increasingly important as systems scale upward and large numbers of individuals are enrolled into various systems and schemes. If the failure of an automated biometric identity verification check results in denial of service, a proportion of individuals are likely to find themselves disenfranchised in this context. The impact of this from a societal perspective will depend upon how well such factors have been considered in advance, together with the nature and practical delivery of associated exception handling processes [2].

Impact Upon the Trust Model Between Citizen and State

This is a very important point, especially when viewed in the context of modern history (the last 100 years). In many countries that would consider themselves civilized and perhaps of a democratic nature, the trust between citizen and state plays a key role. Citizens offer their trust to government and, in doing so, empower government to manage national and international affairs on their behalf. If this trust breaks down, a breeding ground is created where a variety of situations might develop, from underground economies to outright challenges to government and civil unrest. In many countries, part of this trust is inherent in the concept of being considered innocent until proved guilty and in enjoying personal privacy and anonymity. These fundamental concepts of trust seem now to be challenged by certain governmental aspirations. There is a risk that the emphasis changes to ordinary citizens being almost treated as criminal suspects and the right to privacy and anonymity are being withdrawn [2].

The issue is exacerbated when the administrations of foreign countries have an undue influence on a given country's procedures. It may be true that, in the short term, citizens simply go with the flow and accept what many of them will see as the sacrifice of personal freedoms in order to support policies that, they have been led to believe, will create a more secure world. However, in the medium and longer terms, the reality of the situation (such as it may be) may become self-evident and, depending upon popular perception, this may lead to an erosion of trust that will not be in the interest of government. This is a very serious issue that should be taken fully into consideration with respect to current aspirations. You should have no doubt that you are tampering with the very fabric of society, and you should treat this fabric with the care and respect it deserves [2].

Economic Aspects of Biometric Technologies

Economic transactions require trust. The secure provision of identity can help build the needed trust by clarifying the assignment of legal liability and any necessary recourse to the courts. In addition, identifying oneself can signal goodwill. Moreover, personalized data tied to identities provides convenient summaries that may help firms to tailor their goods and services or to offer customers the most appropriate choices, improving the efficiency of the market. More generally, identity indexes transaction history or other data [2].

Identity also serves as a capital asset (credit ratings), formed through investment and subject to depreciation. Ownership of identity capital may be split or diffused (credit rating agencies with different accounts and amounts of information). This increases the need to attach the data to the person seeking credit [2].

These functions of identity have been known in economics for a long time, but identification was not really an economic issue—face-to-face or closed-system transactions lacked significant misidentification risk, and identity fixation in remote transactions or open systems tended to be a legal matter. The value of identity was also approached obliquely, primarily via analysis of reputations. Recent changes in technology and practice call for fresh economic perspectives. Increasingly, "virtual" transactions (where parties may never be able directly to verify each other's identity) have increased the value of identity and made identity theft [9] a more pressing concern. Technical "solutions" offer identification of differing strengths; their interoperability affects the compartmentalization of economic identity and its externalities [2].

Economic Aspects

In this part of the chapter, the impact of biometrics on economic outcomes will be discussed. This includes optimal and actual identity, the emergence of standards, and costs and benefits. This part of the chapter also surveys the present state and likely evolution of market demand and supply. The issues that policy makers need to address as well as the means to address these issues are also explored [2].

Optimal Identity

In cash transactions, parties need not be identified; it is only necessary to verify the right to exchange goods and services for money. However, uncertain or contingent transactions may need more. Buyers may need to prove creditworthiness or certify how purchases will be used; sellers may need to establish provenance or certify quality, origin, and so on via retrospective (professional qualification) or prospective (seller warranty) identity. Sometimes it suffices to prove membership of a specified class (adults, physicians); other cases require identification of specific individuals or their legal representatives [2].

Even if biometrics provides more certain identification, it is not necessarily cost-effective or "optimal," because its additional costs may exceed the benefits of increased certainty of identification. The quality of a particular implementation may be too high for at least one party. Some (regardless of monetary cost) may be too strong for the purpose for which they are employed due to privacy concerns or legal restraints on information collection. Permissible accuracy may be limited—for example, it is essential to establish that voters are eligible and have not already voted, but equally essential not to identify them further. Unless the means and degree of biometric identification are included in negotiations, there is no reason to expect the level of identification to be optimal; there may be too much or too little identification or use of secure channels [2].

Generalized use of one or several large and widely used "strong identification" systems provides an enormous installed base to cover (security and RTD costs and scope for data mining to detect fraud), thus lowering costs and increasing security. It also limits identity compartmentalization to control risks. However, even apart from increased data protection concerns, its very strength makes errors harder to correct. Hardening outer boundaries may reduce overall security if internal precautions are relaxed. Identity theft may be less frequent, but more severe; and identity theft may give way to outright denial-of-identity service attacks [2].

Furthermore, to the extent that biometrics provide cheaper, stronger, and/or faster identification, they tilt the playing field against those who cannot or will not participate. If the vast majority migrate to a biometric solution, alternative channels may disappear, excluding or imposing costs on the minority. Those with privacy concerns may be unable freely to opt out without losing access to goods, services, or societal interactions to which they are entitled, harming those on the inside as well. Due to network effects, any system whose benefits depend on user interactions will be damaged by changes that raise barriers among users [2].

The Emergence of Standards

Biometric implementations have technical and dynamic efficiency effects common to network technologies. Identity is complementary to economic transactions, so equilibria may be unstable or nonexistent. Economies of scale and interoperability favor winner-takes-all ("tipping") equilibria. This works by three channels:

- Market adoption depends on expectations, a technology expected to become a standard is likely to do so.

- Competitive forces are likely to produce a single (or unified) standard approach, especially with greater interconnection among sectors and participants, so early leads are difficult to overcome.

- Sunk costs of adopting standards can strand those making the wrong choice with obsolete investments and reduced benefits. This risk makes firms wait to adopt, particularly where value depends on availability of interoperating and complementary database, communication, sensing, and payment systems. This in turn inhibits investment in developing such complements, and partially accounts for private sector reluctance to adopt biometrics despite falling direct costs [2].

This tendency to tipping is reinforced by pressures for compromise solutions. If interoperability were irrelevant, it would be possible to match each application to that biometric offering the best combination of costs, accuracy, and so forth. But even closed identity management systems need to interoperate, and multiple identity systems impose substantial burdens. Even when optimal biometric solutions differ by application area, there are strong pressures to adopt imperfect compromise solutions [2].

Another mechanism that might damage competition could be the strategic use of intellectual property rights (IPR). A firm holding key patents need fear no competition; if it chooses to allow competitors to license its technology, it can do better by encouraging entry of efficient rivals and extracting further rents from their innovations. Ultimately, such strategies are self-defeating; they encourage bypass competition and antitrust action, keep prices high and limit market growth, and prevent the medicine of competition from driving costs further down. But, as recent iris scan algorithm patenting disputes show, such self-defeating tactics persist. Further ramifications include patent thickets and clusters to deter innovative rivals [2].

There are two alternatives to the emergence of de facto (proprietary) standards as a result of tipping, IPR, or accident: voluntary industry agreements (typically open), and mandated national or international standards. Open standards are more likely to solve the coordination problem and enhance competition by lowering entry barriers and stimulating innovation of complementary products. However, they may take longer to achieve and can mask collusion. Mandated standards can be established quickly—perhaps too quickly if they are based on uncertain assessments (ISDN) or forestall price and quality competition. Regulators may be captured by better-informed industry players by amplifying the anticompetitive effect of proprietary standards [2].

Costs and Benefits

Decisions about biometrics rest on estimates of costs and benefits, relative to alternative means of identification, that offer both advantages (ease of issue or revocation, no problem of template aging, low entry barriers) and disadvantages (vulnerability, hidden cost of lost or multiple passwords). Early adopters have high direct costs, but enjoy increased chances of winning the standardization race, incentives for further development and IPR, and learning curve reduction of future costs, including indirect costs. On the benefit side, available data tend to fall into three categories:

1. Costs of problems biometrics should solve: Annual UK costs for identity theft are estimated by industry analysts at $10 billion (10% of all fraud, and growing). In the United States, where it quadruples annually, identity theft affected 51 million citizens and cost $555 billion in 2006. However, the degree to which biometrics reduces theft and the possible displacement of fraud remain uncertain.

2. Cost savings from immediate deployment: Such data are often proprietary or commercial. They should be presented as lifetime cost of ownership and adjusted for changes in financial, physical, IT, and human capital and impacts on internal processes.

3. Estimates of willingness-to-pay: These estimates provide a lower bound on consumer surplus from biometrics. Better functionality is accompanied by falling costs: The two effects offset in terms

of price, but should be added to estimate welfare gains. Biometrics also let risk-averse consumers save on costly hedging or insurance or make use of more secure or competitive channels [2].

The Biometric Market: Demand

In the recent past, three applications have constituted the bulk of the biometrics demand. First, physical access control has been the dominant application since the advent of biometrics, but is rapidly being supplanted by IT applications. According to industry analysts, it had 97% of the biometric market in 2006, and had been dwindling in 2000 but has revived strongly since 9/11 (see sidebar, "Was 9/11 an Inside Job?"). The dominant trend is expansion to monitor time, attendance, or physical location. IT applications had the second largest share of the market, growing with biometrics' inclusion in laptops, the development of specific interface standards, and biometric implementations in converged computing/communications equipment. The third largest area for biometrics was financial services, which is likely to grow due to changes in fraud types, financial identity management, and banking itself [2].

Was 9/11 an Inside Job?

Sibel Edmonds, a 32-year-old Turkish-American, was hired as a translator by the FBI shortly after the terrorist attacks of September 11, 2001, because of her knowledge of Middle Eastern languages. She was fired less than a year later in March 2002 for reporting shoddy work and security breaches to her supervisors that could have prevented those attacks.

Edmonds, the foremost 9/11 whistleblower, recently indicated from the evidence she has seen that the 9/11 attack was possibly an inside job. Edmonds agrees the weight of evidence leans toward criminal complicity.

Edmonds indicated that the preponderance of evidence, plus the outright cover-up surrounding 9/11, suggests that criminal elements at the very apex of the U.S. military-industrial complex had a direct hand in carrying out the attack. A recent survey by industry analysts clearly shows that 90% of the U.S. Islamic community agrees with her, as well as 80% of Europeans. There is a growing feeling across the United States that something was strangely afoot on 9/11, other than the attacks. The attacks were too easily carried out: the first clue that it could have been an inside job.

To make matters worse, the Bush administration has gone to great lengths to silence a $8,000-a-year University of Wisconsin-Madison teacher, Kevin Barrett, for his Cheney-9/11 theory. The teacher is now under fire for thinking the most likely theory about the 9/11 plot is that it was an "inside job" organized and commanded by Vice President Dick Cheney.

Wisconsin state legislators demanded that the college bar him from teaching an Islamic Studies course. More than 60 state lawmakers are urging the University of Wisconsin-Madison to fire Barrett, who has argued that the U.S. government orchestrated the September 11 terrorist attacks.

A letter sent and signed by 52 assembly representatives and nine state senators condemns a decision to let Kevin Barrett teach an introductory class on Islam in the fall of 2006. U.W.-Madison Provost Pat Farrell launched a review after Barrett spoke in June 2006 on a talk show about his views that the terrorist attacks were the result of a government conspiracy to spark war in the Middle East with the invasion of Iraq, and to steal its oil. The nuclear saber rattling of Iran, the invasion of Lebanon by Israel, and the possibility of an expanded war in the Middle East has resulted in higher oil prices (higher gas prices in the United States). Some conspiracy theorists believe that the plan (part of a much larger globalization plan) was nothing more than an oil grab, in order to sell it on the black market and to China. Not one drop of this oil has ever reached the United States, or paid for the war in Iraq, as U.S. citizens were promised by the administration.

Barrett, 47, also described how some news organizations (the French daily newspaper *Figaro* and Radio France International, in fact) had reported that an agent from the Central Intelligence Agency visited with Osama bin Laden two months before the attacks. What could the CIA and bin Laden have been talking about? Was bin Laden on the CIA's payroll? Hmmm, I wonder!

He also indicated that fires could not have caused the collapse of the World Trade Center towers at free-fall speed, as reported by the special Sept. 11 commission. He believes that the 9/11 report will be universally reviled as a sham and a cover-up very soon.

Note: *The views presented in this sidebar are those of this author, and do not represent the views of the publisher. They are presented here for the reader to make up his or her own mind about what the real truth is. The truth is out there!*

However, the demand for biometrics is rapidly shifting, due to new implementations. Government and other public sector applications will be leading the sector in volume, new technology adoption, project scale, and prominence. After September 11, 2001, transport and immigration (biometric passports) have become key issues, with an emphasis on international interoperability. The public sector is also a leading client in healthcare, where biometrics is increasingly used to prove entitlement and link patients to electronic health records. Other sectors likely to emerge as significant parts of the market are retail and other payments (already being trailed in a wide range of applications), telecommunications services (integrated with other services and linked to individual data), and transport (including private transport) [2].

Supply

The biometric sector follows the experience curve: a few leading firms, many subsequent entrants, and consolidation to a few survivors. The shake-up is well under way; despite strong demand growth, mergers and bankruptcies dominate recent market reports. The cycle is more advanced in fingerprint technology, while newer technologies (iris) still have many small firms pursuing diverse approaches (albeit with tight control of key patents). Concentration is high even during expansion, leading to persistence of dominant firms with specific national and/or sectoral attachments and possible distortion of biometric development [2].

The tendency to concentration is reinforced by specific factors. First, as eventual uses of the technologies are unclear, fixed testing costs are fairly high, which raises entry barriers. Second, early public or private customers seek assurance, which favors incumbents and firms with a large installed base. The key role currently played by very large public procurements can generate an enormous installed user base, which encourages subsequent clients and suppliers of complements to standardize on the incumbent firm and/or approach. Third, the threat to competition is enhanced by the layered structure—hardware, middleware, application, all of which must work with each other. Market power in one layer can extend to others [2].

State of the Market

The industry began and is thriving in the United States, but Europe's share is growing rapidly, particularly in banking. Recent European government initiatives will boost demand even more. Available data suggest consistent dominance by fingerprint, with hand geometry and voice recognition dwindling and iris growing [2].

Supporting these data are overall growth and the growing non-U.S. market (where hand recognition is rarely used). Strong revenue growth in fingerprint is likely to continue as cheaper scanners are bundled with computers, but other biometrics such as facial recognition and iris are also showing strong growth (see sidebar, "Implementing Biometric Systems") [2].

Implementing Biometric Systems

Biometrics is slowly making its way into the universe of verification products for the enterprise. However, based on some new products, it's not just about fingerprint scanners and face recognition systems anymore. Biometrics is evolving into a range of science fiction-like systems that measure esoteric physical characteristics, like typing speed and electrophysiological signals, to name two of many [3].

While biometric products are better and more finely tuned than they used to be, and the classic problems they used to have of false readings and high error rates are diminishing, it still requires careful consideration and planning to implement. It's not magical protection for your network. Like any other verification tool, there are best practices and pitfalls to watch out for [3].

Biometric systems can be costly and are more complicated to implement than other effective traditional two-factor verification systems (tokens, smart cards, and one-time passwords). Also, the market is splintered. There are fingerprint readers, iris scanners, and face recognition systems among the hundreds of biometric products available, and each is different and requires different implementation. So it's not easy to compare them, which leaves IT purchasing managers without a single focal point when evaluating the different products coming across their desks. This doesn't mean biometrics should be ruled out, but it requires more careful planning up front before deployment than other traditional verification systems with longer track records [3].

Most of the new products include the conventional, like fingerprint readers, and the off-beat, like the device that builds a physiological profile of the user and another that captures the user's typing speed. The following is a sampling of some of these offerings:

Aladdin, better known for its AV software, has come out with a prototype of the BioDynamic Reader. This consists of a mouse with two tiny pads (one for each of two fingers) that the user touches to register and gain access. The device builds a profile based on electrophysiological signals captured from the user. The BioDynamic Reader is scheduled for release sometime in 2008 [3].

Another unusual product, the BioPassword, measures the user's keystroke speed and typing style. A new user has to type in their password about a dozen times to build a keystroke profile. After that, the user just types in their user ID and password and the system "knows" who is typing by their keystroke style. If it's someone other than the registered user, access is denied. The BioPassword can be fine-tuned and adjusted by a system administrator, as needed [3].

Traditional fingerprint scanners, some on USB thumb drives, others embedded into laptops, are more the norm among other biometric products. Two examples were ClipDrive Bio from Memory Experts International and the BioPass 3000 from Feitan. Another fingerprint scanner company, BIO-key International, developed a neat software interface that builds the scanned fingerprint on the screen of the user's laptop as they are logging in. The software requires a fingerprint reader, either a USB token or a built-in reader on the laptop. Here are some best practices and things to consider for implementing biometric systems:

- Do a thorough risk analysis of your systems. In some cases, biometrics may be overkill; in others, it may be just what you need to access systems with sensitive customer data or that process high-risk transactions. Only consider using biometrics if the level of risk warrants it.

- Consider customer acceptance when used for logging on to company Web sites. Most home users aren't quite ready to install biometrics on their home computers to do their online banking.

- Be mindful of where the digital data or templates generated by biometric devices will be stored. All raw biometric data from any reader (whether a face recognition system or a keystroke profiler) is analog and must be digitized for consumption by a computer. This data needs to be protected on a dedicated and secure server to prevent it from being stolen and replayed against the system for malicious access.

- Ensure secure transmission of biometric data from the reader, such as a USB token. Encrypt all data to prevent its theft in transit between the reader and the data store [7].

- Just like any other verification data, biometric data needs a home. Therefore, ensure interoperability with existing databases for storing verification data, such as Active Directory or LDAP [3].

The market for biometric products is still growing and, as with any product in its infancy, hasn't succumbed yet to consolidation. With that in mind, keep shopping and don't settle for the latest cool product or fad. Think long term and about what verification systems fit best into your particular network before opening the corporate checkbook [3].

Over time, hardware will become cheaper, interoperable, and commoditized. Algorithms will remain proprietary and distinctive and continue to improve, so IPR will remain profitable. Middleware, which mediates functionality and interoperability, is likely to be convergent, less profitable, and ultimately provided by open-source and/or compatible free software [2].

Application service providers will dominate the growth phase—initially providing solutions but ultimately supporting users and intermediary layers, possibly before acquisition by integrators. Value-added resellers and original equipment manufacturers provide important transitional competition, but the market is likely ultimately to belong to specialized security or diversified ICT integrators. Relationships are likely to be strategic and/or collusive partnerships. Ultimately, biometrics may be wholly subsumed by technology (PCs), integrated ICT, and/or security markets [2].

Policy

Six major issues that might require action by policy makers emerge from the preceding analysis. In a second step, this part of the chapter presents the levers that policy makers have at their disposal to address these issues [2].

Issues

The first is possible market failure (competition may be undermined by tipping or capture) of a single market layer or a set of connected segments. This applies to biometrics per se and to broader IT, transportation, health informatics, market segments, in many of which strong network, interoperability, and complementarity effects can lead to some dominance. The consequences are those usual to competitive failure: allocational inefficiency, retarded or distorted RTD, and associated spillover effects on employment, competitiveness, and so on [2].

A second, somewhat narrower, concern is the development and competitiveness of biometrics and the identity industry. Biometrics shares many characteristics with other high-tech industries (risk, possible slow take-up, limited capital access, threatened obsolescence, high-tech skill dependence, critical importance to other rapidly growing sectors), but stands out because of its importance to security, e-government and other public objectives [2].

The third concern is the tension between standards lock-in and diversity. Market competition on its own may fail to produce timely and appropriate levels of standardization or may get stuck in an inferior standard. Fourth, intellectual property rights (IPR) are obviously important to the competitive health of the market, but pose particular problems relating to interoperability and network effects. Compatibility requirements may reward IPR holders with market power even without beneficial innovation—especially when customers value stability, assurance, and compatibility above other characteristics. The first product to be adopted may well become the de facto industry standard. On the other hand, IPR may encourage beneficial bypass innovation [2].

A fifth point is that biometrics is a key element of government security policy. Yet governments have poor records in managing large IT procurements, and political sensitivities combined with rapid technology development and the importance of international interoperability make value for money even harder to ensure. For instance, it is not obvious who (if anyone) owns liability for flaws in a technology or its implementation. On the basis of empirical evidence, open-source systems seem to be at least as secure as proprietary systems and sometimes much more secure [2].

Finally, the use of one's identity itself is changing from a private good belonging to the individual and useful in a limited range of close interactions, to a form of social capital used in a vast range of poorly observed and uncontrolled interactions and based on data scattered throughout many networks. Difficulties in preventing access to one's identity and its possible abuse in ways that are not immediately obvious makes identity a public good—not least because protection of individual rights and freedoms may require public provision of strong identity [2].

Policy Levers

The preceding issues can be addressed by several policy levers. The first is procurement policy (see Table 27-2) [2]. Large government contracts are often the first major demand component, underwriting private financing and creating industry leaders in a short space of time. Thus, they drive new technologies. The advent of mass-market biometrics coincides with security, e-government, and e-participation initiatives. However, the public sector's "launching customer" role is extremely difficult; it requires appropriate specification, smart contracting, and active partnerships with suppliers in the face of untested technology. Because biometrics is intimately connected with sensitive policy areas, it may challenge the two pillars of European public procurement: equal treatment and transparency. Tools include pre-competitive engagement multiple-sourcing, design competitions, IPR options in contracts, open standards requirements, and insistence on open and transparent supply chain management. Interoperability generally makes it impossible to divide procurement among many firms in advance of open standards, but procurement can be structured to leave even losers with valuable IPR and to provide opportunities for integrators, licensees, and so on to participate in future development [2].

Table 27-2   *Summary of the Interaction Between Issues and Levers*

		Policy Domains			
		Procurement	Standards	Competition	IPR
Issues	Market failure, sector health	X		X	X
	Standards	X	X		X
	IPR	X	X	X	X
	Security	X			
	Public identity	X			

A second policy lever is the standardization policy. There is a potential role for mandated open standards with protection for equivalent alternatives or for incorporation of open standards requirements in procurement, licensing, and other policy decisions [2].

As a third lever, the competition policy must take into account both of the tipping tendencies and the need for innovation. In general, incompatibility makes product innovation "too fast." Another danger is foreclosure (when an integrated provider deliberately makes its equipment incompatible with rival offerings or when the holder of a key patent effectively controls all those who use it). The competition policy can act via a merger and access pricing regulation. The treatment of industry standards consortia is also important; they might manipulate standards, exchange cost information, or refuse to license to outsiders [2].

The fourth policy domain is intellectual property rights (IPR). There is an obvious scope to use mutual recognition and compulsory licensing to control adverse effects or private IPR. A more radical alternative would be a public goods route (general public license) supporting an open source RTD policy, where access to research results is open, usage rights are granted freely, and even derivative innovations may be bound to the public domain. Economic returns may be sought in selling related goods and services or in selling enhanced versions [2].

Legal Aspects of Biometric Technologies

With computer systems recognizing fingerprints or voice, you have gained a powerful tool to verify the identity of an individual and thus ensure essential levels of security. The technique to use human characteristics in identification processes is often referred to as biometric recognition. Biometric technology is no longer an embryonic development, but has become the core of national and international security and immigration policies and is gaining importance as a product for the private sphere [2].

Is Europe Ready for Biometrics?

With the exception of DNA analysis, blood and breath sampling regulated in traffic bills, and (to a lesser degree) fingerprint sampling, there is relatively little legislation in Europe with regard to biometrics. Biometrics' use in private transactions is based on consent. Governmental use of biometrics is only starting; and when biometric enrollment becomes obligatory, for instance in the context of identification schemes such as electronic passports and identity cards, new legislation will be needed [2].

Analysis of the current human rights framework and the data protection framework shows a flexible legal environment that allows for much discretion for public and private actors implementing biometric schemes. Biometric deployment does not threaten procedural rights, such as the presumption of innocence, stated in Article 6 subsection 2 of the European Convention on Human Rights. Also, the sampling of biometrical data respects the right not to incriminate oneself as defined in the European case law. According to the European Court of Human Rights, the right not to incriminate oneself (which is regarded as an aspect of the general right to a fair trial enshrined in Article 6 subsection 1) means that a suspect cannot be forced to supply evidence for his or her conviction. Consequently, the prosecuting authority has to collect evidence without exploiting evidence obtained by force or pressure. Taking bodily samples, even against the will of a suspect, is not considered a limitation of this right [2].

Also important is privacy, a fundamental right included in Article 8 of the European Convention for the Protection of Human Rights and Fundamental Freedoms. Interference by the executive power on the rights and freedoms of the individual should not be permitted unless there is a clear legal basis to do so. The requisite in Article 8.2 of the Convention that a law restricting privacy must be necessary in a democratic society brings us to the difficult relationship between individual rights and collective interests. Because, with most biometric technologies, no penetration of the body's surface is required, it is assumed that the use of these technologies will not be deemed unreasonably intrusive when properly motivated (and based on a legal regulation) or based on consent. Therefore, every application (such as the choices of the EU legislator for two biometrics in the passport and visa system) must provide a satisfactory balance on four criteria: reliability, proportionality, the presence of a fallback option, and prior knowledge or consent. Even if arguments against current EU legislation can be found, when these four criteria are met, decisions will submit to the European Convention on Human Rights [2].

The text of the Constitution of the European Union and, previously, the European Union's Chapter of Fundamental Rights include next to privacy protection the rights to data protection and human dignity, which are not covered in the European Convention. Although the data protection framework has some important consequences for the way biometrics are implemented, fundamental choices such as the choice for centralized biometrical databases are seemingly left untouched by it. Data protection lacks normative content. It is in the first place designed to channel the application of new technologies. However, certain technical problems with the data protection framework are identified, such as the question of whether templates are considered to be personal data, whether biometrical data is sensitive data, and in general, problems with the application of Article 15 of the Directive 95/46 on Privacy Protection already in force [2].

Fundamental Concerns about Human Rights and Power Remain

The deployment of biometrics by public and private factors raises numerous concerns that are not adequately addressed by the current human rights framework and the data protection framework. For instance, this includes concerns of power accumulation; concerns about further use of existing data; concerns on specific threats proper to biometrics; concerns related to the use of the technology in the private sector; concerns about the failure to protect individuals from their inclination to trade their own privacy; and concerns for costs [2].

These concerns are genuine. Policy makers and civil society demand decisions that are well-informed and based on careful consideration of reality. However, there are no empirical data about the current performance of the existing systems, as there are no precise data about why new systems and facilities are needed [2].

The concerns are genuine because European policy makers and civil society know that the longer a technology is used, the more entrenched in life it becomes. They feel that the current (legal) system gives too much leeway to new technological developments that are conceived without proper regard to a human rights perspective. They also feel the U.S. pressure. And they also know about America's mass installation of surveillance technologies (metal detectors, scanners, CCTVs, iris recognition systems, alarms, locks, intercoms, and other forms of surveillance, detection, access control, and biometric equipment) in schools, government premises, stores, offices, workplaces, recreation areas, streets, homes, and other public places. All of this is known without understanding all the purposes behind this security build-up. Common sense has pushed people to adopt a critical attitude (that regrettably is hardly echoed in the current legal framework), refusing to accept simple answers about safety and protection when there is little evidence that security technology actually makes one safer. They have heard about the paradox of technology. They realize that law enforcement often uses new technological security tools on poor and nonwhite people, and fear social outrage about discriminating practices [2].

When allowing biometric images to be processed, one gives up complete control over information that maps distinctively onto one's physical person. Should someone's biometric data become available on public networks (unauthorized release) or be distributed or exchanged commercially (misuse), further risks emerge, to the point where it is difficult to imagine any proportionate gains in security or comfort [2].

This ethical assessment leaves no room for the view on what data protection will do for biometrics. Next to privacy and data protection, the right to have human dignity protected should be taken into consideration. Applying data protection principles implies the presumption that biometrics can be processed or that biometric data can be made available to others (even commercially). Some U.S. firms present their customers the option of making a commodity of their fingerprints in exchange for faster acquisition of cheeseburgers. The choice is portrayed as a casual decision with little or no moral impact, and customers are not encouraged to consciously consider its repercussions. It is easy to imagine people providing biometric samples under time pressure, without precaution. The example of the European dancing club that uses biometrics for access control demonstrates that monetary or other rewards can have a similar effect in making biometric enrollment look trivial [2].

The answers to such concerns must be formulated with reference to the basic features of the democratic constitutional state. From this perspective, opacity/privacy (prohibiting) rules should guarantee those aspects of an individual's life that embody the conditions for his or her autonomy (or self-determination, or freedom, or "personal sovereignty"). Privacy and human dignity must preserve the roots of the individual's autonomy against outside steering or against disproportionate power balances in vertical and horizontal power relations. This is because such interference and unbalanced power relations are not only threatening individual freedom, they are also threatening the very nature of society [2].

The fundamental task should be first to consider whether biometrics should be allowed and when. Developing concepts such as biometrical anonymity or a right to property on biometrical data might be instrumental to achieve this objective. Defining specific biometric prohibitions may be another, more familiar approach. Some possible options are incriminations for theft and unauthorized use of biometric data, and prohibitions. For instance, this includes forbidding the nonencrypted processing and transfer of biometric data, prohibiting the use of biometrics that generate sensitive data when alternatives exist, the use of financial rewards to promote participation in biometric identification programs, or centrally storing easy-to-misuse full, raw images [2].

Legitimate use must be identified by the legislator (the first task). This includes enhancing available transparency tools that need to be considered (the second task). Only after having identified legitimate forms of biometrical processing, should one define the rules and conditions that any allowed use of biometrics should respect. With regard to this second task, there is a need to establish both common principles and language of privacy for biometrics. This includes principles such as equality of access to the network; absolute accuracy of targeting by surveillance systems; systems to ensure the accuracy of the data held within the surveillance systems; mechanisms for amending the false, inaccurate, or modified data; and, systems to protect individuals from their inclination to trade their own privacy. This biometric framework should be established based on appropriate risk assessment that distinguishes between legitimate and illegitimate use of biometrics [2].

Procedures Based on Biometric Evidence Shall Be Unfavorably Received

Biometric evidence is likely to be accepted without too much resistance in European courts. Notwithstanding some differences, all systems in Europe tend to include most forms of evidence. Also, although the principle is elaborated in a different way, the rules governing evidence in all European countries have a tendency to ban only categorically unreliable or illegal (illegally obtained) evidence. In countries belonging to different traditions, some form of corroboration is required as a limit on the freedom of the judge. In the Netherlands, for instance, one confession is not sufficient (art. 341 Code of Criminal Procedure) for a conviction. This evidence has to be corroborated by other evidence [2].

However, some assess critically the impact of DNA analysis on legal systems that employ the rule of free assessment of evidence. Within such systems all means of evidence are equal; the judge can thus choose freely what kind of evidence is relevant to help assess the possible guilt of the defendant. Since DNA analysis offers stronger security and more reliability than older evidential techniques (which may be flawed by subjective elements), there is the danger that judges within such systems of freedom of evidence will be tempted to attach the increased role to DNA evidence (obviously when properly obtained and processed by certified institutions). This might be detrimental to the system of free evaluation of proof based on a possible intimate conviction of the judge [2].

This warning can be generalized to all biometric technologies and to all systems of evidence in Europe. Whenever investigations become complex and the methods of investigation become formalized, the outcome will be harder to evaluate by the court and the defense. To prevent experts taking over the position of the judges, the legal recognition of an automatic right to counterexpertise is needed and, like in civil cases all over Europe, parties should have the right to meet the expert and be heard [2].

Technical Aspects of Biometric Technologies

For a long time, the use of biometrics was limited to forensic applications [4]. Recently, however, it has become possible to digitize, store, and retrieve biometric patterns and have them processed by computers. Large-scale deployment can thus be envisaged in, for example, passports, voter ID cards, national ID cards, and driver's licenses, which will reduce waiting time at border controls or for welfare disbursement. Biometrics provides a challenging solution to increased security needs, as it bases verification on aspects that are specific to each individual. However, biometrics is only one element of a larger system that involves the use of sensors to acquire a biometric sample; the transmission of this data from the sensor to a computer; the access to a database of stored templates in order to find a match; and the decision and subsequent action. Biometrics should not be considered alone, but as part of a global system that must be designed and evaluated in its entirety [2].

Different Well-Known Modalities

Different modalities can be considered; fingerprint and iris scans are currently the most reliable methods, but users often consider them intrusive. Users are more familiar with methods using face, voice, or handwritten signatures, but these are not yet sufficiently efficient for use on a large scale. In view of this, combining several methods would seem more appropriate, but this has still to be validated. There will always be a compromise between the level of accuracy you can obtain through a given modality (as biometric systems will always produce a certain level of error) and the level of constraints you can impose on the user, especially during the enrollment phase. Indeed, the more constraining the acquisition of the patterns, the more accurate the results of the biometric system. Of course it is the application's purpose that mostly impacts user acceptability; requirements to ensure safe air travel need not be the same as those used to access an office or a home [2].

Iris

Of all existing biometric techniques, the one encoding the iris patterns is the most precise one, possibly at the expense of a rather constraining sample acquisition process (the camera must be infrared, and the eyes must be at a very precise distance and angle from the camera). These elements provide a very good quality initial image, which is necessary to ensure such a high level of performance [2].

On the other hand, they may make enrollment time-consuming and call for user training. This method is also relatively expensive and unavoidably involves the scanning of the eye, which can initially prove disconcerting to users. Its reliability, however, means it can be successfully used both for identification and verification, an advantage that few other techniques can offer [2].

Fingerprinting

Fingerprinting is currently the method that offers the best compromise between price, acceptability, and accuracy; and a lot of systems based on this modality are already operational. However, the latest evaluation results show that their performance relies heavily on the quality of the acquired images, particularly during the enrollment phase. It seems that a few percentages of the population cannot be enrolled through fingerprinting (manual workers, people with too wet or too dry hands, etc.), though this can be reduced with the use of prints from two or more fingers and adequate specific enrollment processes for people who have problems. While the existence of a great number of different sensors associated with various technologies is in general beneficial to performance, due to the coupling of sensor and algorithms that is optimized by the designer of the biometric system, it also induces interoperability problems. Fingerprinting is, in general, fairly well accepted, even if it has some forensic connotations, and it allows both identification and verification [2].

Face Recognition

Currently, face recognition is considered to be relatively inaccurate due to the presence of a lot of variability (from 1.39% to more than 13% EER). This is due to changes that occur to people over time, like aging, or simply related to external environmental conditions (poses, facial expressions, illumination, textured background). Therefore this method's performance varies considerably, depending on the recording conditions and the context of application (static images or video, with or without a uniform background, or constant lighting conditions) [2].

Face recognition is not efficient enough at this time to deal with large-scale identification; but it can be useful in the context of verification or limited access control with constraining acquisition conditions (during enrollment the background must be uniform and the user must face the camera at a fixed distance).

With regards to a sample acquisition using a video camera, no system can be considered as sufficiently developed, but there are promising technological innovations that use 3D modeling to cope with the problem of pose. This obviously means an increase of the cost of the global system (use of sophisticated 3D scanners in place of standard medium-cost cameras). However, due to the fact that this modality is well accepted by the user, and that it has been introduced as a standard in travel documents by the ICAO, a lot of research is being conducted to improve the system's accuracy. A big increase in performance can be expected in the next five years, but this modality can never be expected to be as accurate as fingerprinting or iris scanning due to its intrinsic variability and behavioral character. Nevertheless, for convenience, applications (like physical access control or personalization of environment) which impose limited FAR constraints, the use of face recognition is still very interesting as it can be transparent. It would, however, have to be used in association with other methods, in order to reduce error rates or be used against a pre-selected database (trained to use) [2].

DNA

Except for identical twins, each person's DNA is unique. It can thus be considered a perfect modality for identity verification. DNA identification techniques look at specific areas within the long human DNA sequence, which are known to vary widely between people. The accuracy of this technique is thus very high, and allows for both identification and verification. Enrollment can be done from any cell that contains a nucleus; for instance, taken from blood, semen, saliva or hair samples (which is considered intrusive by many users). However, DNA as a biometric for identification uses a very small amount of noncoding genetic information that does not allow deciphering a person's initial genetic heritage. At present, DNA analysis is performed in specialized laboratories and is expensive and time-consuming (roughly four to five hours for the whole procedure). Moreover, the complete lack of standardization means interoperable systems are a long way off. DNA techniques are currently being used by law enforcement. Thus, any wider deployment of DNA-based biometric techniques in the future, if these do indeed become quicker and cheaper, will always face acceptability problems [2].

It seems that it will be a long time before DNA printing becomes a real-time biometric verification method. However, a Canadian laboratory recently announced a proprietary DNA extraction process that takes only 15 minutes and needs only simple equipment. In other words, DNA analysis could be done in real time. Future technical improvements will be of two types: first, more automation and more accuracy in the existing processes; and second, the building of new systems that only require very small amounts of material to provide an identification [2].

Evaluation of Biometric Systems

At first, comparing the error rates of the different systems in each modality and in a restricted number of environments per application, by using estimates of FAR and FRR, one may reach conclusions as to performance. In fact, the performance of the systems is highly dependent on the test conditions (laboratory conditions with a small database and relatively good-quality data). Fair evaluation should include forgeries (natural or simulated) in the database, and this is very rarely done. Fingerprinting and face recognition are subjected to independent international evaluation annually, which now aims at testing more operational situations. Unfortunately, no openly available evaluation on iris recognition is being conducted. Table 27-3 gives what is considered to be the most accurate information available on biometric performance (the least order of magnitude estimates of the performance of the state-of-the-art systems) [2].

More generally, in the evaluation of operational biometric systems, criteria other than performance have to be taken into account (such as robustness, acceptability, facility of data acquisition, ergonomic aspects of the interface, enrollment, and identification time). When choosing a practical fingerprint system, for example, the robustness of the sensor, the possibility of wrong or clumsy manipulation, and dirtiness must be considered. It should also be remembered that a relatively large part of the population will be unable to enroll with any chosen method. Alternative processes will always have to be found for any specific application [2].

Table 27-3 *Selected Technology Error Rates*

Biometric	Face	Finger	Iris
FTE % (Failure to enroll)	n/a	4	7
FNMR % rejection rates	4	2.5	6
FMR1 % verification match error rate	10	<0.01	<0.001
FMR2 % identification error rates for dB size > 1 m	40	0.1	N/A
FMR3 % screening match error rate for dB sizes = 500	12	<1	N/A

Note: Typical biometric accuracy performance numbers reported in large third-party tests. FNMR (also ERR) and FMR (also FAR). N/A is nonavailable data.

Challenges and Limitations

Fraudulent reproduction of biometric data is possible; this depends heavily on the modality, application, and resources being considered and availability of the data to be reproduced. Different questions should be considered when deciding whether a biometric system can be fooled. Is it technologically possible to reproduce biometric data artificially? How easily available is the data? Is the person's cooperation needed? Is it possible to design biometric sensors that can detect impostors [2]?

Resistance of the System to Forgeries

While it is not easy to, for example, get a good three-dimensional image of the finger, it is relatively easy (using a dentist's kit) to get latent fingerprints left by a person on different surfaces and objects and use them to reconstruct a fake finger (still not very reliable). There are also behavioral tests of liveness; some rely only on software, but some require special hardware that distinguishes by physical means living from dead tissue. Nonetheless, a fake finger that would fool all the vitality detectors in a fingerprint sensor could still be built, given sufficient resources [2].

Biometric Data Storage

Biometric data may be stored on portable media such as smart cards if they will be used in verification mode [8]. This ensures that the data cannot be used without the user's own authorization, contrary to what happens with data stored in a central database. Biometric verification/identification can also be realized through remote access, by transmission of the biometric image or template through a network to the device that will process the decision step. This requires a highly secure connection. Watermarking could be used in this case to ensure that the transmitted data have not been corrupted [2].

Of course, smart cards can be lost or stolen. For this reason, the data they contain must be encrypted and backed up. However, if the information is stolen, it is necessary to be able to revoke it and to produce another template that could be used for further identification. Revocation is easy when dealing with PIN codes or passwords, but not with biometric traits, as you cannot change your irises or your fingerprints [2].

Cancelable biometrics is a new research field, and some preliminary propositions have been made. It is possible to generate new facial images for a person by filtering the original image. The coefficients of the filter are randomly generated thanks to a PIN code. Changing the PIN code means changing the filter, and therefore, changing the facial image generated. It has been demonstrated that for face recognition this process does not affect the result of recognition if the matching algorithm relies on correlations. More research is needed to confirm these results on other face recognition methods. The use of such filtering is not straightforward for fingerprints or iris recognition, because it affects the quality of the images and the accuracy of the minutiae detection (fingerprint) or texture analysis (iris). For iris recognition, one solution is to extract a shorter code from the 2,048-bit length code and to use only this information in the matching process [2].

Biometrics as a Way to Increase Privacy, Anonymity, and Security

Biometrics, depending on the way they are deployed, could enhance the security and the privacy of the users. Biometric encryption can thus be used. The fingerprint of one person can be used to produce a PIN which, for example, allows access to a bank ATM. The coded PIN has no connection whatsoever to the finger pattern. The finger pattern only acts as the coding key of that PIN [6], any PIN. What is stored in the bank's database is only the coded PIN. The fingerprint pattern, encrypted or otherwise, is not stored anywhere during the process. Moreover, the successful decoding of a PIN confirms a person's eligibility for a service without having to reveal any personal identifiers; since only the user can decode the PIN (indicating also physical presence), the transaction can go ahead. There is an indirect benefit to privacy. A user can continue to have a multitude of PINs and passwords, and thereby achieve safety through numbers, rather than having one single identification that links everything. However, there are technical problems with biometric encryption. Some solutions have been already proposed and some patents applied for, but further research is needed. The fact that biometric patterns are never exactly the same from one data acquisition to another renders the production of a private key, which has to be similar at each stage, very difficult [2].

Multimodality

The use of several modalities can be considered in order to improve the efficiency of the overall system and provide alternative paths, thus enhancing system flexibility.

Improve the Efficiency of the Overall System

A single-modality biometric system can be subject to a high level of errors. Some errors can be due to noise associated with the acquired data, or to intra-class variability (from one data acquisition to another). In addition, biometric systems may be attacked with forged data or genuine data of a dead person. Using several different modalities together aids in dealing with such unimodal problems, especially when complementary biometrics such as behavioral and physical, which may be discriminative or not, are used. Indeed, multimodality has a clear impact on performance and attacks by impostors. For instance, by combining a fingerprint with a hand shape or face, the use of fake fingerprints may be circumvented, since faces and hands are more difficult to fake than fingers [2].

Provide Alternative Paths, Thus Enhancing System Flexibility

Different modalities can be used in parallel, thus allowing the use of the system for different objectives. For example, a biometric system built for both fingerprint and face recognition could use the face in verification mode if the user has a problem enrolling a fingerprint. Moreover, in case some biometric trait is temporarily unavailable, the other one could be used to allow access. If the user has, for example, a temporary eye problem that makes the iris scan impossible, in a multimodal system, fingerprints could be used instead (see sidebar, "Multimodal Biometric Systems"). The same would apply in cases where people refuse to use a specific modality (for religious or health purposes, for instance). A multimodal system therefore allows flexibility by providing alternatives in the identification process [2].

Multimodal Biometric Systems

Biometric systems relying on a single technology are currently being deployed with various levels of success in many different application contexts (airports, passports, physical and logical access control, etc.). However, by combining more than one modality, enhanced performance reliability and even increased user acceptance could be achieved. Combining less-reliable technologies in sequence could strengthen the overall system performance, and combining them in parallel could increase the flexibility of the system by providing alternative modes for the verification/identification process [2].

Using Multimodality to Achieve Improved Efficiency

Unimodal biometric systems can be subject to many types of errors. Studying the source of such errors will help the design of multimodal systems that can achieve improved performance characteristics.

Some errors may be due to noise associated with the acquired data. Noise may be produced in different ways:

- By sensor performance (image out of focus);

- By poor ambient conditions (reflected light during facial image acquisition);

- By user behavior/status (an incorrectly placed finger) [2].

As a consequence, the biometric input may be incorrectly matched and the user falsely rejected. By combining appropriate technologies together, such noise may be minimized and the end result could be fewer false rejects [2].

Another type of error relates to intra-class variability. Biometric data will naturally vary from one data acquisition to another. This intra-class variability may be stronger for some individuals, especially when monitoring behavioral biometric features such as signature, voice, or gait. This usually results in a variation between the data acquired and enrolled data, which affects the matching process and may lead to system failure. Again, combining technologies with mixed intra-class variability could result in systems that exhibit overall better performance characteristics [2].

Other types of errors relate to the distinctiveness of individual biometric features. By combining two less distinct features, improved overall performance may be achieved. Another error effect that multimodal system design can minimize relates to forging and liveness attacks (fake fingerprint). In this case, combining biometric technologies in sequence is likely to counter such attacks since a lot more effort will be required to spoof the combined system. As a result, multimodality could significantly enhance the performance of verification systems, compared to unimodal systems [2].

Using Multimodality to Enhance the Usability of Systems

Two (or more) modalities could be combined in parallel to produce a system that would allow more flexible use. For example, biometric systems built for both fingerprint and face recognition could allow the use of only the facial image for verification when users have problems enrolling their fingerprints and vice versa. This procedure could prove extremely useful to those users who have temporarily lost the ability to provide one of their biometric traits (for example, a temporary eye problem that rules out an iris scan). The same could apply in cases where people refuse to use a specific modality for religious or health purposes, for instance. A multimodal system therefore allows enhanced flexibility by providing alternatives for the identification process. As such, it also has the potential to be more socially inclusive. In brief, when designing a multimodal system, the following choices must be addressed: Which modalities are going to be combined; and at which stage should technologies be combined [2]?

Which Modalities Are Going to Be Combined?

The choice once again is mainly driven by the application requirements. In addition to the need to enhance performance or usability of the system, other factors such as available resources (including necessary processing power) and costs (of the combined technologies) should be considered. For example, if a mobile platform with a camera (a smart phone) is used, voice and face may be the natural combination [2].

At Which Stage Should Technologies Be Combined?

When the modalities are combined in sequence, the fusion of the information provided by the different modalities can be done at different levels:

- At the feature level, by combining the features extracted in a single input.

- At the decision level, by combining the decisions of separate biometric systems. The last option may be problematic if the systems disagree. In this case, it may lead to further errors (the "bad" performance of a system will degrade the combined multimodal system).

- At the score level, by combining scores generated by the different systems. Fusion at the score level is more widely used. In this case, the combination considers the scores produced by the system before making a final decision. Overall performance is increased, provided that the fusion scheme is adequately chosen. In some cases, the two modalities that are combined may be correlated (for example, lip movement and voice recorded together when a person is speaking, minimizing the possibility of fraud). In such cases, it is interesting to fuse the information at an even earlier stage, namely just after feature extraction, and to build a unique system taking as input a combination of these features [2].

Independent of the procedure chosen to design and develop efficient multimodal systems, it is essential that further research on such systems is conducted. Several research projects (see note) are evaluating multimodal biometric systems, but a major problem is the lack of available multimodal test data [2].

Note: *Two EC-funded research projects on multimodality are in progress. The two projects both involve mobile handheld platforms, which is a new, promising, but also complex orientation in the use of multimodal biometrics. Mobility introduces more noise in captured data and lower quality of data because of cheaper sensors, as well as increased intra-class variability due to changes in capture environments. FP6 IST project SecurePhone "Secure Contracts Signed by Mobile Phone" explores face, voice, and signature simultaneously; and "Multimodal Face and Speaker Identification" research project explores multimodal biometrics combining face and voice on a handheld device [2].*

There are few multimodal databases available; M2VTS, XM2VTS, BANCA, DAVID, and SMARTKOM, most of which are the outcome of past European projects. Most of these databases contain few biometric modalities, usually face and voice, and it has only been recently that a database (BIOMET) including five biometric traits has been built. Developing multimodal databases is more complicated, time-consuming, and expensive than developing unimodal ones; as a result, such databases contain the data of only a few hundred individuals. This in turn makes it difficult to extrapolate the success or failure of a multimodal algorithm or method that is tested to be used in large-scale deployment (thousands or millions of people). Furthermore, current data protection legislation limits the cross-border sharing of such data [2].

> In addition, there is currently no independent evaluation of multimodal systems available. One of the aims, however, of the BIOSECURE European Network of Excellence is to carry out such an evaluation [2].

Application Issues

"Mass identification" applications (border control, national ID cards, visas, etc.) that demand a high level of security (very low FAR) must be distinguished from domestic or personal applications (personal access to PCs) for which the constraints are low FRR and friendly interfaces. Mass identification involves:

- Storage of the data on a central database;
- High accuracy level;
- User constraints for high-quality enrollment [2].

The size of the population may be a problem when considering access times to a database and the fluidity of the entire process. Interoperability is another issue: If a border control system is to be used in several Schengen area entry points, either the same system has to be used by all Schengen states, or the different systems must be interoperable (which means that software and hardware on multiple machines from multiple vendors must be able to communicate). Interoperability between different systems is achieved by using common standards and specifications. At the moment, the standardization of the data formats (for face recognition and fingerprints) is rapidly becoming an important concern with the ISO-SC37 commission. It seems that standardization constraints are essentially suitable for verification systems (1:1), but they increase the processing time of large-scale identification, which can be detrimental to the systems. Very few tests have been conducted so far dealing with real interoperability issues, which thus remain a fundamental concern [2].

In the second type of applications, the focus is on transparency and comfort for the user. In this case, nonintrusive biometrics may be used such as video recording, from which a sequence of images can be obtained, providing different types of correlated information such as gait or voice in correlation with the face images. None of these modalities is efficient enough to be used alone. However, the complementary aspect of the information that the joint use would provide will be an important tool to ensure final reliability in the identification of people [2].

Summary/Conclusion

This chapter dealt with the implementation of social, economic, legal, and technological aspects of biometric and verification systems. Any biometric system has drawbacks and cannot warranty 100% identification rates, nor 0% false acceptance and rejection ratios. One way to overcome the limitations is through the implementation of a combination of different biometric verification systems. In addition, a multimodal biometric recognition is more difficult to fool than a single biometric system, because it is more unlikely to defeat two or three biometric systems than one. This chapter also summarized the different data fusion levels, and how they must be performed in order to improve the results of each combined system on its own.

Finally, this chapter also discussed the impact of biometrics on economic outcomes, such as optimal and actual identity, the emergence of standards, and costs and benefits. The chapter also surveyed the present state and likely evolution of market demand and supply. Issues that policy makers need to address as well as the means to address these issues were also explored.

References

1. Marcos Faundez-Zanuy, "Data Fusion in Biometrics," *IEEE A&E Systems Magazine* [Escola Universitaria Politecnica de Mataro, Avda. Puig i Cadafalch 101–111, 08303 Mataro (Barcelona) Spain]. IEEE Corporate Office, 3 Park Avenue, 17th Floor, New York, NY 10016-5997, vol. 20, no. 1, pp. 34–38, January 2005. Copyright © 2003, IEEE. All Rights Reserved.

2. "Biometrics at the Frontiers: Assessing the Impact on Society." For the European Parliament Committee on Citizens' Freedoms and Rights, Justice and Home Affairs (LIBE), European Commission, Joint Research Center (DG JRC), Institute for Prospective Technological Studies, 2005. Copyright © 2005 European Communities.

3. Joel Dubin, "Biometrics: Best Practices, Future Trends," c/o TechTarget, 117 Kendrick Street, Needham, MA 02494, April 6, 2006. All Rights Reserved, Copyright © 2000–2006, TechTarget, SearchSecurity.com.

4. John R. Vacca, *Computer Forensics: Computer Crime Scene Investigation, 2nd ed.*, Charles River Media (2005).

5. John R. Vacca, *Net Privacy: A Guide to Developing and Implementing an Ironclad eBusiness Privacy Plan*, McGraw-Hill (2001).

6. John R. Vacca, *Public Key Infrastructure: Building Trusted Applications and Web Services*, CRC Press (2005).

7. John R. Vacca, *Satellite Encryption*, Academic Press (1999).

8. John R. Vacca, *The Essentials Guide to Storage Area Networks*, Prentice Hall, Inc., Professional Technical Reference, Pearson Education (2001).

9. John R. Vacca, *Identity Theft*, Pearson Education (2002).

Part 9: Biometric Solutions and Future Directions

28

How Mapping-the-Body Technology Works

Research into tracking and recognizing human movement has so far been mostly limited to gait or frontal posing. This chapter presents a continuous human movement recognition (CHMR) framework that forms a basis for the general biometric analysis of the continuous mapping of the human body in motion, as demonstrated through tracking and recognition of hundreds of skills, from gait to twisting saltos (see Figure 28-1) [1]. CHMR applications to the biometric verification of gait, anthropometric data, human activities, and movement disorders will also be presented. Furthermore, in this chapter, a novel 3D color clone-body-model will be discussed, which is dynamically sized and texture mapped to each person for more robust tracking of both edges and textured regions. Tracking is further stabilized by estimating the joint angles for the next frame using a forward smoothing particle filter with the search space optimized by utilizing feedback from the CHMR system. A new paradigm defines an alphabet of dynemes, units of full-body movement skills, to enable recognition of diverse skills. Using multiple hidden Markov models (HMMs), the CHMR system attempts to infer the human movement skill that could have produced the observed sequence of dynemes. The novel clone-body-model and dyneme paradigm presented in this chapter enable the CHMR system to track and recognize hundreds of full-body movement skills, thus laying the basis for effective biometric verification associated with full-body motion and body proportions [1].

As discussed, biometric verification depends on significant measurable diversity of a particular physical characteristic, such as iris, fingerprint, signature, or gait. The more dimensions and larger between-person variability for each dimension, the better the biometric. The goal is to resolve the apparent conflict of enhancing between individual variations while minimizing within individual variations. With 249 degrees of freedom and good discrimination entropy, the iris biometric is well ahead of others by reliably recognizing 9 million with no false positives [1] and with projections to 1 in 10 billion—more than the population of this planet. Other biometrics such as face and gait are orders of magnitude away from iris recognition accuracy; and, unlike iris, gait and face are affected by age, clothes, and accessories, leaving many problems yet

Figure 28-1   *Overview of the continuous human movement recognition framework. (Source: Adapted with permission from the University of Canterbury.)*

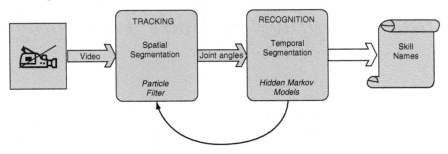

to be solved. However, biometrics that engage the whole body, such as gait, have a place for less proximal biometric verification, where identification is possible without any awareness of the subject to minimize risk of an identity being "faked [1]."

Human movement is commercially tracked by requiring subjects to wear joint markers/identifiers, an approach with the disadvantage of significant set-up time. Such an invasive approach to tracking has barely changed since it was developed in the 1970s. Using a less invasive approach free of markers, computer vision research into tracking and recognizing full-body human motion has so far been mainly limited to gait or frontal posing [1]. Various approaches for tracking the whole body have been proposed by using a variety of 2D and 3D shape models and image models, as listed in Table 28-1 [1].

These approaches determine body-part orientation by tracking only the edges or same color regions. To improve tracking accuracy and robustness by also tracking the textured colors within regions is the goal of the clone-body-model. This model is dynamically sized and texture mapped to each person, enabling both edge and region tracking. No previous approaches use such a method, as can be seen in Table 28-1 [1]. The prediction of joint angles for the next frame is cast as an estimation problem, which is solved using a particle filter with forward smoothing. This approach optimizes the huge search space related to calculating so many particles for these 32 degrees of freedom (DOF) by utilizing feedback from the recognition process [1].

Security systems will become increasingly effective as computers more accurately recognize and understand full-body movement in terms of every-day activities. Stokoe began recognizing human movement in the 1970s by constructing sign language gestures (signs) from hand location, shape, and movement and assumed that these three components occur concurrently with

Table 28-1   *Comparison of Different Human Body Models*

Creators	Shape Model	Image Model
Hogg	Cylinders	Edge
Rohr	Cylinders	Edge
Gavrila & Davis	Superquadrics	Edge
Drummond & Cipolla	Conics	Edge
Goncalves et al.	Cones	Edge
Kakadiaris & Metaxas	Deformable	Edge
Wren & Pentland	2D color blobs	Skin color blobs
Ju et al.	Patches (2D)	Flow
Bregler & Malik	Cylinders	Flow
Wang et al.	Cylinders	Flow
Cham & Rehg	Patches (2D)	Template
Wachter & Nagel	Cones	Flow + Edge
Plänkers & Fua	Deformable	Silhouette + Disparity
Deutscher et al.	Cones	Edge + Silhouette
Brand	Outline	Silhouette moments
Rosales & Sclaroff	Outline	Silhouette moments
Liebowitz & Carlsson	Outline	Hand-marked joints
Taylor	Outline	Hand-marked joints
Leventon & Freeman	Outline	Hand-marked joints

no sequential contrast (independent variation of these components within a single sign). Ten years later, Liddel and Johnson used sequential contrast and introduced the movement-hold model. In the early 1990s, Yamato et al. began using HMMs to recognize tennis strokes. Recognition accuracy rose as high as 99.2% in Starner and Pentland's work in 1996. Constituent components of movement have been named cheremes [1], phonemes [1], and movemes [1].

As can be seen from Table 28-2, most movement recognition research has been limited to frontal posing of a constrained range of partial-body motion [1].

Table 28-2 *Human Movement Recognition Research*

Creator	Recognition Approach
Stokoe 1978	Transcription system => sign = location (tab) + hand shape (dez) + movement (sig)
Tamura and Kawasaki 1988	Cheremes to recognize 20 Japanese signs (gesture signings)
Liddell and Johnson 1989	Use sequences of tab,dez,sig => Movement-Hold model
Yamato, Ohya, and Ishii 1992	HMM recognizes 6 diff tennis strokes for 3 people (25 × 25 pixel window)
Schlenzig, Hunter, and Jain 1994	Recognizes 3 gestures: *hello, goodbye,* and *rotate*
Waldron and Kim 1995	ANN recognizes small set of signs
Kadous 1996	Recognizes 95 Auslan signs with data gloves—80% accuracy
Grobel and Assam 1997	ANN recognizes finger spelling—242 sign vocab with colored gloves—91.3% accuracy
Starner and Pentland 1996	HMM recognizes 40 signs in 2D with constrained grammar—99.2% accuracy
Nam and Wohn 1996	HMM very small set of gestures in 3D—movement primes to construct sequences
Liang and Ouhyoung 1998	Continuous recognition of 250 Taiwanese signs—segment temporal discontinuities
Vogler & Metaxis 1997	HMM continuous recognition of 53 signs—models transitions between signs
Vogler & Metaxis 1998	HMM continuous recognition of 53 signs—word context with CV geometrics
Vogler & Metaxis 1999	Define tab,dez,sig as phonemes—22 signs—magnetic tracking one hand—91.82%

By contrast, this chapter describes a computer vision–based framework that recognizes continuous full-body motion of hundreds of different movement skills (see Figure 28-2) [1]. The full-body movement skills in this study are constructed from an alphabet of 35 dynemes—the smallest contrastive dynamic units of human movement. Using a novel framework of multiple HMMs, the recognition process attempts to infer the human movement skill that could

Figure 28-2
CHMR system tracking and recognizing a sequence of movement skills. (Source: Reproduced with permission from the University of Canterbury.)

have produced the observed sequence of dynemes. This dyneme approach has been inspired by the paradigm of the phoneme as used by the continuous speech recognition research community; pronunciation of the English language is constructed from approximately 50 phonemes, which are the smallest contrastive phonetic units of human speech [1].

Tracking

Various approaches for tracking the whole body have been proposed. They can be distinguished by the representation of the body as a stick figure, 2D contour, or volumetric model and by their dimensionality being 2D or 3D. Volumetric 3D models have the advantage of being more generally valid, with self-occlusions more easily resolved. They also allow 3D joint angles to be able to be more directly estimated by mapping 3D body models onto a given 2D image. Most volumetric approaches model body parts using generalized cylinders [1] or super-quadratics [1]. Some extract features [1], and others fit the projected model directly to the image [1].

Clone-Body-Model

Cylindrical, quadratic, and ellipsoidal [1] body models of previous studies do not contour accurately to the body, thus decreasing tracking stability.

To overcome this problem, in this research on the 3D clone-body-model, regions are sized and texture mapped from each body part by extracting features during the initialization phase. This clone-body-model has a number of advantages over previous body models:

- It allows for a larger variation of somatotype (from ectomorph to endomorph), gender (cylindrical trunks do not allow for breasts or pregnancy), and age (from baby to adult).

- Exact sizing of clone-body-parts enables greater accuracy in tracking edges, rather than the nearest best fit of a cylinder.

- Texture mapping of clone-body-parts increases region tracking and orientation accuracy over the many other models that assume a uniform color for each body part.

- Region patterns, such as the ear, elbow, and knee patterns, assist in accurately fixing orientation of clone-body-parts [1].

The clone-body-model proposed in this chapter consists of a set of clone-body-parts, connected by joints [1]. Clone-body-parts include the head, clavicle, trunk, upper arms, forearms, hands, thighs, calves, and feet. Degrees of freedom are modeled for gross full-body motion (see Table 28-3) [1]. Degrees of freedom supporting finer resolution movements are not yet modeled, including

Table 28-3   *Degrees of Freedom Associated with Each Joint*

Joint	DOF
Neck (atlantoaxial)	3
Shoulder	3*
Clavicle	1*
Vertebrae	3
Hip	3*
Elbow	1*
Wrist	2*
Knee	1*
Ankle	2*
	32 total

*Double for left and right.

the radioulnar (forearm rotation), interphalangeal (toe), metacarpophalangeal (finger), and carpometacarpal (thumb) joint motions [1].

Each clone-body-part consists of a rigid spine with pixels radiating out (see Figure 28-3) [1]. Each pixel represents a point on the surface of a clone-body-part. Associated with each pixel is radius or thickness of the clone-body-part at that point; color as in hue, saturation, and intensity; accuracy of the color and radius; and the elasticity inherent in the body part at that point. Although each point on a clone-body-part is defined by cylindrical coordinates, the radius varies in a cross-section to exactly follow the contour of the body, as shown in Figure 28-4 [1].

Automated initialization assumes only one person is walking upright in front of a static background initially, with gait being a known movement model.

Figure 28-3   *Clone-body-model consisting of clone-body-parts that have a cylindrical coordinate system of surface points b() and up to three DOF for each joint linking the clone-body-parts. Each surface point is a vector b with cylindrical coordinates, color (h, s, i), accuracy of radius (a_r), accuracy of color (a_{hsi}), and elasticity of radius (e_r). (Source: Reproduced with permission from the University of Canterbury.)*

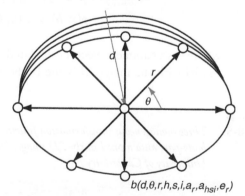

$$b(d,\theta,r,h,s,i,a_r,a_{hsi},e_r)$$

Figure 28-4   *Clone-body-model example rotating through 360 degrees. (Source: Reproduced with permission from the University of Canterbury.)*

Anthropometric data [1] is used as a Gaussian prior for initializing the clone-body-part proportions, with left-right symmetry of the body used as a stabilizing guide from 50th percentile proportions. Such constraints on the relative size of clone-body-parts and on limits and neutral positions of joints help to stabilize initializations. Initially a low accuracy is set for each clone-body-part, with the accuracy increasing as structure from motion resolves the relative proportions. For example, a low color and high radius accuracy is initially set for pixels near the edge of a clone-body-part, high color and low radius accuracy for other near-side pixels, and a low color and low radius accuracy for far-side pixels. The ongoing temporal resolution following self-occlusions enables increasing radius and color accuracy. Breathing, muscle flexion, and other normal variations of body part radius are accounted for by the radius elasticity parameter [1].

Kinematic Model

The kinematic model tracking the position and orientation of a person relative to the camera entails projecting 3D clone-body-model parts onto a 2D image using three chained homogeneous transformation matrices, as illustrated in Figure 28-5 [1]:

$$p(x, b) = I_i(x, C_i(x, B_i(x, b)))$$

where x is a parameter vector calculated for optimum alignment of the projected model with the image, B is the Body frame of reference transformation, C is

Figure 28-5 *Three homogeneous transformation functions B( ), C( ), I( ) project a point from a clone-body-part onto a pixel in the 2D image. (Source: Reproduced with permission from the University of Canterbury.)*

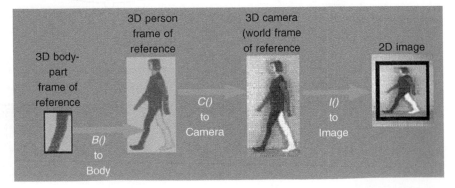

the Camera frame of reference transformation, I is the Image frame of reference transformation, b is a body-part surface point, and p is a pixel in 2D frame of video [1].

Joint angles are used to track the location and orientation of each body part, with the range of joint angles being constrained by limiting the DOF associated with each joint. A simple motion model of constant angular velocity for joint angles is used in the kinematical model. Each DOF is constrained by anatomical joint-angle limits, body-part interpenetration avoidance, and joint-angle equilibrium positions modeled with Gaussian stabilizers around their equilibria. To stabilize tracking, the joint angles are predicted for the next frame. The calculation of joint angles, for the next frame, is cast as an estimation problem, which is solved using a particle filter (condensation algorithm) [1].

Particle Filter

The particle filter was developed to address the problem of tracking contour outlines through heavy image clutter [1]. The filter's output at a given time-step, rather than being a single estimate of position and covariance as in a Kalman filter, is an approximation of an entire probability distribution of likely joint angles. This allows the filter to maintain multiple hypotheses and thus be robust to distracting clutter [1].

With about 32 DOFs for joint angles to be determined for each frame, there is the potential for exponential complexity when evaluating such a high-dimensional search space. Partitioned sampling and layered sampling is proposed here to reduce the search space by partitioning it for more efficient particle filtering [1]. Although annealed particle filtering [1] is an even more general and robust solution, it struggles with efficiency, which improves with partitioned annealed particle filtering [1].

The particle filter is a considerably simpler algorithm than the Kalman filter. Moreover, despite its use of random sampling, which is often thought to be computationally inefficient, the particle filter can run in real time. This is because tracking over time maintains relatively tight distributions for shape at successive time steps, and particularly so given the availability of accurate learned models of shape and motion from the human-movement-recognition (CHMR) system. Here, the particle filter has:

- Three probability distributions in problem specification:
  1. Prior density $p(x)$ for the state x
     - Joint angles x in previous frame

2.   Process density $p(x_t|x_{t-1})$
- Kinematic and clone-body-models (x_{t-1}: previous frame, x_t: next frame)

3.   Observation density $p(z|x)$
- Image z in previous frame

- One probability distribution in solution specification:

1.   State density $p(x_t|Z_t)$
- Where x_t is the joint angles in next frame Z_t [1]

Feedback from the CHMR system utilizes the large training set of skills to achieve an even larger reduction of the search space. In practice, human movement is found to be highly efficient, with minimal DOFs rotating at any one time. The equilibrium positions and physical limits of each DOF further stabilize and minimize the dimensional space. With so few DOFs to track at any one time, a minimal number of particles are required, significantly raising the efficiency of the tracking process. Such highly constrained movement results in a sparse domain of motion projected by each motion vector [1].

Because the temporal variation of related joints and other parameters contains information that helps the recognition process infer dynemes, the system computes and appends the temporal derivatives and second derivatives of these features to form the final motion vector. Hence the motion vector includes joint angles (32 DOFs), body location and orientation (6 DOFs), center of mass (3 DOFs), and principal axis (2 DOFs) all with first and second derivatives [1].

Recognition

To simplify the design, it is assumed that the CHMR system contains a limited set of possible human movement skills. This approach restricts the search for possible skill sequences to those skills listed in the skill model, which lists the candidate skills and provides dynemes (an alphabet of granules of human motion) for the composition of each skill. The current skill model contains hundreds of skills where the length of the skill sequence being performed is unknown. If M represents the number of human movement skills in the skill model, the CHMR system could hypothesize M^N possible skill sequences for a skill sequence of length N. However, these skill sequences are not equally likely to occur due to the biomechanical constraints of human motion. For example, the skill sequence stand-jump-lie is much more likely than stand-lie-jump (as it is difficult to jump while lying down) [1].

This approach applies Bayes' law and ignores the denominator term to maximize the product of two terms: the probability of the motion vectors given the skill sequence and the probability of the skill sequence itself. The CHMR framework described by this equation is illustrated in Figure 28-6 where, using motion vectors from the tracking process, the recognition process uses the dyneme, skill, context, and activity models to construct a hypothesis for interpreting a video sequence [1].

In the tracking process, motion vectors are extracted from the videostream. In the recognition process, the search hypothesizes a probable movement skill sequence using four models:

- The dyneme model models the relationship between the motion vectors and the dynemes.

- The skill model defines the possible movement skills that the search can hypothesize, representing each movement skill as a linear sequence of dynemes.

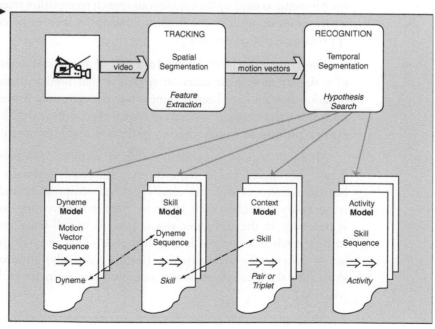

Figure 28-6
Continuous human movement recognition system. The dyneme, skill, context, and activity models construct a hypothesis for interpreting a video sequence. (Source: Reproduced with permission from the University of Canterbury.)

- The context model models the semantic structure of movement by modeling the probability of sequences of skills simplified to only triplets or pairs of skills, as discussed later in the chapter.

- The activity model defines the possible human movement activities that the search can hypothesize, representing each activity as a linear sequence of skills (not limited to only triplets or pairs as in the context model) [1].

Three principal components comprise the basic hypothesis search: a dyneme model, a skill model, and a context model.

Dyneme Model

As the phoneme is a phonetic unit of human speech, so the dyneme is a dynamic unit of human motion. The word dyneme is derived from the Greek *dynamikos* "powerful," from *dynamis* "power," from *dynasthai* "to be able," and in this context refers to motion. This is similar to the phoneme being derived from *phono* meaning "sound" and with *eme* inferring the smallest contrastive unit. Thus dyneme is the smallest contrastive unit of movement. The movement skills in this research are constructed from an alphabet of 35 dynemes that HMMs use to recognize the skills. This approach has been inspired by the paradigm of the phoneme as used by the continuous speech recognition research community where pronunciation of the English language is seen to be constructed from approximately 50 phonemes [1].

The dyneme can be understood as a type of movement notation. An example of a similar movement notation system is that used in dance. Many dance notation systems have been designed over the centuries. Since 1928, there has been an average of one new notation system every four years [1]. Currently, there are two prominent dance notation systems in use: Labanotation and Benesh.

Although manual movement notation systems have been developed for dance, computer vision requires an automated approach in which each human movement skill has clearly defined temporal boundaries. Just as it is necessary to isolate each letter in cursive handwriting recognition, so it is necessary in the computer vision analysis of full-body human movement to define when a dyneme begins and ends. This research defined an alphabet of dynemes by deconstructing (mostly manually) hundreds of movement skills into their correlated lowest common denominator of basic movement patterns [1].

Although there are potentially an infinite number of movements the human body could accomplish, there are a finite number of ways to achieve motion in any direction. For simplicity, consider only *xy* motion occurring in a frontoparallel plane:

- The human body *x* translation caused by:

  - Min-max of hip flexion/extension => (gait, crawl)
  - Min-max of hip abduction/adduction or lateral flexion of spine => cartwheel
  - Min-max of shoulder flexion => (walk on hands, drag-crawl)
  - Rotation about the transverse (forward roll) or antero-posterior (cartwheel) axes
  - Min-max foot rotation => (isolated feet-based translation)
  - Min-max waist angle => (inchworm)

- The human body *y* translation caused by:

  - Min-max center of mass (COM) => (jump up, crouch down)

- The human body no *x* or *y* translation:

  - Motion of only one joint angle => (head turn)
  - Twist–rotation about the longitudinal axis => (pirouette) [1]

The number of dynemes depends on the spatial-temporal resolution threshold. A dyneme typically encapsulates diverse fine granules of motion. A gait step dyneme, for example, encompasses diverse arm motions (shoulder and elbow angular displacements, velocities, and accelerations) where some arm movements have a higher probability of occurring during the step dyneme than others [1].

A hidden Markov model offers a natural choice for modeling human movement's stochastic aspects. HMMs function as probabilistic finite state machines: The model consists of a set of states, and its topology specifies the allowed transitions between them. At every time frame, a HMM makes a probabilistic transition from one state to another and emits a motion vector with each transition [1].

Figure 28-7 shows a HMM for a dyneme [1]. A set of state transition probabilities ($p1$, $p2$, and $p3$) governs the possible transitions between states. They specify the probability of going from one state at time t to another state at time $t + 1$. The motion vectors emitted while making a particular transition represent the characteristics for the human movement at that point, which vary corresponding to different executions of the dyneme. A probability distribution or probability density function models this variation. The functions ($p(y|1)$, $p(y|2)$, and $p(y|3)$) can be different for different transitions. These distributions are modeled as parametric distributions—a mixture of multidimensional Gaussians [1].

Figure 28-7   *Hidden Markov model for a dyneme. State transition probabilities p1, p2, p3 govern the possible transitions between states. (Source: Adapted with permission from the University of Canterbury.)*

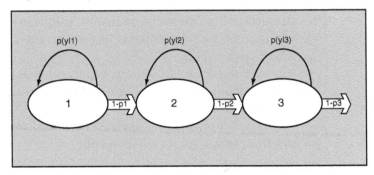

The HMM shown in Figure 28-7 consists of three states. The dyneme's execution starts from the first state and makes a sequence of transitions to eventually arrive at the third state. The duration of the dyneme equals the number of video frames required to complete the transition sequence. The three transition probabilities implicitly specify a probability distribution that governs this duration. If any of these transitions exhibit high self-loop probabilities, the model spends more time in the same state, consequently taking longer to go from the first to the third state. The probability density functions associated with the three transitions govern the sequence of output motion vectors [1].

A fundamental operation is the computation of the likelihood that a HMM produces a given sequence of motion vectors. For example, assume that the system extracted T motion vectors from human movement corresponding to the execution of a single dyneme and that the system seeks to infer which dyneme from a set of 35 was performed. The procedure for inferring the dyneme assumes that the ith dyneme was executed and finds the likelihood that the HMM for this dyneme produced the observed motion vectors [1].

If the sequence of HMM states is known, the probability of a sequence of motion vectors can be easily computed. In this case, the system computes the likelihood of the tth motion vector, y_t, using the probability density function for the HMM state at time t. The likelihood of the complete set of T motion vectors is the product of all these individual likelihoods. However, because the actual sequence of transitions is not known, the likelihood computation process sums all possible state sequences. Given that all HMM dependencies are local, efficient formulas can be derived for performing these calculations recursively [1].

With various dynemes overlapping, a hierarchy of dynemes is required to clearly define the boundary of each granule of motion and so define a high-level movement skill as the construction of a set of dynemes. For example, a somersault with a full-twist rotates 360° about the transverse axis in the somersault and 360° about the longitudinal axis in the full-twist. This twisting salto is then an overlap of two different rotational dynemes. Whole-body rotation is more significant than a wrist flexion when recognizing a skill involving full-body movement. To this end, motion vectors are divided into parallel streams with different weights in the dyneme model to support the following descending hierarchy of five dyneme categories:

- Full-body rotation
- COM motion (including flight)
- Static pose
- Weight transfer
- Hierarchy of DOFs [1]

Each category of motion is delineated by a pause, min, max, or full, half, quarter rotations. For example, a COM category of dyneme is illustrated in Figure 28-8(a) where each running step is delimited by COM minima [1]. A full 360° rotation of the principal axis during a cartwheel in Figure 28-8(b) illustrates a rotation dyneme category [1].

A dyneme model computes the probability of motion vector sequences under the assumption that a particular skill produced the vectors. Given the inherently

Figure 28-8   *COM parameters during running and principal-axis parameters through a cartwheel. (Source: Reproduced with permission from the University of Canterbury.)*

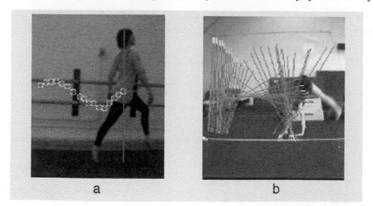

a   b

stochastic nature of human movement, individuals do not usually perform a skill in exactly the same way twice. The variation in a dyneme's execution manifests itself in three ways: duration, amplitude, and phase variations. Dynemes in the surrounding context can cause variations in a particular dyneme's duration, amplitude, and phase relationships, a phenomenon referred to in this chapter as coexecution. Hence, in some cases the dynemes in the surrounding context affect a particular dyneme's motion vector sequence. This coexecution phenomenon is particularly prevalent in poorly executed movement skills. The system models coexecution by assuming that the density of the observations depends on both the specific dyneme and the surrounding dynemes. However, modeling every dyneme in every possible context generates a prohibitively large number of densities to be modeled. For example, if the dyneme alphabet consists of 35 dynemes, and the system models every dyneme in the context of its immediately surrounding neighbors, it would need to model 42,875 densities. Consequently, the approach taken here clusters the surrounding dynemes into a few equivalence classes of categories, thus reducing the densities that require modeling [1].

Skill Model

The typical skill model shown in Table 28-4 lists each skill's possible executions, constructed from dynemes [1]. An individual movement skill can have multiple forms of execution, which complicates recognition.

The system chooses the skill model on a task-dependent basis, trading off skill model size with skill coverage. Although a search through many videos can easily find dyneme sequences representing commonly used skills in various sources, unusual skills in highly specific situations may require manual specification of

Table 28-4 *Typical Minimal Dyneme Skill Model (with the skill walk having two alternative executions)*

Movement	Skill Dyneme
Walk	step (right), step (left)
Walk	step (left), step (right)
Handstand from stand	step, rotate-fwd (180°)
Jump	knee-extension, COM-flight
Backward salto	knee-extension, COM-flight, rotate-bwd (360°)

the dyneme sequence. In fact, the initial definition of skills in terms of dynemes involved extensive manual specification in this research [1].

Training

Before using a HMM to compute the likelihood values of motion vector sequences, the HMMs must be trained to estimate the model's parameters. This process assumes the availability of a large amount of training data, which consists of the executed skill sequences and corresponding motion vectors extracted from the videostream. The maximum likelihood (ML) estimation process training paradigm is used for this task. Given a skill sequence and corresponding motion vector sequence, the ML estimation process tries to choose the HMM parameters that maximize the training of motion vectors' likelihood of computation by using the HMM for the correct skill sequence [1].

The system begins the training process by constructing a HMM for the correct skill sequence. First, it constructs the HMMs for each skill by concatenating the HMMs for the dynemes that compose that skill. Then it concatenates the skill HMMs to form the HMM for the complete skill sequence, where the transitional probabilities between connecting states for these HMMs are set to one and those between nonconnecting states are set to zero. For example, the HMM for the sequence "skip" would be the concatenation of the two dynemes, "step, hop" [1].

Hypothesis Search

The hypothesis search seeks the skill sequence with the highest likelihood given the model's input features and parameters [1]. Because the number of skill sequences increases exponentially with the skill sequence's length, the search might seem at first to be an intractable problem for anything other than short skill sequences from a small lexicon of skills. However, because the model has only local probabilistic dependencies, the system can incrementally search through the hypothesis in a left-to-right fashion and discard most candidates with no loss in optimality [1].

Although the number of states in the context model can theoretically grow as the square of the number of skills in the skill model, many skill triplets never actually occur in the training data. The smoothing operation backs off to skill pair and single skill estimators, substantially reducing size. To speed up the recursive process, the system conducts a beam search, which makes additional approximations such as retaining only hypotheses that fall within the threshold of the maximum score in any time frame [1].

Given a time series, the Viterbi algorithm computes the most probable hidden state sequence; the forward-backward algorithm computes the data likelihood and expected sufficient statistics of hidden events such as state transitions and occupancies. These statistics are used in the Baum-Welch parameter re-estimation to maximize the likelihood of the model given the data. The expectation-maximization (EM) algorithm for HMMs consists of forward-backward analysis and the Baum-Welch re-estimation iterated to convergence at a local likelihood maximum [1].

Brand [1] replaced the Baum-Welch formula with parameter estimators that minimize entropy to avoid the local optima. However, with hundreds of movement skill samples, it is felt that the research presented in this chapter avoided this pitfall with a sufficiently large sample size. The Viterbi alignment is applied to the training data, followed by the Baum-Welch re-estimation. Rather than the rule-based grammar model that is common in speech processing, a context model is trained from the movement skill data set. The hidden Markov model tool kit1 (HTK) is used to support these dyneme, skill, and context models [1].

The HTK is a portable tool kit for building and manipulating hidden Markov models. HTK is primarily used for speech recognition research, although it has been used for numerous other applications including research into speech synthesis, character recognition, gesture recognition, and DNA sequencing. HTK is in use at hundreds of sites worldwide. HTK consists of a set of library modules and tools available in C source form. The tools provide sophisticated facilities for speech analysis, HMM training, testing, and results analysis. The software supports HMMs using both continuous density mixture Gaussians and discrete distributions that can be used to build complex HMM systems [1].

Performance

Hundreds of skills were tracked and classified using a 1.8 GHz, 640 MB RAM Pentium IV platform processing 24-bit color within the Microsoft DirectX 8.1 environment under Windows XP. The video sequences were captured with a JVC DVL-9800 digital video camera at 30 fps, 720 × 480 pixel resolution. Each person moved in front of a stationary camera with a static background and static lighting conditions. Only one person was in frame at any one time. Tracking began when the whole body was visible, which enabled initialization of the clone-body-model [1].

The skill error rate quantifies CHMR system performance by expressing, as a percentage, the ratio of the number of skill errors to the number of skills in the reference training set. Depending on the task, the CHMR system skill error rates can vary by an order of magnitude. The CHMR system results are based on a set of a total of 840 movement patterns, from walking to twisting saltos. From this, an independent test set of 200 skills were selected, leaving 640 in the training set. Training and testing skills were performed by the same subjects. These were successfully tracked, recognized, and evaluated with their respective biomechanical components quantified, where a skill error rate of 4.5% was achieved [1].

Recognition was processed using the (Microsoft-owned) Cambridge University Engineering Department HMM tool kit (HTK), with 96.8% recognition accuracy on the training set alone. Also included was a more meaningful 95.5% recognition accuracy for the independent test set where $H = 194$, $D = 7$, $S = 9$, $I = 3$, $N = 200$ ($H =$ correct, $D =$ Deletion, $S =$ Substitution, $I =$ Insertion, $N =$ test set, Accuracy $= (H - I)/N$). Thus, 3.5% of the skills were ignored (deletion errors) and 4.5% were incorrectly recognized as other skills (substitution errors). There was only about 1.5% insertion errors—that is, incorrectly inserting or recognizing a skill between other skills [1].

The HTK performed Viterbi alignment on the training data followed by Baum-Welch re-estimation, with a context model for the movement skills. Although the recognition itself was faster than real time at about 120 fps, the tracking of 32 DOF with particle filtering was computationally expensive, using up to 16 seconds per frame [1].

Figure 28-9 illustrates the CHMR system recognizing the sequence of skills: stretch and step, cartwheel, and step and step from continuous movement [1]. As each skill is recognized, a snapshot of the corresponding pose is displayed in the fourth tile. Below each snapshot is a stick figure representing an internal identification of the recognized skill. Notice that the cartwheel is not recognized after the first-quarter rotation. Only after the second-quarter rotation is the skill identified as probably a cartwheel [1].

Motion blurring lasted about 10 frames on average, with the effect of perturbing joint angles within the blur envelope (see Figure 28-10) [1]. Given a reasonably accurate angular velocity, it was possible to sufficiently de-blur the image. There was a minimal motion blur arising from rotation about the longitudinal axis during a double twisting salto due to a low surface velocity tangential to this axis from minimal radius with limbs held close to a straight body shape. This can be seen in Figure 28-11, where the arms exhibit no blurring from twisting rotation, contrasted with motion blurred legs due to a higher tangential velocity of the salto rotation [1].

Figure 28-9   *CHMR system recognizing stretching into a cartwheel followed by gait steps. In each picture, four tiles display CHMR processing steps: Tile 1: Principal axis through the body; Tile 2: Body frame of reference (normalized to the vertical); Tile 3: Motion vector trace (subset displayed); Tile 4: Recognizing step, stretch, and cartwheel indicated by stick figures with respective snapshots of the skills. (Source: Reproduced with permission from the University of Canterbury.)*

stretch & step	quarter rotation (unknown skill)	half rotation (cartwheel identified)
finished cartwheel	right step	left step

Figure 28-10   *(a): CHMR system tracking through motion blur of right calf and foot segments during a flic-flac (back-handspring); (b): Alternative particles (knee angles) for the right calf location; (c): Expected value of the distribution. (Source: Reproduced with permission from the University of Canterbury.)*

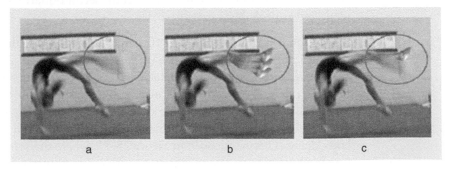

Figure 28-11
Minimal motion blur arising from rotation about the longitudinal axis during a double twisting salto. (Source: Reproduced with permission from the University of Canterbury.)

An elongated trunk with disproportionately short legs is the body-model consequence of the presence of a skirt. The clone-body model failed to initialize for tracking due to the variance of body-part proportions exceeding an acceptable threshold. The CHMR system also failed for loose clothing. Even with smoothing, the joint angles were surrounded by baggy clothes permutated through unexpected angles within an envelope sufficiently large as to invalidate the tracking and recognition [1].

Let's take a look at the biometric verification of gait, anthropometric data, human activities, and movement disorders through the use of the continuous human movement recognition (CHMR) framework, which was introduced earlier in the chapter. A novel biometric verification of anthropometric data is presented here based on the realization that no one is average-sized in as many as 10 dimensions. These body-part dimensions are quantified using the CHMR body model. Gait signatures are then evaluated using motion vectors and temporally segmented by-gait dynemes, and projected into a gait space for an eigengait-based biometric verification. Left-right asymmetry of gait is evaluated using robust CHMR left-right labeling of gait strides. Accuracy of the gait signature is further enhanced by incorporating the knee-hip angle-angle relationship popular in biomechanics gait research together with other gait parameters. These gait and anthropometric biometrics are fused to further improve accuracy. The next biometric identifies human activities, which requires a robust segmentation of the many skills encompassed. For this reason, the CHMR activity model is used to identify various activities from making coffee to using a computer. Finally, human movement disorders were evaluated by studying patients with dopa-responsive Parkinsonism and age-matched normals who were videotaped during several gait cycles to determine a robust metric for classifying movement disorders. The results suggest that the CHMR system enabled successful biometric verification of anthropometric data, gait signatures, human activities, and movement disorders [2].

Biometric Verification of Anthropometric Data, Gait Signatures, Human Activities, and Human Movement Disorders

Although there is a large body of work describing computer vision systems for modeling and tracking human bodies, the vision research community has only recently begun to investigate gait as a biometric. Identifying humans from their gait has become an extremely active area of computer vision [2]. This section describes a robust gait metric with a novel left-step-right-step vector of spatial-temporal parameters to capture the left-right gait asymmetry of the population.

Researchers recently combined body shape and gait into a single biometric applied to the gait databases from CMU (25 subjects), U Maryland (55 subjects), U Southampton (28 subjects), and MIT (25 subjects). Researchers also reported a good 73% recognition rate on a larger sample of 74 subjects [2]. Instead of using a 2D shape-based pose [2], this research employs a novel application of anthropometric dimensions from a 3D body used to uniquely identify individuals from the variability of physical proportions. Although previous work has been done on body-model acquisition from multiple cameras [2], the clone-body-model was sized by the monocular CHMR system described earlier.

The biometric verification of anthropometric data, gait signatures, human activities, and human movement disorders depends on accurately quantifying and recognizing human body movement using a precise model of the body being tracked. This biometric verification process is enabled with data from the CHMR system described earlier, which is used to noninvasively quantify and temporally segment continuous human motion in monocular video sequences. Relative dimensions from the CHMR body model support biometric identification from a library of anthropometric signatures. Gait signatures are correlated using dyneme segmented left-step-right-step motion vector arrays. General human movement activity identification is demonstrated using the CHMR activity model discussed previously [2].

Video image analysis is also able to provide quantitative data on postural and movement abnormalities and thus has an important application in neurological diagnosis and management. This section describes an approach to classifying the gait of Parkinsonian patients and normal subjects using video image analysis results from the CHMR system [2].

Anthropometric Biometrics

Vitruvius in 1st century BC Rome assumed all men were identically proportioned [2], as did Leonardo da Vinci with his famous drawing of the human figure, based on the Vitruvian man (see Figure 28-12) [2]. More than 2,000 years after Vitruvius wrote his 10 books on architecture, Le Corbusier [2] revived interest in the Vitruvian norm with his mapping of human proportions (see Figure 28-13) onto the Golden Section (developed by Euclid in 300 BC Greece, where Euclid had named the extreme and mean ratio) [2].

However, this "average-sized human" model assumed by Vitruvius, da Vinci, and Le Corbusier is a fallacy, as there is no average-sized person. A human with average proportions does not exist. More recent anthropometric data [2] shows that people who are average in two dimensions constitute only about 7% of the population; those in three, only about 3%; those in four, less than 2%. Since there is no one who is average in 10 dimensions [2], a ten-dimensional space of physical proportions can be used as a reasonably accurate biometric. What is not clear from anthropometric data is the natural asymmetry of the human body, which can also be utilized to further improve the accuracy of anthropometric verification. This anthropometric asymmetry becomes apparent with one foot fitting a pair of shoes better than the other foot. This novel biometric promises maximal between-person variability while supporting minimal within-person variability across time within the adult population [2].

Figure 28-12
The Vitruvian man by Leonardo da Vinci. (Source: Reproduced with permission from the University of Canterbury.)

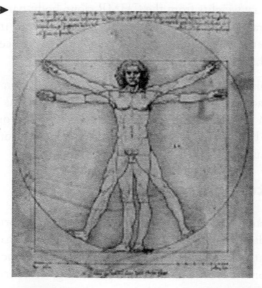

Figure 28-13
*Le Modulor man
by Le Corbusier.
(Source:
Reproduced with
permission from
the University of
Canterbury.)*

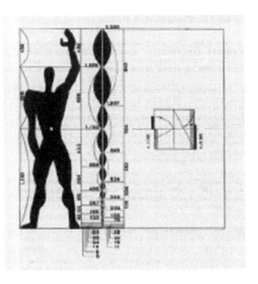

From the initial 50th percentile anthropometric proportions (see Figure 28-14), the body model used in the tracking process is automatically normalized and sized to the relative proportions of the person being tracked, providing at least a ten-dimensional space of physical proportions for this biometric measure [2]. Anthropometric data [2] is used to threshold variance from average body-part proportions, thus allowing for age, race, and gender. Each individual is represented by a normalized vector of physical proportions, with associated accuracy weights from the CHMR clone-body-part averaged radius accuracy. The tracking process maps each person into the training set with this anthropometric vector. The weights enable a confidence measure to be calculated and thresholded for a match [2].

Angular displacement of DOFs during gait enable ongoing improvement in body model accuracy as joint locations; body-part lengths become further revealed through the temporal resolution of self-occlusions. For example, turning 180° to pace back significantly improves the accuracy of frontal dimensions, as shown in Figure 28-15 [2].

Gait Signature

Approaches to gait recognition can be divided into two categories: model-based and holistic. Holistic methods [2] derive statistical information directly from the gait image and attempt to correlate various features for biometric verification. Initial results from holistic approaches are promising, with recognition rates as

Figure 28-14
Front and side views of the 50th percentile proportions. Drawings from H. Dreyfuss, The Measure of Man, 1978. (Source: Reproduced with permission from the University of Canterbury.)

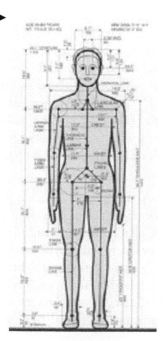
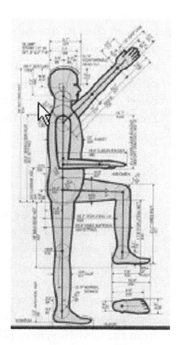

Figure 28-15
Turning and gait step overlaid, 58 frames apart. (Source: Reproduced with permission from the University of Canterbury.)

high as 100% for small databases of hundreds of subjects. However, no research has been done to establish if these high recognition rates will translate to larger databases with thousands of subjects as in face recognition or even millions of subjects as in iris recognition. Model-based approaches rely on a model being fitted to the image data. Researchers have proposed a method for gait recognition based on moving feature analysis. The gait signature was extracted by using a Fourier series to describe the motion of the leg and temporally correlate this to determine the dynamic model from a sequence of images. Performance of this technique was also promising, with recognition rates of up to 90%; however, the test sample was small [2].

Engaging a model-based approach, the CHMR system is used to temporally segment step dynemes for which data from the motion vectors are analyzed to determine a unique gait signature. Similar to the static anthropometric left-right asymmetry of the body is the dynamic left-right asymmetry of gait. Accurate temporal segmentation and identification of the left and right steps is required to fully exploit this asymmetrical parametric diversity of gait populations. In a new approach to biometric verification of gait, this asymmetry is quantified using the motion vectors given the differentiated left and right step segmentation from the CHMR system [2].

A gait pattern classifier takes a temporally normalized sequence of gait delimited motion vectors as the input feature vector—essentially two alternate step dynemes. An eigengait approach [2] is employed, in which a similarity plot is treated the same way that a face is recognized in the eigenface approach and with a similar novel eigenspike approach applied successfully to identify epileptic spikes [2]. The motion vectors of left-right step dyneme pairs are found to be the principal components of the distribution of the feature space. This is followed by the standard pattern classification of new feature vectors in the lower-dimensional space spanned by the principal components [2].

Normalized left and right stride-dyneme motion vectors are concatenated into one single vector. The right stride is appended to the left stride to form a single gait vector g for each person. For recognition, a gait vector is projected into a reduced set of basis vectors. These basis vectors are the global eigenvectors associated with the largest eigenvalues of a covariance matrix of the training set of N people ($g_1 \ldots g_N$) found by the eigenvalue decomposition of their covariance matrix [2].

Of particular interest is the knee-hip angle-angle relationship popular in biomechanics gait research, especially since the minimum possible gait DOFs would include only the hip and knee flexions. In this research, left-right gait asymmetry as a gait feature is explored by using the robust CHMR left-right labeling of gait strides to enable a robust phase alignment of the alternating steps, further enhancing the accuracy of this metric [2].

Gait and anthropometry have the advantage over other biometrics such as fingerprint and iris in that they are noninvasive to the extent that the subject may not even know he is being recognized in security and surveillance applications. The gait and anthropometric biometrics also have a proximity advantage over face detection, since they can operate on a lower-resolution image. With the CHMR approach, it is possible to fuse the gait and anthropometric biometrics to improve accuracy [2].

Activity Identification

Research into human activities generally represents an activity as a single skill, such as walk, run, turn, sit, and stand [2]. This is problematic since human activities are often more complex and consist of a sequence of many possible skills. An activity can be more accurately defined as a sequence of one or more core skills. This research seeks to broaden the distinction between activity and skill. The CHMR activity model discussed earlier in the chapter defines possible human movement activities that the search can hypothesize, representing each activity as a sequence of one or more core skills [2].

For example, making coffee consists of the minimum sequence "spoon-coffee, pour-water." Many other potential skills exist in the make-coffee sequence, with pre-skills such as "boil-water, get-cup, get-spoon" and post-skills such as "stir-coffee, carry-cup." Therefore a set of zero or more related pre- and post-skills are associated with each activity to enable the temporal grouping of skills relating to a particular activity. In this way not only are a sequence of motion vectors temporally segmented into a skill, but a sequence of skills can be temporally segmented into an activity. Five activities were performed, each by three people:

1. **Coffee:** Making coffee

2. **Computer:** Entering an office and using a computer

3. **Tidy:** Picking an object off the floor and placing it on a desk

4. **Snoop:** Entering an office, looking in a specific direction, and exiting

5. **Break:** Standing up, walking around, sitting down [2]

Although an attempt was made to track lifting a coffee pot, carried objects are not recognized as separate from the human body. This research does not cover models beyond a human body model. Consequently, holding large objects such as a coffee pot destabilizes the tracking due to the body part holding the object being dimensioned beyond an acceptable anthropometric threshold. The activities that were defined in this chapter did not involve carrying objects larger than a small coffee cup [2].

The CHMR system is utilized to recognize various activities, from making coffee to using a computer. The CHMR activity model defines the possible human movement activities that the search can hypothesize, representing each activity as a linear sequence of skills [2].

Movement Disorders

Patients with neurological disorders frequently show some degree of gait abnormality. A typical example is Parkinson's disease (PD). Common motor symptoms of PD include rhythmic shaking of one or occasionally more limbs (tremor), slowness in movement (bradykinesia), stiffness of joints (rigidity), slightly bent and flexed posture, and failure of the arms to swing freely when walking [2].

Walking is a highly refined, remarkable, and automatic human skill that is easily taken for granted. The basic reflex for walking, which is probably located in the spinal cord, is present at birth. Parents, relatives, and friends are all very pleased, excited, and proud when an infant takes the first steps. At the other end of the time spectrum, abnormalities of gait and falling tend to be problems of the elderly. Disorders of gait and mobility are second only to impaired mental function as the most frequent neurological effects of aging. Normal gait, stance, and balance require precise input from proprioceptive (position sense), vestibular (inner ear mechanisms and their connections within the brain stem), and visual pathways as well as auditory and tactile information. Two of the three major afferent systems (proprioceptive, vestibular, and visual) must be intact to maintain balance. Afferent data must be integrated in the brain stem and brain through motor (pyramidal and extrapyramidal) and cerebellar pathways, which then serve as the efferent arc of the important skill of walking. Dysfunction in the afferent or efferent systems or in the central integrating centers can lead to gait problems. Gait disorders in the elderly are frequently heterogeneous and often multifactorial in origin [2].

The function of the extrapyramidal system is to modulate posture, right reactions, and associated movements. The Parkinsonian gait is characterized by a flexed posture, diminishing arm swing, and rigid, small-stepped, shuffling gait.

Rising from a sitting position may be slow or impossible. Patients often have difficulty with initiation of movement and turns. Disturbances of balance are often present (impairment of postural reflexes). The legs are stiff and bent at the knee and hips. As the patient walks, the upper part of the body gets ahead of the lower part and the steps become smaller and more rapid (festination). Turning is accomplished with multiple unsteady steps, with the body turning as a single unit (en bloc) [2].

The clinical approach to gait analysis is heavily dependent on subjective observation of the patient's gait. Although the reliability of subjective observation may be improved by systematic procedures and rating scales, the asynchronous series of changes in the complex articulated assembly of the human body presents such a maze of data that few persons could assimilate them all. This limitation may be minimized by quantitative documentation of the patient's performance with reliable instrumentation to provide a permanent record of fact. Quantitative gait analysis is an important clinical tool for quantifying normal and pathological patterns of locomotion and has been shown to be useful for prescription of treatment as well as in the evaluation of the results of such treatment [2].

Commercial quantitative video analysis techniques require patients to be videotaped while wearing joint markers in a highly structured laboratory environment with extensive set-up procedures. This limits the usefulness of video-based analysis in routine clinical practice, so it is rarely used in this capacity. Current video analysis would also be unable to analyze existing videotape libraries. Based on the CHMR model discussed previously, a video analysis system is presented here, free of markers and set-up procedures, that quantitatively identifies gait abnormalities in real time. The aim in this research is to develop a system that is able to meet the needs of a busy movement disorders clinic in both on-line and off-line analysis and diagnosis [2].

Performance

Gait sequences and activity skills were tracked and classified using a 1.8 GHz, 640 MB RAM Pentium IV platform processing 24-bit color within the Microsoft DirectX 8.1 environment under Windows XP. The video sequences were captured with a JVC DVL-9800 digital video camera at 30 fps, 720 × 480 pixel resolution [2].

Each person moved in front of a static blue-screen background with constant lighting conditions and no foreground object occlusion. Only one person was in the frame at any one time. Tracking began when the whole body was visible, which enabled initialization of the body model. Each person walked parallel to the image plane in front of a stationary camera, and then turned to walk back again, repeating this sequence five times on average. The body model accuracy was significantly improved by the first turn [2].

The first turn also enabled accurate texture mapping of the occluded side, and the varying perspectives of the body enabled radii to be more accurately determined, as shown in Figure 28-16 [2]. The large number of frames available in a single turn is of considerable benefit to accurately dimensioning the body model [2].

For completeness, biometric results are less ambiguously quantified using five categories: correctly recognized (true positive), incorrectly recognized, correctly rejected (true negative), false negatives, and false positives. False negatives represent incorrectly rejected candidates from the training set, and false positives are incorrectly recognized candidates not present in the training set. The performance of the anthropometric and gait biometrics is presented next by using these five categories of results [2].

Anthropometric Biometrics

Training samples of 48 people in tight clothing are represented by vectors of physical proportions with associated accuracy weights. The tracking process also

Figure 28-16
Tracking the combined average clone-body-part radius and color accuracy over 17 frames during a turn phase illustrates more dimensions being revealed to further increase clone-body-model accuracy. (Source: Reproduced with permission from the University of Canterbury.)

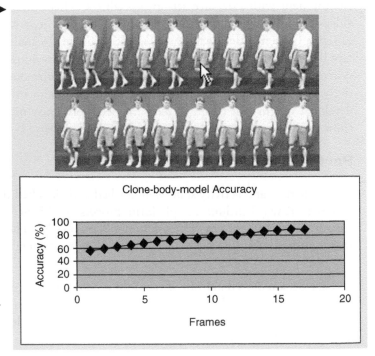

→

Table 28-5 *Biometric Verification of Anthropometric Data*

Correct Recognition	Incorrect Recognition	False Negative	Correct Rejection	False Positive
92%	2%	6%	90%	10%

attempted to recognize 10 people who were not present in the training gallery. The weights enable a confidence measure to be calculated and thresholded for a match [2].

Based on the training data, a recognition rate of 92% was achieved for the anthropometric biometric for a confidence threshold of 99% with one false positive. Dimension inaccuracies were reduced by tight-fitting clothes being worn by the training and test samples. Accuracy of body proportions was significantly improved by the first turn due to varying perspectives of the body that enabled radii to be more accurately determined [2].

In Table 28-5, false negatives represent incorrectly rejected candidates from the training set and false positives are incorrectly recognized as candidates not present in the training set [2]. It was found that dimension inaccuracies were introduced by hair, footwear, and thick clothes such as heavy woolen sweaters. Consequently, some head dimensions were weighted low due to hairstyle-induced inaccuracies. Similarly, foot dimension weights were also low due to adverse footwear influence. It was found that large loose clothes such as coats, skirts, and dresses occluded body parts, causing the body model to fail to initialize for tracking due to the variance of body-part proportions exceeding an acceptable threshold [2].

Gait Signature

A sample of 48 people walking in a sagittal plane became the training gallery, with an additional 10 unknowns. Reasonably tight-fitting clothes were worn by all 58 people with no severe self-occlusions of both legs, which would cause this approach to fail. The two most significant gait biometric predictors were found to be the knee-hip angle-angle relationship and the left-right asymmetry of that relationship, a subset of the left-right step-dyneme vector. Figure 28-17 illustrates the uniqueness of these angle-angle relationships by overlaying the knee-hip diagrams of four different people [2].

Gait-specific features were normalized with respect to the gait cycle. The principal components were found for the distribution of the feature space of gait-step dyneme pairs by standard pattern classification of new feature vectors

Figure 28-17
*Hip-knee
angle-angle
relationships of
four different
people. (Source:
Reproduced with
permission from
the University of
Canterbury.)*

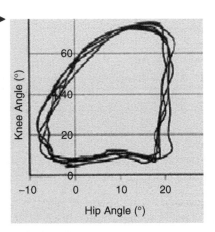

in the lower dimensional space spanned by the principal components. This eigengait analysis yielded the same recognition rate of 88% as the hip-knee angle and asymmetry fusion [2].

In Table 28-6, false negatives represent incorrectly rejected candidates from the training set and false positives are incorrectly recognized candidates not present in the training set [2]. Interestingly, the fusion of both the anthropometric biometric and gait biometric raised the accuracy from 92% and 88% respectively, to 94% [2].

About 30% of subjects moved their arms minimally, causing the far arm to be occluded during the gait cycle, to be accurately measured only when walking

Table 28-6   *Gait Signature Recognition Results*

Gait Feature %	Correct Recognition	Incorrect Recognition	False Negative	Correct Rejection	False Positive
Gait period	48	32	20	60	40
Arm swing amplitude	64	14	22	70	30
Stride amplitude	66	22	12	60	40
Arm swing asymmetry	76	6	18	80	20
Hip-knee angle-angle	82	6	12	90	10
Hip-knee left-right asym.	86	4	10	90	10
Hip-knee angles & asym.	88	2	10	100	0
Eigengait analysis	88	0	12	100	0

back in the opposite direction. It was assumed that the near-side visible arm swung identically when it was occluded. The large number of small arm swing amplitudes accounts for the low 64% gait signature recognition based on arm amplitude and 76% recognition based on arm swing asymmetry. In future studies involving carried objects [2], arm swing will become less relevant to the gait signature.

Activity Identification

The activity error rate quantifies CHMR system performance by expressing, as a percentage, the ratio of the number of activity errors to the number of activities in the reference training set. The CHMR system was tested on a training set of five activities with an activity error rate of 0%. However, the sample size is too small for this result to be significant [2].

Results for the following activities are detailed in Table 28-7 [2]:

- Coffee: Making coffee
- Computer: Entering an office and using a computer
- Tidy: Picking an object (pen) off the floor and placing it on a desk
- Snoop: Entering an office, looking in a specific direction and exiting
- Break: Standing up, walking around, sitting down [2]

With such a small sample of activities, the activity recognition results reflect the skill recognition results of 4.5% skill error rate as explained earlier in the chapter [2].

Although carrying a spoon in "coffee" and a pen in "tidy" caused no tracking problems, attempts to carry objects such as a large mug caused the arm to permutate through unexpected angles within an envelope sufficiently large as

Table 28-7 *Activity Recognition Results*

Activity %	Recognition	False Negative
Coffee	100	0
Computer	100	0
Tidy	100	0
Snoop	100	0
Break	100	0

to invalidate the tracking and recognition. Carrying even larger objects such as a briefcase caused the body model to fail. With no valid body to track, tracking and recognition did not proceed. It is intended to extend the tracking process to recognize carried objects as separate from the human body for a more useful activity recognition biometric [2].

Movement Disorders

The gaits of 20 patients with dopa-responsive Parkinsonism (PD) and 15 age-matched normals were tracked and classified. The PD video data analyzed in this chapter were validated in a previous study [2].

A number of gait parameters were analyzed to determine their significance to the correlation of PD gait (see Figure 28-18) [2]. These features included leg swing, arm swing, gait period, and shape of the gait cycle limb swing waveform. The PD limb swing amplitude was generally less than that of normals, but it was found to vary among both PDs and age-matched normals enough to result in about 11% false positives and so was not a useful feature (refer to Table 28-8) [2]. The period of the gait was also unable to reliably classify PD gait. The most useful feature proved to be the left-right asymmetry of waveform shape due to a significant asymmetry in the PD gait arising from the deterioration of one side more quickly than the other. By using this feature, the system correctly classified 95% of subjects with one false negative [2].

The two graphs in Figure 28-19 illustrate this PD asymmetry by contrasting an irregular PD gait with the regular leg swing of a normal gait [2]. PD patients in Figure 28-18 show either no arm swing (subjects C, G, and L) or the significant asymmetry typical of PD gait (subjects E, F, and I) [2]. Also visible was

Figure 28-18   *PD gait samples illustrating characteristic body flexion with asymmetrical or minimal arm swing. (Source: Reproduced with permission from the University of Canterbury.)*

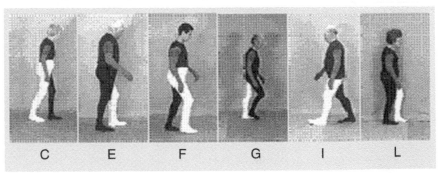

C   E   F   G   I   L

Table 28-8 *Correlation of Limb Swing Amplitude, Period, and Left-Right Asymmetry*

%	Correct PD	False Negative	Correct Normal	False Positive
Amplitude: leg	90	10	80	20
Arm	85	15	73	27
Gait period	55	45	88	12
Gait asymmetry	95	5	93	7

Figure 28-19 *Graphs contrasting an asymmetrical leg swing typical of PD gait with a normal symmetrical gait. (Source: Reproduced with permission from the University of Canterbury.)*

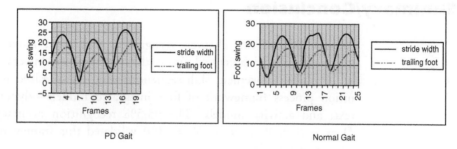

PD Gait   Normal Gait

the flexed body and limb shape (subjects C, E, F, and G) common in PD. This contrasts with the somatotype and age-matched normals in Figure 28-20 [2]. The degree to which body and limbs were flexed was not addressed by this study.

The single PD gait sample not detected by this system had a gait similar to normal, but some tremor was visible. However, due to the low-resolution

Figure 28-20
Age-matched normal gait samples. (Source: Reproduced with permission from the University of Canterbury.)

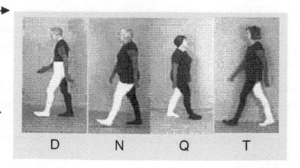

images and low frame rate, the tremor was not able to be analyzed. Another problem arose from the minimal arm swing common in PD gaits. With minimal arm swing in many PD gaits, tracking the far arm caused problems because it was occluded during the entire gait passes. To stabilize tracking in this case, the location of the far arm was assumed to be near vertical or similar to the location of the near arm [2].

Finally, the subjects used in this study had been used in a previous study and were therefore established as known PDs and normals. The classification accuracy was improved in this study by analyzing entire gait cycles rather than a static gait snapshot of each subject as in the previous study [2].

Summary/Conclusion

As described in this chapter, recognition of human movement skills has been successfully processed using the Cambridge University HMM tool kit. Probable movement skill sequences were hypothesized using the recognition process framework of four integrated models—dyneme, skill, context, and activity models. The 95.5% recognition accuracy ($H = 194$, $D = 7$, $S = 9$, $I = 3$, $N = 200$) validated this framework and the dyneme paradigm [1].

However, the 4.5% error rate attained in this research is not yet evaluating a natural-world environment, nor is this a real-time system, with up to 16 seconds to process each frame. The CHMR system did achieve 95.5% recognition accuracy for the independent test set of 200 skills, which encompassed a much larger diversity of full-body movement than any previous study. Although this 95.5% recognition rate was not as high as the 99.2% accuracy achieved by recognizing 40 signs, a larger test sample of 200 skills were evaluated in this chapter [1].

With a larger training set, lower error rates are expected. Generalization to a user-independent system encompassing partial body movement domains such as sign language should be attainable. To progress toward this goal, the following improvements seem most important:

- Expand the dyneme model to improve discrimination of more subtle movements in partial-body domains. This could be achieved by either expanding the dyneme alphabet or having domain-dependent dyneme alphabets layered hierarchically below the full-body movement dynemes.

- Expand the clone-body-model to include a complete hand-model for enabling even more subtle movement domains such as finger signing and to better stabilize the hand position during tracking.

- Use a multicamera or multimodal vision system such as infrared and visual spectrum combinations to better disambiguate the body parts in 3D and track the body in 3D.

- More accurately calibrate all movement skills with multiple subjects performing all skills on an accurate commercial tracking system by recording multiple camera angles to improve on depth-of-field ambiguities. Such calibration would also remedy the qualitative nature of tracking results from computer vision research in general.

- Enhance tracking granularity using cameras with higher resolution, frame rate, and lux sensitivity [1].

So far, movement domains with exclusively partial-body motion such as sign language have been ignored. Incorporating partial-body movement domains into the full-body skill recognition system is an interesting challenge. Can the dyneme model simply be extended to incorporate a larger alphabet of dynemes, or is there a need for subdomain dyneme models for maximum discrimination within each domain? The answers to such questions may be the key to developing a general purpose of an unconstrained skill recognition system [1].

The results suggest that this approach has the potential to enable the biometric verification of a general human movement utilizing a noninvasive biomechanical analysis. This chapter also presented a general robust and efficient biometric analysis by applying it to anthropometric data, gait signatures, various human activities, and movement disorders [1].

Future Research

The research in this chapter has demonstrated that the proposed CHMR system has not only tracked and recognized hundreds of skills, but also successfully applied biometric verification to anthropometric data, gait signatures, human activities, and movement disorders. These biometrics were recognized free of joint markers, set-up procedures, and hand-initialization. The CHMR body model data was successfully applied as a biometric for body proportions and gait dyneme segmented motion vectors were successfully supported as a biometric for gait signatures [2].

A novel biometric verifying anthropometric data was presented in this chapter, based on a maximal between-person variability of about 10 dimensions of body proportions, with a promising minimal within-person variability across time. A recognition rate of 92% was achieved with one false positive, supporting this anthropometric signature as a valid biometric [2].

Biometric verification of gait signatures achieved 88% recognition with no false positives using the eigengait approach. Although this is better than the 73% reported by researchers [2], it is not as good as others have achieved for smaller sample sizes [2]. The most significant gait feature was found to be the left and right hip-knee angle-angle relationship encompassing a left-right asymmetry. The fusion of anthropometric and gait biometrics raised the accuracy from 92% and 88% respectively, to 94%. These results indicate that applying a fused anthropometric-gait biometric verification could form the basis for a security application. Future studies will extend to fast and slow gaits of each subject and include carried items [2].

Human movement activities were identified with no activity error. Various activities from using a computer to making coffee were successfully tracked and recognized. However, the number of activities in the sample were too small for this result to be conclusive [2].

It was also demonstrated that this approach was able to successfully track and classify gait to detect PD with a success rate of 95% with one false positive. The results suggest that this approach has the potential to guide clinicians on the relative sensitivity of specific postural/gait features in diagnosis and quantifying progress. However, detecting the small rapid motion of a tremor would necessitate a higher frame rate and resolution than was used in this study [2].

Future studies aim to extend the skill, semantic, and activity models and also to improve the robustness and accuracy of the system, especially the poorly observable depth DOFs, by applying to the particle filter inflated posteriors and dynamics for sample generation and then reweighing the results. Future research will adopt the receiver operating characteristic (ROC) methodology by using ROC curves to present results for more clarity [2].

Finally, loose clothing and carried items that occluded body parts reduced the effectiveness of these biometrics. An improvement can be achieved by modeling the draping of loose clothing to more fully reveal the true body shape [2]. The body model also needs to be extended to allow for the wide variety of loose clothing encountered in general situations. Tracking stability can be increased by enhancing the body model to include degrees of freedom supporting radioulnar (forearm rotation), interphalangeal (toe), metacarpophalangeal (finger), and carpometacarpal (thumb) joints to further stabilize the hand and feet positions. Future studies aim to further improve the accuracy of the biometric verifications presented in this chapter by increasing the sample sizes and both the spatial and temporal resolutions [2].

References

1. R. D. Green and L. Guan, "Quantifying and Recognizing Human Movement Patterns from Monocular Video Images—Part I: A New Framework for Modeling Human Motion" [R. D. Green is with the Human Interface Technology Lab, University of Canterbury, Christchurch, New Zealand; Prof. L. Guan is with the Department of Electrical and Computer Engineering, Ryerson University, Toronto, ON M5B 2K3, Canada], 2005.

2. R. D. Green and L. Guan, "Quantifying and Recognizing Human Movement Patterns from Monocular Video Images—Part II: Applications to Biometrics" [R. D. Green is with the Human Interface Technology Lab, University of Canterbury, Christchurch, New Zealand; Prof. L. Guan is with the Department of Electrical and Computer Engineering, Ryerson University, Toronto, ON M5B 2K3, Canada], 2005.

29

Selecting Biometric Solutions

As organizations search for more secure verification solutions for user access, e-commerce [10], and other security applications, biometrics is gaining increasing attention. But should your company use biometrics? And, if so, which ones should you use, and how do you choose them? There is no one best biometric technology. Different applications require different biometrics [1].

To select the right biometric solutions for your situation, you will need to navigate through some complex vendor products and keep an eye on future developments in technology and standards. Your options have never been more diverse. After years of research and development, vendors now have several products to offer. Some are relatively immature, having only recently become commercially available, but even these can substantially improve your company's information security posture. This chapter briefly describes some emerging biometric technologies to help guide your decision making. The security field uses three different types of verification:

- **Something you know:** A password, PIN, or piece of personal information (such as your mother's maiden name).

- **Something you have:** A card key, smart card, or token (like a SecurID card).

- **Something you are:** A biometric [1].

Of these, a biometric is the most secure and convenient verification tool. It can't be borrowed, stolen, or forgotten, and forging one is practically impossible.

Note: Replacement-part surgery, by the way, is outside the scope of this chapter.

Biometrics measure individuals' unique physical or behavioral characteristics to recognize or authenticate their identity. Common physical biometrics include fingerprints; hand or palm geometry; and retina, iris, or facial characteristics [1].

Figure 29-1   *How a biometric system works. (1) Capture the chosen biometric; (2) process the biometric and extract and enroll the biometric template; (3) store the template in a local repository, a central repository, or a portable token such as a smart card; (4) live-scan the chosen biometric; (5) process the biometric and extract the biometric template; (6) match the scanned biometric against stored templates; (7) provide a matching score to business applications; (8) record a secure audit trail with respect to system use. (Source: Adapted with permission from TopickZ Inc.)*

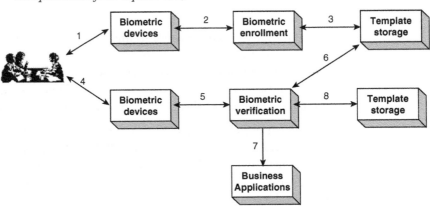

Behavioral characters include signature, voice (which also has a physical component), keystroke pattern, and gait (see sidebar, "The Gait Recognition Solution"). Of this class of biometrics, technologies for signature and voice are the most developed. Figure 29-1 illustrates the process involved in using a biometric system for security [1].

The Gait Recognition Solution

This sidebar systematically analyzes different components of human gait for the purpose of human identi-fication. Dynamic features are investigated by researchers, such as the swing of the hands/legs, the sway of the upper body, and static features like height, in both frontal and side views. Both probabilistic and non-probabilistic techniques are used for matching the features. Various combination strategies may be used, depending upon the gait features being combined. Three simple rules are discussed: the Sum, Product, and MIN rules that are relevant to feature sets. Experiments using four different data sets demonstrate that fusion can be used as an effective strategy in recognition [2].

Biometrics, such as face, voice/speech, iris, fingerprints, gait, and so on have come to occupy an increasingly important role in human identification due primarily to their universality and uniqueness.

Face recognition systems have good performance with canonical views at high resolution and good lighting conditions. Current iris recognition systems are designed to work when the subjects are placed at relatively close distances from the imaging system. A possible alternative is gait or, simply, the way a person walks. While medical studies [2] have shown that gait is indeed a unique signature of humans, psychophysical evidence [2] also points to the viability of gait recognition. Gait, a nonintrusive biometric, can be captured by cameras placed at a distance. Illumination changes are not a cause for serious concern. In particular, gait analysis might even be attempted in nighttime conditions using IR imagery. The potential applications of gait analysis/recognition systems include access control, surveillance, activity monitoring, and kinesiology.

Researchers know that gait and posture provide you with cues to recognize people. Consider a familiar person walking at a sufficiently far distance so that the face is not clearly visible to the naked eye. To recognize the person, you may try to combine information such as posture, arm/leg swing, hip/upper body sway, or some unique characteristic of that person. Generally speaking, information may be fused in two ways. The data available may be fused and a decision can be made based on the fused data, or each signal/feature can be matched separately, using possibly different techniques and the decisions made may be fused. The former is called data fusion while the latter is decision fusion. Researchers [2] have shown that decision fusion is a special case of data fusion.

Note: *The converse relationship need not be true. Consequently, data fusion, which tends to be more complex to implement, need not be a bottleneck.*

Researchers investigate different techniques to combine classification results of multiple measurements extracted from the gait sequences and demonstrate the improvement in recognition performance. Three different sets of features are extracted from the sequence of binarized images of the walking person. First, researchers have investigated the swing in the hands and legs. Since gait is not completely symmetric, in that the extent of forward swing of hands and legs is not equal to the extent of the backward swing, researchers have built the left and right projection vectors. To match these time varying signals, dynamic time warping is employed. Second, fusion of leg dynamics and height combines results from dynamic and static sources. A hidden Markov model is used to represent the leg dynamics [2]. While the preceding two components consider the side view, the third case explores frontal gait. Researchers also characterize the performance of the recognition system using the cumulative match scores [2] computed using the aforesaid matrix of similarity scores. As in any recognition system, researchers would like to obtain the best possible performance in terms of recognition rates. The combination of evidence obtained is not only logical but also statistically meaningful. Researchers can show that combining the evidence using simple strategies such as Sum, Product, and MIN rules improves the overall performance [2].

Methodology

Researchers assume that, within the field of view of the stationary camera, only one person is present. This simplifies the task of tracking. Background subtraction [2] is used to convert the video sequence into a sequence of binarized images in which a bounding box encapsulates the walking subject.

All the features of interest are extracted from the aforesaid sequence of binarized images. Three aspects of gait are discussed here: motion of the hands and legs, dynamics of the legs alone, and frontal gait. The researchers address the issue of foot dominance as well. Different strategies such as Sum, Product, and MIN rules [2], as applicable in each of the cases, are used.

The left and right projection vectors are constructed from the image sequence to study the motion of the hands and legs. Dynamic time warping is used to match the two vector sequences separately. The overall similarity score is taken to be the sum of the two scores. Next, the truncated width vector captures the leg dynamics. The hidden Markov model is used to describe the motion of the leg within a walk cycle. In the evaluation phase, the absolute value of the forward log probability is recorded as the similarity score. These scores are weighted by a factor that depends on the height of the subject. Then, frontal gait sequences are represented using the width vector, suitably normalized for apparent changes in the height as the subject approaches the camera. A set of width vectors are built for the side view and the two are matched, separately, using DTW. The Sum rule is used to combine the two similarity scores [2].

Motion of the Arms and Legs

In the four-limb system, the researchers seek to find a consistent pattern by systematically analyzing all the four limbs and a pair of limbs. If the degree of coupling between, say, the legs is significantly more than the coupling between the right leg and left hand, then the researchers would assign a higher weight to the similarity score obtained by comparing the leg motion in the reference and test pattern. The researchers must first consider the arms and legs of the subject. While it is tempting to assume that gait is a symmetric activity, there exists an asymmetry between the forward and backward swing of the limbs. By maintaining this dichotomy, the researchers build the left and right projection vectors as follows. Given a binarized image, they first align the box so that the subject is in the center of the bounding box. The left and right projection vectors are computed as illustrated in Figure 29-2(a) and (b), respectively [1].

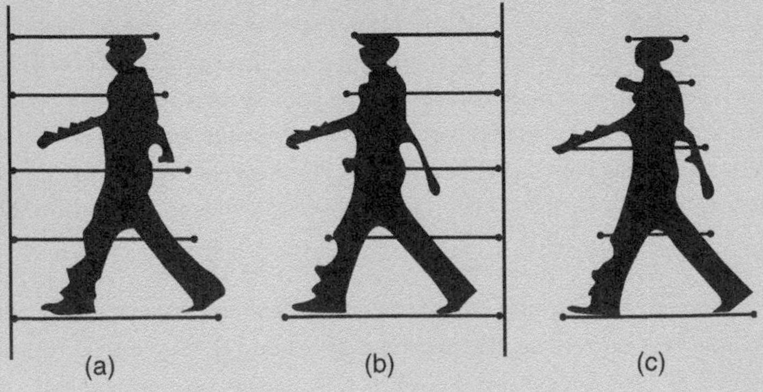

Figure 29-2
Illustrating the generation of (a) left projection vector, (b) right projection vector, and (c) width vector. (Source: Reproduced with permission from TopickZ Inc.)

(a)   (b)   (c)

After feature selection and extraction, the next logical step is matching. Direct frame-by-frame matching is not a realistic scheme since humans may slightly alter the speed and style of walking with time. Instead of restricting the frames of possible matches, it would be prudent to allow a search region at each time instant during evaluation. Dynamic time warping (DTW) provides for such a mathematical framework [2] in that it allows for nonlinear time normalization. The researchers then form two matrices of similarity scores by matching the left and right projection vectors in the gallery (reference/training) with those in the probe (testing) set separately.

Like hand dominance (right/left handedness), foot dominance (right/left leggedness) also exists. While matching therefore, the researchers may assume that improperly aligned (right/left leg forward) reference and test sequences affect the performance. This is an issue because it is not possible to distinguish between the left/right limbs from 2D binarized silhouettes. Suppose there are five (half-) cycles in both the gallery and probe sequences for a particular subject. To account for foot dominance, the researchers match the first four half-cycles of the two sequences and generate a matrix of similarity scores. Then, they match the gallery sequence with a phase-shifted probe sequence to generate another matrix of similarity scores. Of the two phase-shifted test sequences, only one can provide a match that is in-phase unless the subject does not exhibit foot dominance. Without loss of generality, the researchers may assume that foot dominance exists in all subjects. Then one of the two test sequences is a better match unless corrupted by noise. Therefore, the two similarity scores are combined using the MIN rule [2].

Leg Dynamics

Previously, both the hands and legs were considered while selecting the features. If the movement of the hands is restricted (if the subject is carrying an object in his or her hands) or if the sequence is excessively noisy in the torso region due to a systematic failure in background subtraction, then leg dynamics carries information about the subject's gait. The researchers construct a width vector (width of the outer contour of the binarized silhouette) of size $N \times 1$ from each of the images of size $N \times M$ in the sequence, as illustrated in Figure 29-2(c) [2]. Resistance to noise is provided in two stages. While part of the noise is removed during the computation of the width vector using the spatial correlation of pixels, eigen decomposition and width vector reconstruction utilizes the temporal nature of the data. The sequence of width vectors (matrix of width vectors) $W = \{W_k, k = 1, 2, \ldots, F\}$ representing the width vector of size $N \times 1$, at time $t = k$, is standardized and the scatter matrix computed. Eigen decomposition yields the eigen vectors, the largest K of which are retained. The projections of the width vectors on the K—the largest eigen vectors—yield coefficients that are in turn used to reconstruct the gait sequence by summing the appropriately weighted K. Figure 29-3 illustrates the effect of eigen-smoothing on the gait sequence [2].

A cursory examination of the width vectors suggests that the leg region may exhibit a more consistent pattern compared to other parts of the body such as the arms. At the same time, the gross structure of the body and height are also useful in discriminating between subjects. While leg dynamics concentrate on the variation of the width vector in the horizontal direction of the leg region alone, the height of the

Figure 29-3   *Effect of eigen decomposition and reconstruction on the width vectors. (a) Over-lapped raw width vectors, (b) smoothed width vectors. (Source: Reproduced with permission from the University of Maryland at College Park.)*

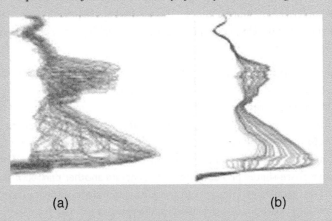

(a)   (b)

subject varies in an orthogonal direction. The width vector is truncated so that only the information about the leg is retained. This sequence of truncated width vectors is the first feature set, say set A. The researchers estimate the height of the subject from the image sequence using robust statistics. The estimated height of the individuals forms the second feature set, say set ß. Euclidean distance is used to compare the feature set ß of the estimated height of the subjects in the probe and gallery sets [2].

To compare the truncated width vectors that contain the information about leg dynamics, the researchers use the hidden Markov model (HMM) [2], which is a generalization of the DTW framework. There exists a Markovian dependence between frames, since the way humans go about the motion of walking has limited degrees of freedom; K-means clustering is used to identify the key frames or stances during a half-cycle. The researchers found that a choice of $k = 5$ is justified by the rate-distortion curve. The researchers thus project the sequence of images on the stance set creating a 5D vector (frame-to-stance distance or *FSD*) representation for each frame and use these samples to train a HMM using the Baum-Welch algorithm [2]. The Viterbi algorithm is used in the evaluation phase to compute the forward probabilities. The absolute values of the log probability values are recorded as the similarity scores [2].

If the decisions made are statistically independent, the researchers may write the final error probability. In practice, however, it is difficult to validate this assumption. Instead, the researchers use the low correlation of decisions across feature sets, as corroboration to the hypothesis that the errors in the two feature sets, the leg dynamics and the height, are uncorrelated. The researchers use the product rule to combine the scores to compute the overall similarity scores [2].

Frontal Gait

Hitherto, the researchers have studied gait in its canonical view, so that the apparent motion of the walking subject is maximal. This does not preclude the possibility of using other views ranging from the frontal view to any arbitrary angle of viewing. Even in the frontal view, where the apparent leg/arm swing is the least, there may be several cues that can be used toward human recognition. More specifically, the head posture, hip sway, or oscillating motion of the upper body, among other features, may pave the way for recognition. As before, to focus the researchers' attention on gait, they extract the outer contour of the subject from the binarized gait sequence in the form of the width vector, suitably normalized for an apparent change in height as the subject approaches the stationary camera [2].

For matching these sequences, the researchers use the DTW technique. When both the frontal and fronto-parallel (side) gait sequences are available, it is natural to combine these two orthogonal views before making the final decision about the identity of the subject. One way to combine multiple views is through the use of 3D models. Currently, 3D models have been built using sequences captured inside the lab under controlled conditions. The researchers therefore adopt the decision-fusion approach and combine the matching scores obtained by matching the frontal and side gait sequences separately using the Sum rule [2].

Experiments

The researchers thus report their experiments using the following data sets:

- CMU Data Set: Consists of 25 subjects walking on a treadmill. Seven cameras are mounted at different angles and the researchers use two of the views for their experiments, viz. the frontal and the side views. The first half of the gait sequence is used for training while the second half is used for testing (see Tables 29-1 and 29-2) [2].

- MIT Data Set: Consists of the side view of the outdoor gait sequences of 25 subjects collected on four different days. Four experiments are designed. Data from three days provides the training data, and data from the fourth day is used as the test sequences (see Table 29-6) [2].

- UMD Data Set: Contains the outdoor gait sequences captured by two cameras (frontal and side views). The 44 subjects are recorded in two sessions. The researchers train with the video data collected from the first session and test with that of the second session (see Tables 29-3, 29-4, and 29-7) [2].

- USF Data Set: Consists of outdoor gait sequences of 71 subjects walking along an elliptical path on two different surfaces (grass and concrete) wearing two different types of footwear (A and B). Two cameras, R and L, capture that data. Seven experiments are set up, as shown in Table 29-5 [2].

Table 29-1 *Cumulative Match Scores at Rank 1 and Rank 5 for the CMU Data Set: Combining Leg Dynamics and Height Using Sum Rule*

Feature	CMS at Rank 1	CMS at Rank 5
Leg dynamics	91	100
Fusion: leg dynamics and height	96	100

Table 29-2 *Cumulative Match Scores at Rank 1 and Rank 5 for the CMU Data Set: Effect of Frontal and Side Gait Fusion*

Feature	CMS at Rank 1	CMS at Rank 5
Leg dynamics	91	95
Fusion: leg dynamics and height	93	95
Frontal and side	96	97

Table 29-3 *Cumulative Match Scores at Rank 1 and Rank 5 for the UMD Data Set: Effect of Frontal and Side Gait Fusion*

Feature	CMS at Rank 1	CMS at Rank 5
Frontal gait	66	86
Side gait	58	74
Frontal and side	85	95

Table 29-4 *Cumulative Match Scores at Rank 1 and Rank 5 for the UMD Data Set: Foot Dominance and Effect of Fusing Evidence from Two Gait Sequences (Each Four Half-Cycles Long), with One Sequence Being Phase-Shifted*

Feature	CMS at Rank 1	CMS at Rank 5
First sequence	68	84
Phase-shifted sequence	70	88
Fusion	77	89

Table 29-5 *USF Data Set: The Seven Probe Sets with the Common Gallery Being G,A,R and Consisting of 71 Subjects*

Experiment	Probe	Difference
A	G,A,L (71)	View
B	G,B,R (41)	Shoe
C	G,B,L (41)	Shoe, view
D	C,A,R (70)	Surface
E	C,B,R (44)	Surface, shoe
F	C,A,L (70)	Surface, view
G	C,B,L (44)	Surface, shoe, view

The numbers in the parentheses are the number of subjects in each probe set.

Table 29-6 *Cumulative Match Scores at Rank 1 and Rank 3 for the MIT Data Set: Combining Leg Dynamics and Height by Adding the Similarity Scores*

Evaluation Scheme	CMS at Rank 1	CMS at Rank 3
Day 1 vs. days 2, 3, 4	29	50
Day 2 vs. days 1, 3, 4	50	100
Day 3 vs. days 1, 2, 4	20	54
Day 4 vs. days 1, 2, 3	30	52

Table 29-7 *Cumulative Match Scores at Rank 1 and Rank 5 for the UMD Data Set: Combining Leg Dynamics and Height Using Sum Rule*

Feature	CMS at Rank 1	CMS at Rank 5
Leg dynamics	31	65
Fusion: leg dynamics and height	49	72

Table 29-1 shows that while the leg dynamics by itself has rich information fusion, it can only improve the performance [2]. Results obtained from using the leg dynamics in the cases of MIT and UMD data sets are shown in Tables 29-6 and 29-7, respectively [2]. Table 29-4 shows that foot dominance is indeed present in certain individuals in the database and that fusing classification results from out-of-phase gait-sequences and serves to increase identification rates. Figure 29-4 suggests that asymmetry about a vertical axis in the side view may be addressed by considering the two halves of the body on either side of the vertical axis [2]. The results of matching the left and right projection vectors separately were combined using the Sum rule. Tables 29-2 and 29-3 show that the performance of frontal gait recognition can be enhanced by using the side view as well [2].

The researchers observe in Figure 29-4 that the right projection vector, which captures the forward swing, outperforms the left projection vector [2]. This suggests that, in this database, the forward swing of the hands and legs tends to have a lesser degree of variability with time between the gallery and probe sequences. The MIT data set, unlike the other data sets, has a low frame rate. Second, errors in background subtraction necessitate frame-dropping. This could be a reason for the poor performance [2].

Figure 29-4   *Identification rates for the USF data base: Effect of fusion of left and right projection vectors. Gallery in all the experiments is sequences from surface: grass, shoe; type: A; camera view: right. (Source: Reproduced with permission from the University of Maryland at College Park.)*

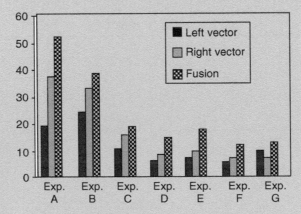

Conclusion

Different features that affect gait, such as the swing of the hands and legs, the sway in the body as observed in frontal gait, and static features like height, were systematically analyzed. Starting with dynamic time warping, which is a variant of template matching, a more generalized scheme, the HMM was chosen for matching [2].

The matrices of similarity scores between the gait sequences in the gallery and probe sets were computed. Sum, Product, and MIN rules were used to combine the decisions made using the separate features. As expected, the overall recognition performance improved due to fusion. Experiments were conducted on four different data sets, and each data set presented different types of challenges [2].

Choosing the Right Biometric Solution

Before choosing a biometric user verification solution, an organization should evaluate its needs carefully. The following list includes items that should be considered—the order of importance depends on the environment and level of security needed.

- Level of security required
- Accuracy
- Cost and implementation time
- User acceptance [1]

Level of Security

Voice and signature recognition techniques are generally considered to be appropriate for many non-PC access authorization uses, but in most cases are not good candidates for PC and network user verification. Biometric techniques that identify physical features are more accurate; therefore, they offer a higher level of security [1].

Accuracy

Retinal scanning and iris identification are both highly accurate ways of identifying individuals; however, they are expensive to implement and most organizations do not need this level of accuracy. Hand, face, and fingerprint verification techniques offer good accuracy for a smaller investment in scanning hardware [1].

Physical changes such as cuts, scars, and aging can affect the accuracy of certain types of biometric verification techniques. However, user identification databases can be updated to overcome most of these problems [1].

Cost and Implementation Time

When implementing a biometric user verification system, an organization should work with its PC vendor to evaluate the cost and time associated with the following factors:

- Researching, purchasing, and installing PC-compatible verification hardware and software;

- Biometric capture hardware (readers, cameras, scanners, and so on) and associated software (see sidebar, "Sweaty Biometric Software Solution");

- Hardware and software to maintain the user information database;

- Time required to integrate the verification hardware and software into the existing environment;

- Training IT staff to manage the new system;

- Training users in the new verification protocol;

- Collecting and maintaining a database of user identification data;

- Updating the database as necessary [1].

Sweaty Biometric Software Solution

Sweaty hands might make you unpopular as a dance partner, but they could someday prevent hackers from getting into your bank account. Researchers at Clarkson University have found that fingerprint readers can be spoofed by fingerprint images lifted with Play-doh or gelatin or a model of a finger molded out of dental plaster. The group even assembled a collection of fingers cut from the hands of cadavers [3].

In a systematic test of more than 60 of the carefully crafted samples, the researchers found that 90% of the fakes could be passed off as the real thing. But when researchers enhanced the reader with an algorithm that looked for evidence of perspiration, the false-verification rate dropped to 10% [3].

The idea of using perspiration is promising as a way to beat hackers because sweating follows a pattern that can be modeled. In live fingers, perspiration starts around the pore and spreads along the ridges, creating a distinct signature of the process. The algorithm detects and accounts for the pattern of perspiration when reading a fingerprint image [3].

Dead Fingers Don't Sweat

Since liveness detection is based on the recognition of physiological activities as signs of life, researchers hypothesized that fingerprint images from live fingers would show a specific changing moisture pattern due to perspiration, but cadaver and spoof fingerprint images would not. The research, funded by a $3.1 million grant from the National Security Agency and conducted in collaboration with other universities, is part of an ongoing effort to improve biometric verification and identification [3].

Other methods are in the works as well. Fingerprint readers essentially take a picture of a fingerprint and match it to a sample in the database. To get around spoofs involving lifted fingerprints, NEC researchers have developed technology that actually takes a picture of the tissue underneath the fingertip to get a three-dimensional image that can be matched against a database sample. Fujitsu has developed a verification technology that looks at vein patterns [3].

Although biometric identification technologies continue to improve, each has its own flaws. Voice verification is fairly accurate and tough to spoof, but it can be affected by a bad phone connection. Iris scans work well, but are commercially impracticable. Face scanning is actually less accurate than most, but consultants for the U.S. State Department indicate that the technology was chosen for electronic passports because that particular identity test seems to make people feel less like criminals [3].

User Acceptance

User verification based on fingerprint recognition does not use a person's complete fingerprint. Instead, the intersections of lines in the finger or thumb print, called minutiae points, are captured and used for identification (see sidebar, "Verification Versus Identification") [1].

Verification Versus Identification

In the biometric industry, a distinction is made among the terms identification, recognition, and verification. Identification and recognition are essentially synonymous terms. In both processes, a sample is presented to the biometric system during enrollment. The system then attempts to find out who the sample belongs to, by comparing the sample with a database of samples in the hope of finding a match (a one-to-many comparison) [1].

Verification is a one-to-one comparison in which the biometric system attempts to verify an individual's identity. In this case, a new biometric sample is captured and compared with the previously stored template. If the two samples match, the biometric system confirms that the applicant is who he or she claims to be [1].

The same four-stage process (capture, extraction, comparison, and match/nonmatch) applies equally to identification, recognition, and verification. Identification and recognition involve matching a sample against a database of many, whereas verification involves matching a sample against a database of one [1].

The key distinction between these two approaches centers on the questions asked by the biometric system and how these fit within a given application. During identification, the biometric system asks, "Who is this?" and establishes whether a biometric record exists, and, if so, the identity of the enrollee whose sample was matched. During verification, the biometric system asks, "Is this person who he or she claims to be?" and attempts to verify the identity of someone who is using, say, a password or smart card [1].

Users generally find less intrusive biometric techniques, such as fingerprint, face, or hand identification, most acceptable. However, some users may be reluctant to have their fingerprints recorded in a database. An organization should provide its employees with information and training on the chosen biometric method, so they have a chance to become familiar with the requirements before the system is implemented [1].

Most of the biometric solutions/methods have been thoroughly discussed in previous chapters throughout the book. Nevertheless, let's briefly take one more look at some of those biometric solutions/methods, in order to have a clear understanding—and be absolutely sure of the best solution for you.

Fingerprints

A fingerprint looks at the patterns found on a fingertip. There are a variety of approaches to fingerprint verification. Some emulate the traditional police method of matching minutiae; others use straight pattern-matching devices; and still others are a bit more unique, including things like fringe patterns and ultrasonics. Some verification approaches can detect when a live finger is presented; some cannot [1].

A greater variety of fingerprint devices is available than for any other biometric. As the prices of these devices and processing costs fall, using fingerprints for user verification is gaining acceptance, despite the common-criminal stigma [1].

Fingerprint verification may be a good choice for in-house systems, where you can give users adequate explanation and training, and where the system operates in a controlled environment. It is not surprising that the workstation access application area seems to be based almost exclusively on fingerprints, due to the relatively low cost, small size, and ease of integration of fingerprint verification devices [1].

Hand Geometry

Hand geometry involves analyzing and measuring the shape of the hand. This biometric offers a good balance of performance characteristics and is relatively easy to use. It might be suitable where there are more users or where users access the system infrequently and are perhaps less disciplined in their approach to the system [1].

Accuracy can be very high if desired, and flexible performance tuning and configuration can accommodate a wide range of applications. Organizations are using hand geometry readers in various scenarios, including time and attendance recording, where they have proved extremely popular. Ease of integration into other systems and processes, coupled with ease of use, makes hand geometry an obvious first step for many biometric projects [1].

Retina

A retina-based biometric involves analyzing the layer of blood vessels situated at the back of the eye. An established technology, this technique involves using a low-intensity light source through an optical coupler [17] to scan the unique patterns of the retina. Retinal scanning can be quite accurate but does require the user to look into a receptacle and focus on a given point. This is not particularly convenient if you wear glasses or are concerned about having close contact with the reading device. For these reasons, retinal scanning is not warmly accepted by all users, even though the technology itself can work well [1].

Iris

An iris-based biometric, on the other hand, involves analyzing features found in the colored ring of tissue that surrounds the pupil. Iris scanning, undoubtedly the less intrusive of the eye-related biometrics, uses a fairly conventional camera element and requires no close contact between the user and the reader. In addition, it has the potential for higher than average template-matching performance. Iris biometrics work with glasses in place and is one of the few devices that can work well in identification mode. Ease of use and system integration have not traditionally been strong points with iris scanning devices, but you can expect improvements in these areas as new products emerge [1].

Face

Face recognition analyzes facial characteristics. It requires a digital camera to develop a facial image of the user for verification. This technique has attracted considerable interest, although many people don't completely understand its capabilities. Some vendors have made extravagant claims (which are very difficult, if not impossible, to substantiate in practice) for facial recognition devices. Because facial scanning needs an extra peripheral not customarily included with basic PCs, it is more of a niche market for network verification. However, the casino industry has capitalized on this technology to create a facial database of scam artists for quick detection by security personnel [1].

Signature

Signature verification analyzes the way a user signs her name. Signing features such as speed, velocity, and pressure are as important as the finished signature's static shape. Signature verification enjoys a synergy with existing processes that other biometrics do not. People are used to signatures as a means of transaction-related identity verification, and most would see nothing unusual in extending this to encompass biometrics. Signature verification devices are reasonably accurate in operation and obviously lend themselves to applications where a signature is an accepted identifier. Surprisingly, relatively few significant signature applications have emerged compared with other biometric methodologies. But if your application fits, it is a technology worth considering [1].

Voice

Voice verification is not based on voice recognition but on voice-to-print verification, where complex technology transforms voice into text. Voice biometrics has the most potential for growth, because it requires no new hardware—most PCs already contain a microphone. However, poor quality and ambient noise can affect verification. In addition, the enrollment procedure has often been more complicated than with other biometrics, leading to the perception that voice verification is not user friendly. Therefore, voice verification software needs improvement. One day, voice may become an additive technology to fingerscan technology. Because many people see fingerscanning as a higher verification form, voice biometrics will most likely be relegated to replacing or enhancing PINs, multiple passwords (see sidebar, "Multiple-Password Solutions"), or account names.

Multiple-Password Solutions

In what might seem like a bit of science fiction, users can now log into GE Fanuc's Proficy iFIX plant-operations system by scanning their fingerprints, facial features, or retinas instead of typing in passwords. Proficy iFIX is a human-machine interface/supervisory control and data acquisition (HMI/SCADA) application. The biometric capabilities, available in the newly released version 4.0 of the package, are supposed to make life easier for manufacturers—particularly biotech and pharmaceutical companies that require workers to enter electronic signatures at various stages of the production process [4].

Pharmaceutical customers see a great value in collecting electronic signatures and creating operator audit trails, but the more you apply them, the more tedious it becomes to enter passwords continually. Biometric support alleviates this problem [4].

Tightly regulated manufacturing environments require electronic signatures for compliance, tracking and tracing, and audit trails. Besides having to enter passwords numerous times, the need to remember strong passwords (combinations of letters and numbers) can slow response times and lead to operator fatigue [4].

In a biometric-enabled HMI/SCADA system, operators submit to fingerprint, hand measurement, or retina scans that authenticate their identities and link them to their passwords. Then the system automatically enters passwords when required to create electronic signatures or audit trails [4].

Proficy iFIX users can choose between an interface to Saflink's SAFsolution Enterprise Edition security software or developer tools to build interfaces to any other verification technology. Users pick their own biometric reader hardware to interface with the system. Current iFIX users can add the capability without making changes to existing HMI/SCADA applications [4].

GE Fanuc will roll out similar biometric enhancements to other products that support 21 CFR Part 11, the regulation requiring electronic signatures for authenticating drug-manufacturing processes. While GE Fanuc may be at the forefront of the biometric-enabled software movement, it is expected to have a lot of company soon. According to New York–based International Biometric Group, the global market for biometrics will grow from $2.6 billion in 2006 to $4.6 billion in 2008 (see Figure 29-5) [4]. In addition to password verification, biometric technology is used for access control and identity management in manufacturing, IT, retail, financial, and government operations [4].

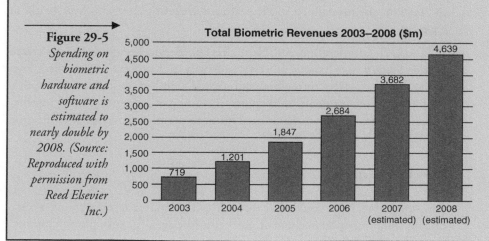

Figure 29-5
Spending on biometric hardware and software is estimated to nearly double by 2008. (Source: Reproduced with permission from Reed Elsevier Inc.)

Uses for Biometrics

Security systems use biometrics for two basic purposes: to verify or to identify users. Identification tends to be the more difficult of the two uses because a system must search a database of enrolled users to find a match (a one-to-many search). The biometric that a security system employs depends in part on what the system is protecting and what it is trying to protect against [1].

Physical Access

For decades, many highly secure environments have used biometric technology for entry access. Today, the primary application of biometrics is in physical security: to control access to secure locations (rooms or buildings). Unlike photo identification cards, which a security guard must verify, biometrics permit unmanned access control. Biometric devices, typically hand geometry readers, are in office buildings, hospitals, casinos, health clubs, and even a Moose Lodge. Biometrics are useful for high-volume access control. For example, biometric-controlled access was conducted on hundreds of thousands of people during the Winter and Summer Olympic Games; Disney World uses a fingerprint scanner to verify season-pass-holders entering the theme park [1].

Engineers are developing several promising prototype biometric applications to support the International Air Transport Association's Simplifying Passenger Travel (SPT) initiatives. One such program is EyeTicket, which Charlotte/Douglas International Airport in North Carolina and Flughafen Frankfurt/Main Airport in Germany are evaluating. EyeTicket links a passenger's frequent-flyer number to an iris scan. After the passenger enrolls in the system, an unmanned kiosk performs ticketing and check-in (without luggage) [1].

The U.S. Immigration and Naturalization Service's Passenger Accelerated Service System uses hand geometry to identify and process pre-enrolled, low-risk frequent travelers through an automated immigration system. Currently deployed in nine international airports, including Washington Dulles International, this system uses an unmanned kiosk to perform citizenship-verification functions [1].

Virtual Access

For a long time, biometric-based network and computer access were areas often discussed but rarely implemented. Recently, however, the unit price of biometric devices has fallen dramatically, and several designs aimed squarely at this application are on the market. Analysts see virtual access as the application that will provide the critical mass to move biometrics for network and computer access from the realm of science-fiction devices to regular system components. At the same time, user demands for virtual access will raise public awareness of the security risks and lower resistance to the use of biometrics [1].

Physical lock-downs can protect hardware, and passwords are currently the most popular way to protect data on a network. Biometrics, however, can increase a company's ability to protect its data by implementing a more secure key than a password. Using biometrics also allows a hierarchical structure of data protection, making the data even more secure: Passwords supply a minimal level of access to network data; biometrics, the next level. You can even layer biometric technologies to enhance security levels [1].

E-commerce Applications

E-commerce developers are exploring the use of biometrics and smart cards to more accurately verify a trading party's identity. For example, many banks are interested in this combination to better authenticate customers and ensure nonrepudiation of online banking, trading, and purchasing transactions. Point-of-sale system vendors are working on the cardholder verification method, which would enlist smart cards and biometrics to replace signature verification. MasterCard estimates that adding smart card–based biometric verification to a POS credit card payment will decrease fraud by 80% [1].

Some companies are using biometrics to obtain secure services over the telephone through voice verification. Developed by Nuance Communications, voice verification systems are currently deployed nationwide by both the Home Shopping Network and Charles Schwab. The latter's marketing catch phrase is "No PIN to remember, no PIN to forget" [1].

Covert Surveillance

One of the more challenging research areas involves using biometrics for covert surveillance. With facial and body recognition technologies, researchers hope to use biometrics to automatically identify known suspects entering buildings or traversing crowded security areas such as airports. The use of biometrics for covert identification as opposed to verification must overcome technical challenges such as simultaneously identifying multiple subjects in a crowd and working with uncooperative subjects. In these situations, devices cannot count on consistency in pose, viewing angle, or distance from the detector [1].

Selecting a Biometric Technology Solution

Biometric technology is one area that no segment of the IT industry can afford to ignore. Biometrics provide security benefits across the spectrum, from IT vendors to end users, and from security system developers to security system users. All these industry sectors must evaluate the costs and benefits of implementing such security measures. A very detailed description of biometric benefits can be found in Chapter 30. Different technologies may be appropriate for different applications, depending on perceived user profiles, the need to interface with other systems or databases, environmental conditions, and a host of other application-specific parameters (see Table 29-8) [1].

Ease of Use

Some biometric devices are not user friendly. For example, users without proper training may experience difficulty aligning their head with a device for enrolling and matching facial templates [1].

Error Incidence

Two primary causes of errors affect biometric data: time and environmental conditions. Biometrics may change as an individual ages. Environmental

Table 29-8   *Different Biometric Technology Solutions*

Characteristic	Fingerprints	Hand Geometry	Retina	Iris	Face	Signature	Voice
Ease of use	High	High	Low	Medium	Medium	High	High
Error incidence	Dryness, dirt, age	Hand injury, age	Glasses	Poor lighting	Lighting, age, glasses, hair	Changing signatures	Noise, colds, weather
Accuracy	High	High	Very high	Very high	High	High	High
Cost	*	*	*	*	*	*	*
User acceptance	Medium	Medium	Medium	Medium	Medium	Medium	High
Required security level	High	Medium	High	Very high	Medium	Medium	Medium
Long-term stability	High	Medium	High	High	Medium	Medium	Medium

*The large number of factors involved makes a simple cost comparison impractical.

conditions may either alter the biometric directly (for example, if a finger is cut and scarred) or interfere with the data collection (for instance, background noise when using a voice biometric) [1].

Accuracy

Vendors often use two different methods to rate biometric accuracy: false-acceptance rate or false-rejection rate. Both methods focus on the system's ability to allow limited entry to authorized users. However, these measures can vary significantly, depending on how you adjust the sensitivity of the mechanism that matches the biometric. For example, you can require a tighter match between the measurements of hand geometry and the user's template (increase the sensitivity). This will probably decrease the false-acceptance rate (FAR), but at the same time can increase the false-rejection rate (FRR). So be careful to understand how vendors arrive at quoted values of FAR and FRR [1].

Because FAR and FRR are interdependent, it is more meaningful to plot them against each other, as shown in Figure 29-6 [1]. Each point on the plot represents a hypothetical system's performance at various sensitivity settings. With such a plot, you can compare these rates to determine the crossover error rate (CER). The lower the CER, the more accurate the system. Generally, physical biometrics are more accurate than behavioral biometrics [1].

Figure 29-6 *Chart showing crossover error rate in biometric accuracy. The crossover error rate attempts to combine two measures of biometric accuracy. (Source: Adapted with permission from TopickZ Inc.)*

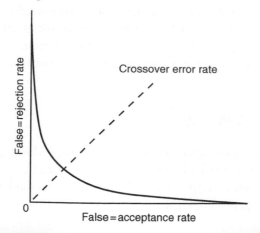

Cost

There are many costs associated with the selection of biometrics. Cost components include the following:

- Biometric capture hardware
- Back-end processing power to maintain the database
- Research and testing of the biometric system
- Installation, including implementation team salaries
- Mounting, installation, connection, and user system integration costs
- User education, often conducted through marketing campaigns
- Exception processing, or handling users who cannot submit readable images because of missing appendages or unreadable prints
- Productivity losses due to the implementation learning curve
- System maintenance [1]

User Acceptance

Generally speaking, the less intrusive the biometric, the more readily it is accepted. However, certain user groups (some religious and civil liberties groups) have rejected biometric technologies because of privacy [11] concerns [1].

Required Security Level

Organizations should determine the level of security needed for the specific application: low, moderate, or high. This decision will greatly impact which biometric is most appropriate. Generally, behavioral biometrics are sufficient for low-to-moderate security applications; physical biometrics, for high-security applications [1].

Long-Term Stability

Organizations should consider a biometric's stability, including maturity of the technology, degree of standardization, level of vendor and government support, market share, and other support factors. Mature and standardized technologies usually have stronger stability [1].

The Future of Biometrics

Although companies are using biometrics for verification in a variety of situations, the industry is still evolving and emerging. To both guide and support the growth of biometrics, the Biometric Consortium formed in December 1995. The recent Biometric Consortium has highlighted two important areas [1].

Standardization

The biometrics industry includes more than 260 separate hardware and software vendors, each with their own proprietary interfaces, algorithms, and data structures. Standards are emerging to provide a common software interface, to allow the sharing of biometric templates, and to permit effective comparison and evaluation of different biometric technologies [1].

The BioAPI standard defines a common method for interfacing with a given biometric application. BioAPI is an open-systems standard developed by a consortium of more than 70 vendors and government agencies. Written in C, it consists of a set of function calls to perform basic actions common to all biometric technologies, such as:

- Enroll user
- Verify asserted identity (verification)
- Discover identity [1]

Not surprising, Microsoft, the original founder of the BioAPI Consortium, dropped out and developed its own BAPI biometric interface standard [1].

Another draft standard is the Common Biometric Exchange File Format, which defines a common means of exchanging and storing templates collected from a variety of biometric devices. The Biometric Consortium has also presented a proposal for the Common Fingerprint Minutia Exchange format, which attempts to provide a level of interoperability for fingerprint technology vendors [1].

Biometric assurance (confidence that a biometric device can achieve the intended level of security) is an active research area. Current metrics for comparing biometric technologies, such as the crossover error rate and the average enrollment time, are limited because they lack a standard test bed on which to base their values. Several groups, including the U.S. Department of Defense's Biometrics Management Office, are developing standard testing methodologies. Much of this work is occurring within the contextual framework of the Common Criteria, a model that the international security community developed to standardize evaluation and comparison of all security products [1].

Hybrid Biometric Technology Solutions Uses

One of the more interesting uses of biometrics involves combining biometrics with smart cards and public-key infrastructure (PKI) [12]. A major problem with biometrics is how and where to store the user's template. Because the template represents the user's personal characters, its storage [14] introduces privacy concerns. Furthermore, storing the template in a centralized database leaves that template subject to attack and compromise. On the other hand, storing the template on a smart card enhances individual privacy and increases protection from attack, because individual users control their own templates [1].

Vendors enhance security by placing more biometric functions directly on the smart card. Some vendors have built a fingerprint sensor directly into the smart card reader, which in turn passes the biometric to the smart card for verification. At least one vendor, Biometric Associates, has designed a smart card that contains a fingerprint sensor directly on the card. This is a more secure architecture because cardholders must authenticate themselves directly to the card [1].

PKI uses public- and private-key cryptography for user identification and verification. It has some advantages over biometrics: It is mathematically more secure, and it can be used across the Internet [16]. The main drawback of PKI is the management of the user's private key. To be secure, the private key must be protected from compromise; to be useful, the private key must be portable. The solution to these problems is to store the private key on a smart card and protect it with a biometric [1].

In the Smart Access common government ID card program, the U.S. General Services Administration is exploring this marriage of biometrics, smart cards, and PKI technology. The government of Finland is considering using these technologies in deploying the Finnish National Electronic ID card [1].

In other words, smart card–enabled biometric solutions are a solid foundation for information and communications security in enterprises, government agencies, and other organizations. They secure access to PCs and applications, buildings, and rooms. They unfold their potential for cutting costs primarily by bundling various security functions (multifunctionality), increasing the level of security (physical and logical protection), and simplifying verification processes (single sign-on, process optimization). The return on investment that can be expected from them is high: According to industry analysts, it's approximately $3.6 million a year at a company with 3,000 employees. With that in mind, let's now take a detailed look at smart card biometric–enabled solutions, and their spin-offs [5].

Smart Card Biometric Technology Solutions

Smart card–enabled solutions for information security are compelling (see Figure 29-7), primarily for two reasons. First, they are very easy to use [5]. A company's employees are convinced right away when they discover that all they need for recurring daily activities, such as access to rooms, verification at their PCs, or buying their lunch in the company cafeteria is a card with a PIN code [5].

Second, they demonstrably cut costs at enterprises. And the cost savings grow with the size of the company. These cards reduce costs mainly due to three factors:

- Simplification of verification processes (single sign-on, process optimization);

- Increase in the level of security (physical and logical protection);

- Bundling of various security functions (multifunctionality) [5].

Figure 29-7
Security smart card. (Source: Reproduced with permission from Siemens AG.)

Something that Brings Progress in Healthcare

Since the pressure to cut costs on government agencies and associations is enormous, the merits of smart card–enabled solutions in the healthcare arena are particularly easy to convey (see Figure 29-8) [5]. The cost savings that can be expected are so persuasive that not only Germany is thinking aloud about introducing a health card. The region of Lombardy in Italy, for example, is counting on a smart card–enabled solution from Siemens for its health service. The new information system improves patient care and cuts costs. Eleven million multifunctional health cards will be in use by the end of 2007 [5].

For the same reasons, universities and hospitals profit from the introduction of the card. They can leverage a far larger range of functions than is currently envisaged for government agencies [5].

Figure 29-8
Healthcare smart card. (Source: Reproduced with permission from Siemens AG.)

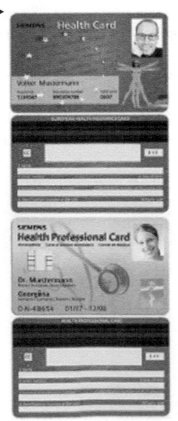

Also Cuts Costs in Enterprises

Each of these organizations (government agencies, hospitals, and universities) makes certain demands of its security infrastructure. Information must be easy to access, but not by everyone, and tailored to traditional workflows and on the basis of differentiated authorization levels (see Figure 29-9) [5]. The security concept should reliably repel threats arising from the use of modern means of telecommunication and IT. Finally, the aim is to simplify and speed up work processes where possible and thus make them cheaper. Those are essentially the same objectives that a company's management has its eye on when the issue of information security is raised. The following discussion describes why smart card–enabled solutions are so successful at many organizations and enterprises [5].

The Smart Card as a Multifunctional Corporate ID

Voice communication has not been perceived to date as an especially critical security issue in enterprises. By contrast, information security is increasingly a focus of interest in IP-based networks, such as the Internet [5].

The communications world is currently undergoing a change, with the result that voice communication will be handled in future over IP-based data networks. Voiceover IP has become a serious alternative to telephoning in a separate network. Yet it will only become a secure alternative if a voiceover IP provider knows the security problems in IP-based data networks, as well as the specific security issues in telecommunications [5].

Figure 29-9
Enterprise smart card. (Source: Reproduced with permission from Siemens AG.)

Competence for All-Round Solutions

A smart card–enabled solution is one building block of an overall security solution. Smart cards can improve network security solutions, for example, as an ideal complement if a virtual private network is to be used to integrate mobile employees. A VPN exploits the possibility of establishing encrypted [13] connections over the Internet. This secure data channel is implemented by means of the IPSec protocol. Siemens sees three means of delivering the basic IPSec functionality for the VPN. The iPSec protocol is either provided by the access router, or it is implemented by means of a software solution at the point of transition between the public and private network, for instance, in a firewall. In addition, dedicated IPSec gateways that operate with a pure hardware solution have become popular. In every case, the smart card is a secure and handy medium in access verification in a network [5].

Yet smart cards can also play a major role in large security infrastructures at the organizational level. In the dynamics of everyday business life, it is essential to establish constantly and beyond all doubt who is allowed to access specific company resources, with what permissions, and when. Enterprises with a large number of customer suppliers, employees, and, as is to be expected, a high level of worker fluctuation, profit greatly from intelligent identity and access management (see Figure 29-10) [5].

The advantage of such a solution lies in automation. Digital identities store the authorization profiles of persons and groups, thereby simplifying

Figure 29-10
Integrated smart card security solution. (Source: Reproduced with permission from Siemens AG.)

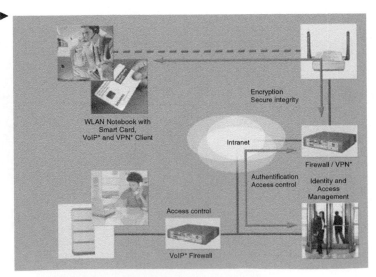

administration changes to permissions that can be applied to groups instead of to every single profile. All of the digital identities are incorporated centrally in a meta directory and distributed automatically to all security-sensitive systems. Applications like e-mail, operating systems, single sign-on (SSO), or public key infrastructures (PKIs) always receive the latest identities from the central directory and can thus check access requests and control them without any delay or manual activity. Smart cards are, for example, ideal for securely storing digital certificates that can be used as proof of identity for verification in a single sign-on solution or to enable secure e-mail [5].

One Card, Many Functions

Smart cards are becoming established in more and more areas of enterprise communications as part of a security policy. That comes as no surprise when you think how many functions such a small card can assume simultaneously and yet just how simple it is to use. The following example shows how a smart card–enabled security solution makes the wide range of very different processes at enterprises simpler, more secure, and thus cheaper day in, day out [5].

Physical Security—Access Control

The typical working day of an employee (let's call him Mr. Jackson) starts at the gate in front of the company parking lot. Mr. Jackson passes his corporate card by the contactless reader and is authorized to enter the grounds. After parking, he gains access to the building in the same way. The smart card–enabled solution logs all access to buildings and rooms, as well as any errors and alarms. All events are easy to reconstruct. Of course, the system only grants Mr. Jackson the defined access rights he needs for his work [5].

Time Data Collection

When Mr. Jackson enters the building, he can also record his time. In addition, he can also forward his time to the central administration department if that is desired [5].

Secure Access to PCs and Applications

At his workstation, Mr. Smith inserts the smart card in the reader next to his PC. He logs on to the local system and authenticates himself to gain access to specific network resources. As well as logging on to the computer by means of a smart card and PIN, a biometric method optionally ensures that the employee is authenticated by means of his fingerprint [5].

File and Hard-Disk Encryption

In addition, Mr. Jackson can encrypt files on the local hard disk, on portable data media, or on shared volumes. The reason for this is so they are protected permanently against being viewed by unauthorized persons [5].

Electronic Signature

The smart card provides the ability to significantly improve business processes by replacing a handwritten signature with an electronic equivalent. As a result, costly and time-consuming paper-based processes can be avoided. Here's an example:

Mr. Jackson prepares a proposal for a customer, signs it electronically using his smart card, sends it to the customer by e-mail, and archives it electronically. Thanks to the electronic signature, the customer can be sure that this is an official offer, and Mr. Jackson can be sure that no one can modify the offer without the change being detected [5].

At the end of the day, Mr. Jackson does his travel expense accounting using a Web-based application in his company's Internet portal. As verification, he signs the data using his smart card. His boss gets a notification, checks the data, and also signs it using his smart card [5].

Secure E-mail

Mr. Jackson checks his mailbox and finds a new e-mail. It contains important contract information, so the mail is encrypted. He can decrypt it using his smart card and read the e-mail. But can he be sure that the e-mail is really from the sender it claims to be from? Once again, a smart card–based technology helps. The sender is identified beyond doubt by means of an electronic signature [5].

Secure and Remote LAN Access

Later, Mr. Jackson has an appointment with a customer. He calls on the customer and discovers during their talks that it would be a good idea to show a recent presentation that he has not stored locally on his laptop. He quickly sets up a secure remote link of the company network over the Internet and downloads the data to his mobile system [5].

Secure Payment Transactions

Mr. Jackson even uses his corporate card in breaks from work. In the cafeteria, at the kiosk, or at various machines, he can make intracompany payments quickly and easily. He can use the smart card reader at his PC to inspect the latest transactions on his account. He can also reload the card at a loading station. The amount drawn can be debited directly from his account [5].

Biometrics—Secure Verification by Smart Card

A state-of-the-art method for physical and logical security is biometrics. An unmistakable digital template that is stored securely on a smart card is created from a person's fingerprint (see Figure 29-11) [5]. If the smart card holder attempts to gain access to a PC, for example, his or her fingerprint is scanned again by a sensor and compared with the reference on the card. The card decides automatically whether the person has authenticated his or her identity or not [5].

The advantage of biometrical methods is that users of a security infrastructure do not have to remember anything. They do not need a PIN or password to prove their identity to applications, at doors, or on PCs. At the same time, their ID card remains forgery-proof and unique [5].

The Argument for Single Sign-On

Smart card–enabled solutions also permit improved forms of access management. The term single sign-on subsumes various ways of simplifying access to networks and applications that range from a simple login with a user name and password to smart cards with certificates. A common feature is that after logging on once to the network, users have access to all the resources for

Figure 29-11
An unmistakable digital template that is stored securely on a smart card is created from a person's fingerprint. (Source: Reproduced with permission from Siemens AG.)

which they have authorization. They only have to remember a single PIN, but can perform various authenticator activities, such as in the company portal or to use a protected application. The most important advantages are the following:

- Added security, because a complex password is used instead of several simple, easily guessed passwords. Login data and passwords can be stored securely and especially conveniently on a smart card with a password management solution.

- Time savings thanks to quicker logging in.

- Fewer inquiries to help desks from people who have forgotten their passwords.

- Simpler and more reliable administration, particularly in conjunction with an identity and access management solutions [5].

The Example of Business Conduct Guidelines— Greater Efficiency Thanks to the Card

A further example of the efficiency of smart card–based electronic communications channels is the business conduct guidelines that must be regularly acknowledged and signed by employees at larger enterprises. If the guidelines are sent in paper form, that means considerable costs for printing and handling from the very beginning of the process [5].

It is necessary to check which guidelines have been signed and returned on time and which have not. A reminder has to be sent to tardy employees. The results of the counts are transferred to the computer so they can be checked. Finally, this mass of paperwork has to be archived [5].

Such discontinuities in the media chain and manual operations are avoided with the end-to-end use of electronic media. The gain in time and security regarding results and confidentiality is enormous [5].

Return on Investment for Smart Card–Enabled Solutions

Enterprises with a workforce of 50 or more can expect significant annual savings if they use smart cards to ensure physical and logical security. The cards protect access to buildings and rooms, as well as PCs and networks. Finally, they enable secure e-mail in a simple way. The precise return on investment that can be expected must be calculated on an individual basis. Costs are cut primarily in the following areas:

- Password inquiries to IT staff
- The time needed for signing on
- The time required for access control
- The personnel costs of access control
- The issue of temporary access permissions [5]

There are also savings potentials in the area of security:

- Fewer violations of IT security policies
- Less system downtime due to sabotage [5]

Now, as part of hybrid biometric technology solutions uses, let's briefly look at radio-frequency ID (RFID) chips. RFID chips will soon be in cash, credit cards, your driver's license, cheap crap at grocery stores, cars, car tires, and even possibly under your skin (see sidebar, "Getting Under Your Skin").

RFID Biometric Solutions

The RFID biometric technology process starts with a tag, which is made up of a microchip with an antenna and a reader with an antenna. The reader sends out radio-frequency waves that form a magnetic field when they join with the antenna on the RFID tag. A passive RFID tag creates power from this magnetic field and uses it to energize the circuits of the RFID chip. The chip in the radio-frequency identification tag sends information back to the reader by means of radio-frequency waves. The RFID reader converts the new waves into digital information. Semi-passive RFID tags use a battery to run the circuits of the chip, but communicate by drawing power from the RFID reader.

Getting Under Your Skin

Forgetting computer passwords is an everyday source of frustration, but a solution may literally be at hand—in the form of computer chip implants. With a wave of his hand, Amal Graafstra, a 29-year-old entrepreneur based in Vancouver, Canada, opens his front door. With another, he logs onto his computer [9].

Tiny radio-frequency identification (RFID) computer chips inserted into Graafstra's hands make it all possible. "I just don't want to be without access to the things that I need to get access to. In the worst case scenario, if I'm in the alley naked, I want to still be able to get in [my house]," Graafstra said in an interview in New York, where he is promoting the technology: "RFID is for me" [9].

The computer chips, which cost about $2, interact with a device installed in computers and other electronics. The chips are activated when they come within three inches of a so-called reader, which scans the data on the chips. The "reader" devices are available for as little as $50. Information about where to buy the chips and readers is available online at the "tagged" forum (http://tagged.kaos.gen.nz/), where enthusiasts of the technology chat and share information. Graafstra indicated at least 20 of his tech-savvy pals have RFID implants [9].

ABRACADABRA

Mikey Sklar, a 28-year-old Brooklyn resident, indicated, "It does give you some sort of power of 'Abra-cadabra,' of making doors open and passwords enter just by a wave of your hand." The RFID chip in Sklar's hand, which is smaller than a grain of rice and can last up to 100 years, was injected by a surgeon in Los Angeles [9].

Tattoo artists and veterinarians also could insert the chips into people, he indicated. For years, veterinarians have been injecting similar chips into pets so the animals can be returned to their owners if they are lost [9].

Graafstra was drawn to RFID tagging to make life easier in this technological age, but Sklar indicated he was more intrigued by the technology's potential in a broader sense. In the future, technological advances will allow people to store, transmit, and access encrypted personal information in an increasing number of wireless [15] ways, Sklar indicated. Wary of privacy issues, Sklar indicated he is developing a fabric "shield" to protect such chips from being read by strangers seeking to steal personal information or identities [9].

One advantage of the RFID chip, Graafstra indicated, is that it cannot get lost or stolen. And the chip can always be removed from a person's body [9].

"It's kind of a gadget thing, and it's not so impressive to have it on your key chain as it is to have it in you," Sklar indicated. "But it's not for everyone" [9].

RFID Applications

Radio-frequency identification is used in many different applications. Like other forms of automated identification, such as barcodes and scanning, it is very application specific.

RFID has been used globally for over a decade and is a respected form of automatic identification. RFID systems share a common foundation: They utilize the electromagnetic wave spectrum to transfer data. Within the family of RFID solutions, there are many different types of technologies, that lend themselves to specific applications, such as retail, security, supply chain, and other valuable application areas. Let's briefly look at some of these from an Orwellian point of view.

Retail RFID Applications

Wal-Mart is mandating RFID adoption by its suppliers, which will force all of corporate America to switch from barcodes, which merely track what kind of product something is, with RFID, which uses an 18-digit number to track which specific product it is. Barcodes track a model of car tire; RFID would track the specific tire—which could then be cross-referenced in the great Homeland Insecurity Totalitarian Information Awareness überdatabase. Simple RFID readers will probably be set up just about everywhere, which will then read all RFID chips in the vicinity. This is far, far more intrusive than the nightmarish vision of George Orwell's *1984*. The movie *Brazil* by Terry Gilliam (1985) was a warning about what type of society these sorts of technological slavery systems would create.

As previously explained, RFID chips don't have their own energy source; they are passive. They emit a signal when specific frequencies of radio energy are used to "paint" them. RFIDs contain tiny antennas that receive that RF energy and then re-radiate their encoded information.

The main problem to their widespread adoption is cost (it's too expensive to put them in every cereal box) and the lack of scanning systems to read them in stores. But with mass production and a few billion dollars from Wal-Mart, the military, Homeland Security grants, and other rulers of the Brave New World Order, these technical obstacles will be overcome soon.

Law Enforcement RFID Applications

The next fashion accessory for some inmates at the Los Angeles County jail will be a radio-frequency identification bracelet. The country's largest jail system has launched a pilot project with Alanco Technologies to track inmates using the technology [6].

The first phase will involve setting up an RFID system in the 1,800-inmate east facility of the Pitchess Detention Center in Castaic, California. If it succeeds, and funding can be obtained, the county will spread the system throughout its prison facilities [6].

In prison networks with such technology, RFID readers are planted throughout a jail in such large numbers that bracelet-wearing inmates can be continually tracked. When an inmate comes within range of a sensor, it detects his or her presence and records the event in a database. Thus, if an assault occurs at night, prison officials can look at the RFID logs and identify who was at the scene at the time of the incident. Tampering with the bracelet sends an alarm to the system. The system can also warn of gang gatherings [6].

Orwellian as tagging sounds, inmate violence has declined in prisons where similar RFID systems have been installed, according to Alanco. Guards also wear RFID tags in these facilities. The primary concern of the sheriff's department is the safety of both their staff and the inmates housed in their facilities [6].

In 2005, there were an estimated six inmate deaths and injuries to 2,853 inmates and 99 jail staff in the seven facilities that make up the L.A. county jail system, according to the county. Alanco estimates that the prison system alone could become a billion-dollar market, while jails could account for $600 million to $800 million in revenue [6].

Airport RFID Applications

Let's not mince words about the stupidy of using biometric IDs to create two classes of travelers (low risk, higher risk) and privatizing the system that does it.

What the Trusted Traveler program does is create two different access paths into the airport: high security and low security. The intent is that only good guys will take the low-security path, and the bad guys will be forced to take the high-security path. The Trusted Traveler program is based on the dangerous myth that terrorists match a particular profile and that you can somehow pick terrorists out of a crowd if you could only identify everyone. That's simply not true. Most of the 9/11 terrorists were unknown and not on any watch list (see sidebar, "9/11 Deception by Pentagon"). Timothy McVeigh was an upstanding American citizen before he blew up the Oklahoma City federal building. What incentive do these for-profit companies have to not sell someone a pass? Who is liable for the mistakes [7]?

9/11 Deception by Pentagon

Some staff members and commissioners of the Sept. 11 panel concluded that the Pentagon's initial story of how it reacted to the 2001 terrorist attacks may have been part of a deliberate effort to mislead the commission and the public, rather than a reflection of the fog of events on that day, according to sources involved in the debate. Suspicion of wrongdoing ran so deep that the 10-member commission, in a secret meeting at the end of its tenure in summer 2004, debated referring the matter to the Justice Department for criminal investigation, according to several commission sources. Staff members and some commissioners thought that e-mails and other evidence provided enough probable cause to believe that military and aviation officials violated the law by making false statements to Congress and to the commission, hoping to hide the bungled response to the hijackings [8].

In the end, the panel agreed to a compromise, turning over the allegations to the inspectors general for the Defense and Transportation Departments, who can make criminal referrals if they believe they are warranted. "We to this day don't know why NORAD [the North American Aerospace Command] told us what they told us," indicated Thomas H. Kean, the former New Jersey Republican governor who led the commission. "It was just so far from the truth. It's one of those loose ends that never got tied" [8].

Although the commission's landmark report made it clear that the Defense Department's early versions of events on the day of the attacks were inaccurate, the revelation that it considered criminal referrals reveals how skeptically those reports were viewed by the panel and provides a glimpse of the tension between it and the Bush administration. A Pentagon spokesman indicated that the inspector general's office will soon release a report addressing whether testimony delivered to the commission was "knowingly false." A separate report, delivered secretly to Congress in May 2005, blamed inaccuracies in part on problems with the way the Defense Department kept its records [8].

A spokesman for the Transportation Department's inspector general's office indicated its investigation is complete and that a final report is being drafted. Laura Brown, a spokeswoman for the Federal Aviation Administration, indicated that she could not comment on the inspector general's inquiry [8].

In an article, *Vanity Fair* magazine reported aspects of the commission debate (though it did not mention the possible criminal referrals) and published lengthy excerpts from military audiotapes recorded on Sept. 11. ABC News aired excerpts [8].

For more than two years after the attacks, officials with NORAD and the FAA provided inaccurate information about the response to the hijackings in testimony and media appearances. Authorities suggested that U.S. air defenses had reacted quickly, that jets had been scrambled in response to the last two hijackings, and that fighters were prepared to shoot down United Airlines Flight 93 if it threatened Washington [8].

In fact, the commission reported a year later, audiotapes from NORAD's northeast headquarters and other evidence showed clearly that the military never had any of the hijacked airliners in its sights and at one point chased a phantom aircraft (American Airlines Flight 11) long after it had crashed into the World Trade Center. Maj. Gen. Larry Arnold and Col. Alan Scott told the commission that NORAD had begun tracking United 93 at 9:16 a.m., but the commission determined that the airliner was not hijacked until 12 minutes later. The military was not aware of the flight until after it had crashed in Pennsylvania [8].

These and other discrepancies did not become clear until the commission, forced to use subpoenas, obtained audiotapes from the FAA and NORAD. The agencies' reluctance to release the tapes (along with e-mails, erroneous public statements, and other evidence) led some of the panel's staff members and commissioners to believe that authorities sought to mislead the commission and the public about what happened on Sept. 11 [8].

"I was shocked at how different the truth was from the way it was described," John Farmer, a former New Jersey attorney general who led the staff inquiry into events on Sept. 11, indicated in a recent interview. "The tapes told a radically different story from what had been told to us and the public for two years. This is not spin. This is not true" [8].

Arnold, who could not be reached for comment, told the commission in 2004 that he did not have all the information unearthed by the panel when he testified earlier. Other military officials also denied any intent to mislead the panel [8].

John F. Lehman, a Republican commission member and former Navy secretary, indicated in a recent interview that he believed the panel may have been lied to but that he did not believe the evidence was sufficient to support a criminal referral. "My view of that was that whether it was willful or just the fog of stupid bureaucracy, I don't know," Lehman indicated. "But in the order of magnitude of things, going after bureaucrats because they misled the commission didn't seem to make sense to me [8]."

Need more reasons why this could be a bad idea? It's not like the credentials accepted by airports today can't be easily forged. Maybe anything that raises the barrier to forgery is better as long as it applies to everyone and it doesn't give anyone (the TSA in particular) a false sense of security. One other point: Soft targets in combination with the potential for a lot of casualties seems to me to be the biggest area of vulnerability. In the bigger scheme of things, between the security in the airports, the security on the planes, and highly sensitized passengers (who, historically, since 9/11 have acted pretty quickly at the first sign of a threat), airliners are not exactly easy prey. No one is saying that the United States should not have the best security possible in airports. It's just that some much softer and very populous targets could use equal if not more attention [7].

Summary/Conclusion

Biometric technology has been around for decades but has mainly been used for highly secretive environments with extreme security measures. The technologies behind biometrics are still emerging. Finally, this chapter gave a snapshot of the dynamics under way in this emerging biometric market, and hopefully it helped you consider all the possible alternatives when selecting and acquiring new biometric technologies.

References

1. Simon Liu and Mark Silverman, "A Practical Guide to Biometric Security Technology" [Simon Liu is director of computer and communications systems at the National Library of Medicine. He is also an adjunct professor at Johns Hopkins University. Mark Silverman is a technical advisor at the Center of Information Technology, National Institutes of Health.], TopickZ Inc., Toronto, Ontario, Canada, 2005.

2. Naresh Cuntoor, Amit Kale and Rama Chellappa, "Combining Multiple Evidences for Gait Recognition," Center for Automation Research, University of Maryland at College Park, College Park, MD 20742, 2005.

3. Michael Kanellos, "New Biometrics Software Looks for Sweat," CNET Networks, Inc., Headquarters, 235 Second Street, San Francisco, CA 94105, December 21, 2005. CNET News.com, ZDNet News, Copyright © 2006 CNET Networks, Inc. All Rights Reserved.

4. "Biometrics Solve Multiple-Password Entry Problem," *Manufacturing Business Technology Magazine*, 2000 Clearwater Dr., Oak Brook, IL 60525, Vol. 24, No. 8, page 36, August 2006. Copyright © 2006 Reed Business Information, a division of Reed Elsevier Inc. All Rights Reserved. Manufacturing Business Technology® is a registered trademark of Reed Elsevier Inc.

5. "HiPath Security—Smartcard-enabled Security Solutions," Siemens AG, Otto-Hahn-Ring 6, 81739 Munchen, Germany, January 1, 2005. Copyright © Siemens AG 2005.

6. Michael Kanellos, "L.A. County Jail Tags Inmates With RFID," CNET Networks, Inc., Headquarters, 235 Second Street, San Francisco, CA 94105, May 17, 2005. CNET News.com, ZDNet News, Copyright © 2006 CNET Networks, Inc. All Rights Reserved.

7. David Berlind, "Dumb Idea?: Biometric Hall Pass for Travelers," CNET Networks, Inc., Headquarters, 235 Second Street, San Francisco, CA 94105, February 1, 2006. CNET News.com, ZDNet News, Copyright © 2006 CNET Networks, Inc. All Rights Reserved.

8. Dan Eggen, "9/11 Panel Suspected Deception by Pentagon," *The Washington Post*, 1150 15 St. NW, Washington, DC 20071, August 2, 2006. Copyright © 2006 The Washington Post Company.

9. Jamie McGeever, "Computer Chips Get Under Skin of Enthusiasts," January 6, 2006. Copyright © 2006 Reuters Ltd. All Rights Reserved.

10. John R. Vacca, *Electronic Commerce, 4th ed*, Charles River Media (2003).

11. John R. Vacca, *Net Privacy: A Guide to Developing and Implementing an Ironclad eBusiness Privacy Plan*, McGraw-Hill (2001).

12. John R. Vacca, *Public Key Infrastructure: Building Trusted Applications and Web Services*, CRC Press (2005).

13. John R. Vacca, *Satellite Encryption*, Academic Press (1999).

14. John R. Vacca, *The Essentials Guide to Storage Area Networks*, Prentice Hall, Professional Technical Reference, Pearson Education (2001).

15. John R. Vacca, *Wireless Data Dymistified*, McGraw-Hill (January 2003).

16. John R. Vacca, *Practical Internet Security*, Springer (2006).

17. John R. Vacca, *Optical Networking Best Practices Handbook*, John Wiley & Sons (2006).

30

Biometric Benefits

Imagine you're Jack Bauer of TV's *24*, and you have to get into a secret laboratory to disarm a deadly biological weapon and save the world for the hundredth time. But first, you have to get past the security system. It requires more than just a key or a password—you need to have the villain's irises, his or her voice, and the shape of his or her hand to get inside [1].

You might also encounter this scenario, minus the deadly biological weapon, during an average day on the job. Airports, hospitals, hotels, grocery stores, and even Disney theme parks increasingly use biometrics for added security [1].

In this chapter, you'll learn about the benefits of using biometric systems that use handwriting, hand geometry, voiceprints, iris structure, and vein structure (see Figure 30-1) [2]. You'll also learn why more businesses and governments use the technology and whether fake contact lenses, a recorded voice, and a silicone hand could really get Jack Bauer into the lab (and let him save the world again) [1].

Note: Biometrics and forensics have a lot in common, but they're not exactly the same. Biometrics uses your physical or behavioral characteristics to determine your identity or to confirm that you are who you claim to be. Forensics uses the same kind of information to establish facts in civil or criminal investigations [1].

The Benefits of Working Biometrics

You take basic security precautions every day—you use a key to get into your house and log on to your computer with a username and password. You've probably also experienced the panic that comes with misplaced keys and forgotten passwords. It isn't just that you can't get what you need—if you lose your keys or jot your password on a piece of paper, someone else can find them and use them as though they were you [1].

Figure 30-1   *Vein scanning is one form of biometric identification. Image courtesy Hitachi Engineering Co. (Source: Reproduced with permission from Hitachi Engineering & Services Co., Ltd.)*

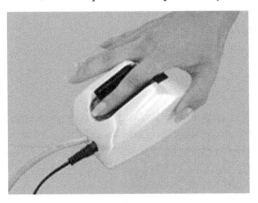

Instead of using something you have (like a key) or something you know (like a password), biometrics uses who you are to identify you. Biometrics can use physical characteristics, like your face, fingerprints, irises (see Figure 30-2), or veins, or behavioral characteristics like your voice, handwriting, or typing rhythm [3]. Unlike keys and passwords, your personal traits are extremely difficult to lose or forget. They can also be very difficult to copy. For this reason, many people consider them to be safer and more secure than keys or

Figure 30-2   *Biometrics uses unique features, like the iris of your eye, to identify you. Photo courtesy Iridian Technologies. (Source: Reproduced with permission from Iridian Technologies, Inc.)*

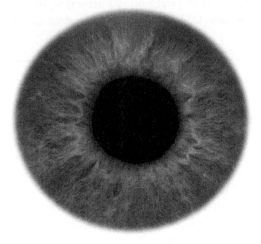

passwords. Biometric systems can seem complicated, but they all use the same three steps:

- **Enrollment:** The first time you use a biometric system, it records basic information about you, like your name or an identification number. It then captures an image or recording of your specific trait.

- **Storage:** Contrary to what you may see in movies, most systems don't store the complete image or recording. They instead analyze your trait and translate it into a code or graph. Some systems record this data onto a smart card that you carry with you.

- **Comparison:** The next time you use the system, it compares the trait you present to the information on file. Then, it either accepts or rejects that you are who you claim to be [1].

Systems also use the same three components:

- A sensor that detects the characteristic being used for identification;

- A computer that reads and stores the information;

- Software that analyzes the characteristic, translates it into a graph or code and performs the actual comparisons [1].

Biometric security systems, like the fingerprint scanner available on the IBM ThinkPad T43 (see Figure 30-3), is becoming more common for home use [1]. Next, let's examine how biometrics provides security using other traits, starting with handwriting.

Figure 30-3
Fingerprint scanner. (Source: Reproduced with permission from HowStuffWorks, Inc.)

fingerprint scanner

Note: Movies and television shows often depict the process of comparing traits in a way that is fun to watch, but not accurate. For example, you may see a whole fingerprint compared to other whole fingerprints until a computer finds a match. This method would be slow and difficult. Instead of comparing actual pictures, biometric systems use various algorithms to analyze and encode information about the trait. This information takes up only a few bits of space [1].

Handwriting

At first glance, using handwriting to identify people might not seem like a good idea. After all, many people can learn to copy other people's handwriting with a little time and practice. It seems like it would be easy to get a copy of someone's signature or the required password and learn to forge it [1].

But, biometric systems don't just look at how you shape each letter; they analyze the act of writing. They examine the pressure you use and the speed and rhythm with which you write. They also record the sequence in which you form letters, like whether you add dots and crosses as you go or after you finish the word [1].

Unlike the simple shapes of the letters, these traits are very difficult to forge. Even if someone else got a copy of your signature and traced it, the system probably wouldn't accept their forgery [1].

A handwriting recognition system's sensors can include a touch-sensitive writing surface or a pen that contains sensors that detect angle, pressure, and direction. The software translates the handwriting into a graph and recognizes the small changes in a person's handwriting from day to day and over time (see sidebar, "Determining Accuracy") [1] as shown in Figure 30-4 [4].

Figure 30-4
This Tablet PC has a signature verification system. Photo courtesy SOFTPRO. (Source: Reproduced with permission from SOFTPRO GmbH.)

Determining Accuracy

All biometric systems use human traits that are, to some degree, unique. Which system is best depends on the necessary level of security, the population who will use the system, and the system's accuracy. Most manufacturers use measurements like these to describe accuracy:

- **False-accept rate (FAR):** How many impostors the system accepts.

- **False-reject rate (FRR):** How many authorized users the system rejects.

- **Failure-to-enroll rate (FTE):** How many people's traits are of insufficient quality for the system to use.

- **Failure-to-acquire rate (FTA):** How many times a user must present the trait before the system correctly accepts or rejects them [1].

Hand and Finger Geometry

People's hands and fingers are unique—but not as unique as other traits, like fingerprints or irises. That's why businesses and schools, rather than high-security facilities, typically use hand and finger geometry readers to verify users, not to identify them. Disney theme parks, for example, use finger geometry readers to grant ticket holders admittance to different parts of the park. Some businesses use hand geometry readers in place of timecards [1].

Systems that measure hand and finger geometry use a digital camera and light. To use one, you simply place your hand on a flat surface, aligning your fingers against several pegs to ensure an accurate reading. Then, a camera takes one or more pictures of your hand and the shadow it casts. It uses this information to determine the length, width, thickness, and curvature of your hand or fingers. It translates that information into a numerical template (see Figure 30-5) [5].

Hand and finger geometry systems have a few strengths and weaknesses. Since hands and fingers are less distinctive than fingerprints or irises, some people are less likely to feel that the system invades their privacy [11]. However, many people's hands change over time due to injury, changes in weight, or arthritis. Some systems update the data to reflect minor changes from day to day [1].

For higher-security applications, biometric systems use more unique characteristics, like voices. Let's look at those next [1].

Figure 30-5 *A hand geometry scanner. Photo courtesy Ingersoll-Rand. (Source: Reproduced with permission from Ingersoll-Rand Company.)*

Note: A biometric system can either authenticate that you are who you say you are, or it can identify you by comparing your information to all of the information on file. Verification is a one-to-one comparison; it compares your characteristic with your stored information. Identification, on the other hand, is a one-to-many comparison [1].

Voiceprints

Your voice is unique because of the shape of your vocal cavities and the way you move your mouth when you speak. To enroll in a voiceprint system, you either say the exact words or phrases that it requires, or you give an extended sample of your speech so that the computer can identify you no matter which words you say [1].

When people think of voiceprints, they often think of the wave pattern they would see on an oscilloscope. But the data used in a voiceprint is a sound spectrogram, not a wave form. A spectrogram is basically a graph that shows a sound's frequency on the vertical axis and time on the horizontal axis. Different speech sounds create different shapes within the graph, as shown in Figure 30-6 [1].

Figure 30-6 *Speaker recognition systems use spectrograms to represent human voices. Photo courtesy Richard Horne. (Source: Reproduced with permission from HowStuffWorks, Inc.)*

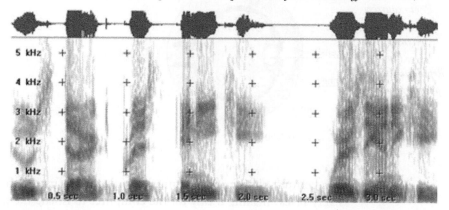

Spectrograms also use colors or shades of gray to represent the acoustical qualities of sound.

Some companies use voiceprint recognition so that people can gain access to information or give authorization without being physically present. Instead of stepping up to an iris scanner or hand geometry reader, someone can give authorization by making a phone call. Unfortunately, people can bypass some systems, particularly those that work by phone, with a simple recording of an authorized person's password. That's why some systems use several randomly chosen voice passwords or use general voiceprints instead of prints for specific words. Others use technology that detects the artifacts created in recording and playback [1].

Other systems are more difficult to bypass. Let's look at some of those next [1].

Note: For some security systems, one method of identification is not enough. Layered systems combine a biometric method with a keycard or PIN. Multimodal systems combine multiple biometric methods, like an iris scanner and a voiceprint system [1].

Iris Scanning

Iris scanning can seem very futuristic, but at the heart of the system is a simple CCD digital camera. It uses both visible and near-infrared light to take a clear,

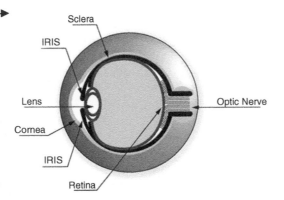

high-contrast picture of a person's iris. With near-infrared light, a person's pupil is very black, making it easy for the computer to isolate the pupil and iris (see Figure 30-7) [3].

When you look into an iris scanner (see Figure 30-8) [3], either the camera focuses automatically or you use a mirror or audible feedback from the system to make sure that you are positioned correctly. Usually, your eye is three to 10 inches from the camera. When the camera takes a picture, the computer locates:

- The center of the pupil
- The edge of the pupil
- The edge of the iris
- The eyelids and eyelashes [1]

It then analyzes the patterns in the iris and translates them into a code. Iris scanners are becoming more common in high-security applications because people's eyes are so unique (the chance of mistaking one iris code for another is 1 in 10 to the 78th power [1]. They also allow more than 200 points of reference for comparison, as opposed to 60 or 70 points in fingerprints [1].

The iris is a visible but protected structure, and it does not usually change over time, making it ideal for biometric identification. Most of the time, people's eyes also remain unchanged after eye surgery, and blind people can use iris scanners as long as their eyes have irises. Eyeglasses and contact lenses typically do not interfere or cause inaccurate readings [1].

Note: Some people confuse iris scans with retinal scans. Retinal scans, however, are an older technology that require a bright light to illuminate a person's retina. The sensor would then take a picture of the blood vessel structure in the back of the person's eye. Some people found retinal scans to be uncomfortable and invasive. People's retinas change as they age, which could lead to inaccurate readings [1].

Vein Geometry

As with irises and fingerprints, a person's veins are completely unique. Twins don't have identical veins, and a person's veins differ between their left and right sides. Many veins are not visible through the skin, making them extremely difficult to counterfeit or tamper with. Their shape also changes very little as a person ages [1].

To use a vein recognition system, you simply place your finger, wrist, palm, or the back of your hand on or near the scanner. A camera takes a digital picture using near-infrared light. The hemoglobin in your blood absorbs the light, so veins appear black in the picture. As with all the other biometric types, the software creates a reference template based on the shape and location of the vein structure [1].

Scanners that analyze vein geometry are completely different from vein scanning tests that happen in hospitals (see Figure 30-9) [1]. Vein scans for medical purposes usually use radioactive particles. Biometric security scans, however, just use light that is similar to the light that comes from

Figure 30-9
Vein scanners use near-infrared light to reveal the patterns in a person's veins. (Source: Reproduced with permission from HowStuffWorks, Inc.)

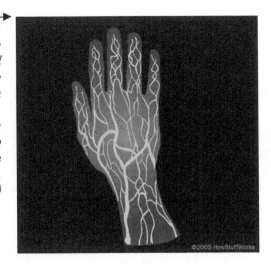

a remote control. NASA has lots more information on taking pictures with infrared light [1].

Next, let's look at some of the concerns about biometric methods. How well do they work?

How Well Do Biometrics Work?

Despite the benefits, can you be absolutely certain that a biometric device will work as claimed (see Figure 30-10) [1]? Will it securely keep the bad guys out, while effortlessly letting the good guys in [6]?

In real life, security versus convenience turns out to be pretty much a non-issue, since the combination of biometric identification plus a keypad code provides virtually unbreakable security. Here's why [6].

Biometric devices can be adjusted to favor security or user convenience. Think of a car alarm. When your car alarm is very sensitive, the probability of the bad guys stealing it is low. Yet the chance of you accidentally setting off the alarm is high. Reduce the sensitivity, and the number of false alarms goes down, but the chance of someone stealing your car increases [6].

The security requirements of a national defense contractor might demand that the device at the front door be adjusted to keep the bad guys out, for example. On the other hand, if hundreds of employees will clock in using a biometric reader at a low-security facility, you'll want to adjust the unit's sensitivity to let the good guys in [6].

Figure 30-10

As seen on TV: Television shows and movies can make it look spectacularly easy or spectacularly difficult to get past biometric security. They usually show people trying to get past the sensors rather than replacing the data in the system with their own or "piggybacking" their way in by following someone with authorization. Here are some of the more common tricks and whether they're likely to work. (Source: Reproduced with permission from HowStuffWorks, Inc.)

Biometric Type	Countermeasure	Could it work?
 Handwriting	Forged signature	Probably not. The system measures the act of writing – not the result.
 Hand geometry	Exact model of the person's hand	Probably, unless the system also performs a "liveness test," like measuring body temperature of pulse.
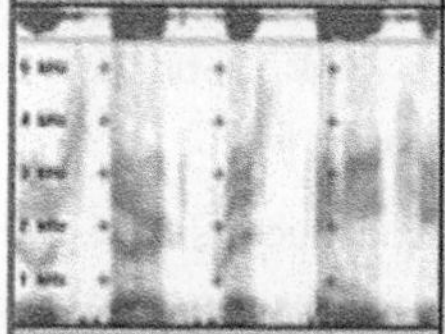 Voiceprint	Recording of the person's voice	May be. Some systems, especially those that are not telephone-based, can detect the noise generated during recording and playback. Others request random passwords instead of one specific phrase.
 Iris Screening	Picture of the person's iris printed on a contact lens	May be. This depends on the quality of the printing and whether the system performs tests that reveal the presence of the fake lens.
 Vein geometry	Model of the person's hand	Probably not. It would be extremely difficult to reconstruct an exact copy of a person's veins using materials that the scanner would identify as real.

People like things that work. If the biometric doesn't allow employees effortless access, frustration will quickly rise and the biometric may never be accepted (see sidebar, "Biometrics for Systems Access Control"). Fortunately, this is extremely unlikely [6].

Biometrics for Systems Access Control

People and passwords—in the long run, they just don't work very effectively together. At least that's what Telesis Community Credit Union, a Chatsworth, Calif.-based financial services provider that manages $3.4 billion in assets, found out. Their team ran a network password cracker as part of an enterprise security audit in 2005 to see if employees were adhering to Telesis' password policies. They weren't [7].

Within 30 seconds, Telesis had identified about 80% of people's passwords, whose group immediately asked employees to create strong passwords that adhered to the security requirements. A few days later, the Telesis team ran the password cracker again: This time, they cracked 70% [7].

Telesis couldn't get employees to maintain strong passwords, and those that did forgot them, so the help desk would have to reset them. Telesis decided to secure network and application access with a biometric system that eliminated the need for user IDs and passwords, opting for the DigitalPersona fingerprint system from DigitalPersona Inc. in Redwood City, California [7].

The use of biometrics (the mathematical analysis of characteristics such as fingerprints, veins in irises and retinas, and voice patterns) as a way to authenticate users' identities has been a topic of discussion for years. Early commercial success stories have largely come from applying biometrics to projects with provable returns on investment: time and attendance, password reduction and reset, and physical access control. Though biometric work remains primarily in the pilot stages, the events of 9/11 pushed emerging commercial products to center stage—a spot some say they weren't ready to claim. Vendor focus shifted from the private sector toward the huge contracts that many expected would be awarded in the public sector [7].

The attacks on 9/11 brought focus to what was going on in biometrics, and vendors switched gears. Where previously they were thinking about biometrics for enterprise access, they decided government contracts were the next gold mine and jumped on that. The problem with this strategy is that commercial biometric systems aren't standardized and haven't been tested in large-scale implementations of the type federal agencies are undertaking, such as the US-VISIT and Transportation Worker Identification Credential projects [7].

The problem, however, was more a lack of public-sector readiness than technology shortfalls. In 2001, the private sector was aggressively researching and testing biometrics, and the public sector had a couple of projects. After September, the biometrics industry reread the whole landscape and decided to gravitate toward the public sector, going after a market that wasn't ready for them. But, there are plenty of smaller stories of biometrics hitting the bottom line in the private sector [7].

Finger on Access

That has been the case for Telesis, which has rolled out fingerprint-based network and systems access technology in its headquarters and credit-union branches. Once Telesis has thoroughly tested the system, the company will deploy it in the offices of Business Partners LLC, its business loan services partner. Users no longer need to remember IDs and passwords because DigitalPersona authenticates enrolled personnel via fingerprint scanners, tying the fingerprints to 256-character passwords that it randomly generates every 45 days [7].

Telesis looked at a single sign-on application, but was uncomfortable with the idea that one verification would provide access to the network and all connected applications. With the current deployment, employees touch their scanners to gain access to each application they use, including homegrown and third-party Web-based applications [7].

The system is already integrated with Microsoft Corp.'s Active Directory for network access, and fingerprint profiles are encrypted [12] and stored directly in Active Directory, relieving worries Telesis had that they might be stored as images that could be compromised. Telesis' IT department is reviewing applications that require ID and password sign-ons and creating profiles for them in the DigitalPersona server [7].

During the deployment's testing phase, the Telesis team encountered a few issues related to mobile workers. For corporate travelers, the company considered equipping laptops with scanners, but most Telesis executives don't carry their laptops unless giving presentations; they prefer to use hotel business centers or Internet cafes [15] to access the corporate intranet. When they do that, they use static but difficult-to-crack passwords [7].

Another segment of Telesis' mobile population ("roaming" tellers) are another concern. Telesis wants to be able to lock down all workstations, so that the Ctrl-Alt-Delete function won't bring up the user ID and password log-in option; but, then roamers wouldn't be able to use the teller workstations they need. Although Telesis feels it's difficult to quantify ROI, it is pleased with the streamlined network access, reduced password-reset requests and the improved security ratings audits it has found since it adopted DigitalPersona [7].

Security or Convenience?

The kind of biometric application Telesis is piloting (user verification for access to computer systems) hasn't thus far seen the adoption rates that many had expected. Telesis doesn't expect to see many more such deployments before 2012 [7].

A lot has been heard about biometrics, but the reality is that most of the projects are still in pilot stages. The most mature applications of biometric technology are in systems that control physical access to facilities and keep records of time and attendance. With time and attendance, companies can use finger-, hand-, or facial-recognition technology; get rid of access cards and mechanical punch-in devices; and it's not a security issue—it's to save money [7].

Though it's not using biometrics for actual system access, Washington-based Marriott International Inc. is using voice verification technology to reset the passwords that enable access to its intranet, Active Directory service, and several nonproprietary applications. The system, Vocent Password Reset from Vocent Solutions Inc. in Mountain View, California, complements existing reset options. Users can also change passwords using PC or Web-based tools, or they can call the help desk. Around a third of the 60,000 Marriott employees who are assigned passwords take advantage of the Vocent option [7].

The system made sense, because it utilizes Marriott's phone system and requires no special hardware. The Vocent application provides two-factor verification, checking a user's voice patterns against a stored voiceprint while simultaneously verifying user information through voice recognition. Marriott captures a voiceprint through a one-time registration, and at the same time, they gather some key information that they use during the password-reset process. Given the costs of manual password resets (industry analysts estimate that they cost $30 to $53 per incident), Marriott's self-service deployment has translated into strong savings, particularly since IT requires that passwords be changed every 90 days. Marriott has a very large user base, with more than 50,000 associates, so you can imagine the amount of human intervention required for manual password resets [7].

Waiting for Standards

The technology behind biometrics represents an emerging commercial market, but adoption of such systems won't really take off until vendors and users agree on standards in areas such as application programming interfaces, common file formats, and data interchange. The scope of massive federal initiatives such as the U.S. Department of Defense's Defense Biometric Identification System demands standardized, interoperable technologies. The DoD is using fingerprint biometrics as part of a verification process for providing personnel and associates (6 million people to date) with smart cards for physical and network access. It's also piloting iris- and facial-recognition technologies [7].

It's key that you have interoperable systems because everybody's mobile; you can't buy a proprietary biometric system that ultimately only works at one base. A recent memo issued by the DoD CIO mandates that the agency's biometric collection practices align with FBI standards so the agencies can share data [7].

When the DoD first became big consumers of smart cards, they knew there weren't perfect standards in place, but they were able to leverage their size and work with other agencies and technology providers to help create standards. Hopefully, the federal agencies will have the same impact in driving biometric standards [7].

False-Accept Rates

The probability that a biometric device will allow a bad guy to pass is called the false-accept rate. This figure must be sufficiently low to present a real deterrent. False-accept rates claimed for today's biometric access systems range from 0.0001% to 0.1%. The biometric hand readers at the front entrances of 60% of U.S. nuclear power plants have a false-accept rate of 0.1% [6].

It's important to remember that the only way a bad guy can get access is if a bad guy tries. Thus, the false-accept rate must be multiplied by the number of attempts by bad guys to determine the number of possible occurrences [6].

False-Reject Rates

For most applications, letting the good guys in is just as important as keeping the bad guys out. The probability that a biometric device won't recognize a good guy is called the false-reject rate [6].

The false-reject rates quoted for current biometric systems range from 0.00066% to 1.0%. A low false-reject rate is very important for most applications, since users will become extremely frustrated if they're denied access by a device that has previously recognized them [6].

An Example May Be Helpful

A company with 100 employees has a biometric device at its front door. Each employee uses the door four times a day, yielding 400 transactions per day [6].

A false-reject rate of 1.0% predicts that every day, four good guys (1% of 400) will be denied access. Over a five-day week, that means 20 problems. Reducing the false-reject rate to 0.1% results in just two problems per week [6].

A low false-reject rate is very important for most applications, since users will become extremely frustrated if they're denied access by a device that has previously recognized them. As mentioned previously, the combination of a low false-reject rate plus a simple keypad code provides virtually unbreakable security [6].

Equal Error Rates

Error curves give a graphical representation of a biometric device's "personality." The point where false accept and false reject curves cross is called the equal error rate. The equal error rate provides a good indicator of the unit's performance. The smaller the equal error rate, the better [6].

Validity of Test Data

Testing biometrics is difficult because of the extremely low error rates involved. To attain any confidence in the statistical results, thousands of transactions must be examined [6].

Some error rates cited by manufacturers are based on theoretical calculations. Other rates are obtained from actual field testing. Field data are usually more reliable. In the case of false-reject rates, only field test data can be considered accurate, since biometric devices require human interaction. For example, if the device is hard to use, false-reject rates will tend to rise. A change in the user's biometric profile could also cause a false-reject (a finger is missing, for example) [6].

None of these conditions can be accurately quantified by purely theoretical calculations. On the other hand, false-accept rates can be calculated with reasonable accuracy from cross-comparison of templates in large template databases [6].

Currently, most field test error rates have been generated by various biometric manufacturers using end-user data. Tests have also been conducted by independent laboratories such as the U.S. Department of Energy's Sandia National Laboratories [6].

It's important to remember that error rates are statistical: They are derived from a series of transactions by a population of users. In general, the larger the population and the greater the number of transactions, the greater the confidence level in the accuracy of the results [6].

If the error rate is reported at 1:100,000, and only 100 transactions were included in the study, the confidence level in the results should be very low. If the same error rate was reported for 1 million transactions, the confidence level would be much higher [6].

The magnitude of the reported results affects the size of the sample required for a reasonable confidence level. If the reported error rate is 1:10, then a sample of 100 transactions may provide a sufficient confidence level. Conversely, a 100-transaction sample would be too small if the error rate was reported as 1:100,000 [6].

Privacy and Other Concerns

Some people object to biometrics for cultural or religious reasons. Others imagine a world in which cameras identify and track them as they walk down the street, following their activities and buying patterns without their consent. They wonder whether companies will sell biometric data the way they sell e-mail addresses and phone numbers. People may also wonder whether a huge database will exist somewhere that contains vital information about everyone in the world, and whether that information would be safe there [1].

At this point, however, biometric systems don't have the capability to store and catalog information about everyone in the world. Most store a minimal amount of information about a relatively small number of users. They don't generally store a recording or real-life representation of a person's traits—they convert the data into a code. Most systems also work only in the one specific place where they're located, like an office building or hospital. The information in one system isn't necessarily compatible with others, although several organizations are trying to standardize biometric data. In addition to the potential for invasions of privacy, critics raise several concerns about biometrics, such as:

- **Overreliance:** The perception that biometric systems are foolproof might lead people to forget about daily, commonsense security practices to protect the system's data.

- **Accessibility:** Some systems can't be adapted for certain populations, like elderly people or people with disabilities.

- **Interoperability:** In emergency situations, agencies using different systems may need to share data, and delays can result if the systems can't communicate with each other [1].

The Future of Biometrics

Biometrics can do a lot more than just determine whether someone has access to walk through a particular door. Some hospitals use biometric systems to make sure mothers take home the right newborns. Experts have also advised people to scan their vital documents, like birth certificates and Social Security cards, and store them in biometrically secured flash memory in the event of a national emergency. Here are some biometric technologies you might see in the future:

- New methods that use DNA, nail bed structure, teeth, ear shapes, body odor, skin patterns, and blood pulses;

- More accurate home-use systems;

- Opt-in club memberships, frequent buyer programs, and rapid checkout systems with biometric security;

- More prevalent biometric systems in place of passports at border crossings and airports [1].

Future Applications: Some Common Ideas

There are many views concerning potential biometric applications. Some of the popular examples are:

- ATM machine use

- Workstation and network access

- Travel and tourism

- Internet transactions
- Telephone transactions
- Public identity cards [8]

ATM Machine Use

Most of the leading banks have been experimenting with biometrics for ATM machine use and as a general means of combating card fraud. Surprisingly, these experiments have rarely consisted of carefully integrated devices into a common process, as could easily be achieved with certain biometric devices. Previous comments in this book concerning user psychology come to mind here, and one wonders why industry analysts in this area have not seen a more professional and carefully considered implementation from this sector. The banks will of course have a view concerning the level of fraud and the cost of combating it via a technology solution such as biometrics. They will also express concern about potentially alienating customers with such an approach. However, it still surprises many in the biometric industry that the banks and financial institutions have so far failed to embrace this technology with any enthusiasm [8].

Workstation and Network Access

For a long time, workstation and network access was an area often discussed but rarely implemented until recent developments saw the unit price of biometric devices fall dramatically and several designs were aimed squarely at this application. In addition, with household names such as Sony, Compaq, KeyTronics, Samsung, and others entering the market, these devices appear almost as a standard computer peripheral. Many are viewing this as the application that will provide critical mass for the biometric industry and create the transition between a sci-fi device to a regular systems component, thus raising public awareness and lowering resistance to the use of biometrics in general [8].

Travel and Tourism

There are many in this industry who have the vision of a multi-application card for travelers that by incorporating a biometric, would enable them to participate in various frequent flyer and border control systems as well as paying for their air ticket, hotel room, rental car, and so on, all with one convenient token. Technically this is eminently possible, but from a political and commercial point of view there are still many issues to resolve, not the least being who would own the card, be responsible for administration, and so on. These may not be insurmountable problems, and perhaps you may see something along these lines emerge. A notable challenge in this respect would be packaging such an initiative in a way that would be truly attractive for users [8].

Internet Transactions

Many immediately think of on-line transactions as being an obvious area for biometrics, although there are some significant issues to consider in this context. Assuming device cost could be brought down to a level whereby a biometric (and perhaps chip card) reader could be easily incorporated into a standard-built PC, you still have the problem of authenticated enrollment and template management, although there are several approaches one could take to that. Of course, if your credit card (see sidebar, "New Credit Card Scam") already has incorporated a biometric, this would simplify things considerably. It is interesting to note that certain device manufacturers have collaborated with key encryption providers to provide an enhancement to their existing services. Perhaps we will see some interesting developments in this area in the near future [8].

New Credit Card Scam

This scam is for real and pretty slick, since they provide you with all the information, except the one piece they want.

Note: *The callers do not ask for your card number; they already have it. This information is worth reading. By understanding how the Visa and MasterCard telephone credit card scam works, you'll be better prepared to protect yourself.*

One employee was called on Wednesday from Visa, and another was called on Thursday from MasterCard.

The scam works like this: A person calling says, "This is (name), and I'm calling from the Security and Fraud Department at Visa. My badge number is 12460. Your card has been flagged for an unusual purchase pattern, and I'm calling to verify. This would be on your Visa card which was issued by (name of bank). Did you purchase an anti-telemarketing device for $497.99 from a marketing company based in Arizona?" When you say "No," the caller continues with, "Then we will be issuing a credit to your account. This is a company we have been watching and the charges range from $297 to $497, just under the $500 purchase pattern that flags most cards. Before your next statement, the credit will be sent to (gives you your address), is that correct?"

You say "yes." The caller continues—"I will be starting a fraud investigation. If you have any questions, you should call the 1-800 number listed on the back of your card (1-800-VISA) and ask for Security.

You will need to refer to this control number." The caller then gives you a six-digit number. "Do you need me to read it again?"

Here's the important part on how the scam works. The caller then says, "I need to verify you are in possession of your card." He'll ask you to "Turn your card over and look for some numbers." There are seven numbers; the first four are part of your card number, the next three are the security numbers that verify you are the possessor of the card. These are the numbers you sometimes use to make Internet purchases to prove you have the card. The caller will ask you to read the three numbers to him. After you tell the caller the three numbers, he'll say, "That is correct, I just needed to verify that the card has not been lost or stolen, and that you still have your card. Do you have any other questions?" After you say "No," the caller then thanks you and states, "Don't hesitate to call back if you do," and hangs up.

You actually say very little, and they never ask for or tell you the card number. But after one family was called recently, they called back within 20 minutes to ask a question. Are they glad they did! The *real* Visa Security Department told them it was a scam and that in the last 15 minutes a new purchase of $497.99 was charged to their card.

Long story? In short, that family made a real fraud report and closed their Visa account. Visa is reissuing them a new number. What the scammers want is the 3-digit PIN number on the back of the card. Don't give it to them. Instead, tell them you'll call Visa or Master Card directly for verification of their conversation. The real Visa told the family that they will never ask for anything on the card as they already know the information since they issued the card! If you give the scammers your three-digit PIN number, you think you're receiving a credit. However, by the time you get your statement, you'll see charges for purchases you didn't make, and by then it's almost too late and/or more difficult to actually file a fraud report.

What makes this more remarkable is that the next day, the same family got a call from a "Jason Richardson of MasterCard" with a word-for-word repeat of the Visa scam. This time the family didn't let him finish. They hung up! They then filed a police report, as instructed by Visa. The police said they are taking several of these reports daily! They also urged the family to tell everybody you know that this scam is happening.

Please pass this on to all your family and friends. By informing each other, you protect each other.

Telephone Transactions

No doubt many telesales and call center managers have pondered the use of biometrics. It is an attractive possibility to consider, especially for automated processes. However, voice verification is a difficult area of biometrics, especially if one does not have direct control over the transducers, as indeed you wouldn't when dealing with the general public. The variability of telephone handsets coupled to the variability of line quality and the variability of user environments presents a significant challenge to voice verification technology, and that is before you even consider the variability of understanding among users [8].

The technology can work well in controlled closed loop conditions, but is extraordinarily difficult to implement on anything approaching a large scale. Designing in the necessary error correction and fallback procedures to automated systems in a user friendly manner is also not a job for the faint-hearted [8].

Perhaps, you will see further developments which will largely overcome these problems. Certainly there is a commercial incentive to do so; and no doubt, there is much research under way in this respect [8].

Public Identity Cards

A biometric incorporated into a multipurpose public ID card would be useful in a number of scenarios, if one could win public support for such a scheme. Unfortunately, in this country as in others, there are huge numbers of individuals who definitely do not want to be identified. This ensures that any such proposal would quickly become a political hot potato and a nightmare for the official concerned. You may consider this a shame or a good thing, depending on your point of view. From a dispassionate technology perspective, it represents something of a lost opportunity, but this is of course nothing new. It's interesting that certain local authorities in the United Kingdom have issued citizen cards with which named cardholders can receive various benefits, including discounts at local stores and on certain services. These do not seem to have been seriously challenged, even though they are in effect an ID card [8].

Now, continuing with the theme of biometrics in the future, let's look at what biometrics might be like in the year 2017. Four scenarios are presented here for your education, as well as entertainment.

Biometrics in 2017: Scenarios and Exercises

This part of the chapter presents four scenarios (see sidebar, "Scenario Methodology"):

- Biometrics in everyday life

- Biometrics in business

- Biometrics in health

- Biometrics at the border

In this part of the chapter, the scenarios are analyzed and placed in context [9].

Scenario Methodology

Scenarios are considered to be one of the main tools for looking at the future, but it is important to clearly situate what their objective is. Normally, their objective is not to predict the future, but to present plausible futures in order to understand what might happen. Scenarios are used to stimulate discussions on the major technological, economic, social, and political factors that are to be taken into account when thinking about possible futures. In theory, the number of possible futures is almost infinite, but scenario exercises usually reduce them to three to five manageable future possibilities [9].

There is no single approach regarding scenarios, but scenario exercises are commonly the outcome of group work, group discussions, and/or scenario workshops. Since there are different types of scenarios, it is important to specify which type of scenario is being developed. The biometric scenarios presented here are trend or reference scenarios. They start from the present and work forward on the basis of expected trends and events. They are intended to be realistic rather than normative or extreme. Normative scenarios, for instance, present a desirable vision of the future and the necessary steps to realize that vision (back-casting). An example of trend scenarios are the MUDIA scenarios on how (on-line) media are expected to evolve in the future [9].

The objective here is to open up the scope of thinking on the future of biometrics, beyond the current passport and visa application plans. One of the themes of this chapter is the so-called diffusion effect (as biometric technologies become better, cheaper, more reliable, and are used more widely for government applications, they will be implemented in everyday life, in businesses, at home, in schools, and in other public sectors). The scenarios, therefore, try to envisage what the results of this diffusion effect might be [9].

The four scenarios, as shown in Figure 30-11, are carefully selected to encompass key environments for the introduction of biometrics [9]. These

Figure 30-11   *Four biometric scenarios. (Source: Adapted with permission from the Institute for Prospective Technological Studies.)*

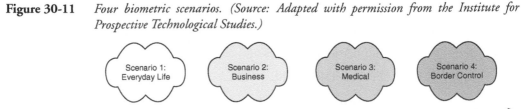

environments differ, for instance, in terms of the role played by governments and public authorities; in fact, they can be placed on a continuum, as shown in the figure, with private actors predominant in the first two scenarios and public actors in the last two. The everyday and business scenarios have limited government involvement. The medical environment, particularly in Europe, is a public or private environment that is carefully regulated, not least as a result of the government's budgetary involvement in health provision. The fourth scenario, biometrics at the border, is not only regulated but also under strong control of public authorities [9].

These differences between the four scenarios can also be viewed with respect to the issues of privacy and security. The use of biometrics at the border has clear security purposes that are likely to take precedence over privacy. This is clearly not the case in the everyday scenario where privacy, particularly in the home, is legally and socially protected. The implementation of biometrics in business will have to take into account privacy and data protection rules. But the protection of personal data may be strongest in the case of the biometrics in health, given the sensitive and thus private nature of medical data. The objective is not to detail all these issues but rather to raise awareness that these differences exist and that they will have an impact on how biometrics can be implemented [9].

These four scenarios thus present different contexts for the use of biometrics. The choice of biometric technology for each situation is based on the analysis outlined in Chapter 2 of this book. Nevertheless, the specific examples should be seen as illustrative rather than a prediction of how and where each technology will be used. The scenarios are neither mutually exclusive nor all-encompassing, but they do present some of the major domains for biometric applications in the future: work, private life, government, and health [9].

Scenario on Biometrics in Everyday Life

The everyday life scenario describes a day in the life of a traditional nuclear family. It is a middle-class dual-income household with two children, a teenager and a toddler. As both parents work, the grandparents provide support in managing the household. The scenario is presented as a diary entry by the teenage son, Spike. He is in trouble at school because he has spoofed the cafeteria's biometric entry system in order to help out a friend. His mother, who is called to the school to discuss this, has a car with a fingerprint scanner to start the engine. The grandmother goes to pick up the youngest son, but the nursery's multi-modal biometric system falsely denies her entry. On the other hand, she has no problem with the face recognition system used on the buses. At home, there is a common digital storage [13] space called the virtual residence, where password access is replaced by an iris scanner. There is also a biometric toy that recognizes registered users. Household appliances can use biometrics to secure access, such as the cooker (which uses hand geometry). Finally, unauthorized use of computer games is made more difficult via biometric verification, in this example, using a fingerprint [9].

Spoofing Physical Access/Entitlements

The scenario shows that spoofing biometric systems is clearly possible. It does not only depend on the biometric technology (though certainly some technologies [iris] are more difficult to circumvent than others), but also on the way the technology is implemented (thresholds and hardware). In the case of the school cafeteria entry system, cheap iris scanners make the system easy to fool. To be able to discover spoofing, systems need to check for irregularities such as double entry attempts (manually or automatically). This is easier to do within a closed system that has a small local database, like the one at the school, compared to a large-scale database containing millions of stored templates [9].

Biometrics to Replace Keys (for Convenience and Security)

The fingerprint scanner in the car is installed to prevent unauthorized use and theft. It is a local system that only needs to verify a limited number of authorized users (and Spike, the son, is not one of them). Enrollment will probably need to be managed by the car owner. The system is bought and installed for security reasons. Insurance companies can stimulate the demand for such systems. For users it is convenient since they always have their keys with them (the finger) but, in the case of breakdown, alternative procedures need to be available. These may, however, take some time, as suggested in the script. It may be the case that spare keys are available at home or at an authorized dealer or garage [9].

Physical Access and Security Thresholds

The biometric technology for access to daycare centers needs to be highly secure. Therefore, the daycare center combines two biometric technologies, in this case, face and voice recognition. Templates will probably be stored in a central database but within a closed system. The threshold for false acceptance is set low at the expense of a higher false rejection rate. This may mean that regular (yearly) enrollment is necessary since people's biometrics may change (slightly) over time. Face recognition seems to be particularly sensitive to this problem but, more generally, regular enrollment is an issue for all technologies. Being falsely rejected may cause user annoyance and user frustration, and as a result, may negatively affect the quality of a submitted biometric trait (Granny's voice), as it is not pleasant to be wrongly rejected by an automated system. In the end, human intervention needs to be available as a fallback procedure [9].

The public transport face recognition system is used to check if people are entitled to use it. (Have they paid the correct ticket?) The threshold is set in favor of convenience (allowing more false positives). In contrast with the daycare center where there is a central database, templates for the public transport system will most likely be stored on a smart card. The less likely alternative would be that buses connect wirelessly [14] in real time with a central database for matching [9].

Digital Access

Biometric access to digital spaces can replace knowledge-based password access. Secure access to a shared digital space also makes personal digital territories possible within that common folder. Another issue here, however, is related to usability. Taking a biometric scan (be that fingerprint, face, or, as in the case of the scenario, iris) requires a clear positioning of the biometric trait on the scanning device for a good result. Scanning devices are not always designed in a user-friendly way (making sure the user knows what to do, where to focus, or how to push), nor are people always in the position to provide the trait in the prescribed way, as illustrated in the scenario (the father is nearsighted). The iris recognition system is bought off the shelf and is installed and managed by the end-user [9].

Biometric Toys

The biometric toy is introduced to illustrate the possibility of alternative uses and business models that are not inspired by security, safety, and convenience. It shows that biometrics can be used in a playful way. Biometric technologies can enable the recognition of people in a natural way. They are part of the repertoire of so-called natural interfaces that envisage human-machine interactions becoming more similar to the way humans interact with each other in the real world (via speech, gesture, touch, look, etc.) [9].

Biometric toys could contribute to the wider acceptance of biometrics in society, not only because children would in this way already be acquainted to them and would learn to use biometrics when they are still young, but also because such localized and off-line applications have less privacy and security concerns. It may be necessary, however, to pay special attention to raising awareness and education because there is a fear that the use of biometrics by children may desensitize them to the data protection risks that they may face as adults through the use of their biometrics [9].

Biometrics for Safety Versus Reluctance to Use Them

The use of the stove is protected by a hand geometry reader to avoid accidents caused by children. The choice of the hand as well as other biometrics that are based on touching (finger) may appear natural in the kitchen but at the same time may be less suitable there, since hand and fingers get dirty while cooking. This also affects the biometric sensors. Contactless biometrics such as face could be more suitable here [9].

The example shows that people can be reluctant to use certain, but not all, biometrics. They may be accustomed to using biometrics and they may not be against them as such, but they just get tired of using them all the time, or rather, of enrolling again and again for every stand-alone application that one can imagine [9].

Biometrics for Digital Rights Management

Biometrics might be useful for digital rights management (DRM) to replace code and/or password-protected files. It can be assumed that people, especially youngsters, will look for possibilities to bypass these systems. The example shows that fingerprint spoofing may be possible, but also that it takes some time to do, especially when taking into account that the newer generation fingerprint sensors have a liveness detection functionality [9].

To summarize, the everyday life scenario illustrates that people can be confronted with biometrics in many different ways in their lives. They are used to secure access (that is to prevent unauthorized access) to both physical and digital places, but also to check entitlements. They can be installed (voluntarily or not) for the protection of both physical (car) and digital goods (DRM). They might be used for safety purposes (stove) but also for toys [9].

It is clear that biometric technologies are never 100% secure. Choices need to be made between different biometrics. But equally important is the implementation. Thresholds need to be set and decisions need to be made, usually in the form of tradeoffs. Finally, some usability and user acceptance issues are raised. People may accept biometrics for certain aspects and reject them for others [9].

Scenario on Biometrics in Business

Biometrics in business encompasses the use of biometrics by companies. This can be for internal and external purposes (with employees internally and with clients, other companies, or third parties externally). This scenario is presented as a memo to senior management of a large multinational supermarket chain that has embraced the use of biometrics but is concerned that it is not reaping the expected benefits. The memo raises several issues, such as a biometric access system to the company premises and secure electronic payments enabled by a third party. Customers also make use of biometrics in order to access stores. The sharing of biometric databases between companies is highlighted as a new use of biometrics to be pursued [9].

Staff Access to Company Premises

Biometric access to company premises may be installed to allow only authorized people to enter, but it can also be used in order to manage people more effectively. In this case, it is used for checking working hours. The memo implies that with the older system of punch-cards, punching-in or punching-out could be done by someone else. With biometric verification, this becomes much more difficult [9].

The staff entrance situation highlights the importance of human factors when using biometrics. Alternative procedures need to be foreseen for the cases where biometric access is refused and these procedures might be neglected, as humans tend to do when it is more convenient for them. The scenario foresees human monitoring of the system to ensure correct use. Another usability issue is raised with the example of sweaty hands, showing that both physical and psychological factors can decrease the performance of biometric applications [9].

Electronic Payments

Electronic payments require strong verification. Biometrics can add an additional layer of security to the process, which is particularly desirable when large amounts of money are concerned. To enable this, banks may want to have biometric verification that is managed by them in order to verify and guarantee correct enrollment and regular re-enrollment. Enrollment may be local while the database is centralized. Adding a biometric to the transaction enables stronger control a posteriori in the case that something goes wrong, since the person who transferred the money can be identified [9].

Companies and Their Customers

The use of biometrics in stores shows that companies will probably need to convince customers to enroll and participate in their biometric systems, especially if it is not clear what the added value for the customer is. For the companies, one of the reasons to invest in biometrics might be to identify and know customers better, so that more products can be sold and logistics can be improved. Companies, however, will have to address bottlenecks in terms of accessibility, privacy, and customer acceptance. Customer reluctance may be tackled by offering a financial benefit (price reductions, enroll and win, promotions) or by providing strong privacy protection (pseudonymous biometric system) [9].

The supermarket chain's initial idea was to use biometrics to provide people with a personal greeting when they entered the shop. But this initiative was withdrawn because it was perceived to be very privacy-invasive. As noted in the memo, customer preferences have been monitored for many years via loyalty cards, but that may be less visible compared to biometric identification. Companies may also need to think about how to deal with customers that cannot provide the biometric feature and, as a result, are excluded from these benefits [9].

Sharing of Enrollment and Databases

The implementation of biometric applications in the business environment might be quite cost-intensive and laborious, and as a result, might make biometrics less feasible for smaller enterprises. To tackle this, it is imaginable that companies will want to collaborate and create virtual networks for sharing biometric investments and biometric applications. Why not share the enrollment process, rather than each company organizing its own enrollment? Why not share biometric databases, rather than each company setting up and maintaining its own database? For customers, this might be interesting since a network of companies can offer a single enrollment. This raises many questions in terms of security, privacy, liability, maintenance, and so forth [9].

It is not explicitly mentioned in the scenario, but there is currently little knowledge on the potential of biometrics in business outside the well-known security and safety schemes. Convenience can be a driver, but it is not clear if it will provide enough reason to invest in biometrics [9].

Scenario on Biometrics in Health

The biometrics in health scenario presents a series of e-mails between doctors in two different countries, describing various applications that exist in each. Zoe Helus, the first doctor, describes how biometrics have been implemented for physical access and network access and mentions an example of an unsuccessful application. Izzi Cornelius replies with a description of an electronic health card and identity checks in the maternity ward. Zoe's second e-mail offers a subjective opinion on the applications and biometrics in general [9].

Prior to discussing the script, a few general points can be made on this scenario. Positive identification is essential in the health sector. Retrieving medical histories, administering medicine, handing out prescriptions, and carrying out medical procedures all rely on the correct identification of the individual. In addition, there is a strong need for privacy which stems from the sensitive nature of medical data. These two requirements make the healthcare sector a likely field for the application of biometrics [9].

Physical Access

In the first situation, biometrics are used in order to limit access to restricted areas to authorized staff. Missing medical supplies are an acknowledged problem faced by hospitals and clinics; therefore, it seems to be a cost incentive to introduce biometrics as a solution. Hospital administrators can estimate the cost of missing supplies and compare this to the cost of introducing a biometric-based system or a nonbiometric-based system. It is therefore possible to evaluate the benefits of introducing such a system. As the application operates within a closed environment with a limited number of users, there are no issues of interoperability, and high performance levels might be achieved. One point to note here is that biometrics are just one part of the overall technological solution; the scenario describes how systems also make use of other elements such as RFID tags and smart cards [9].

Network Access

A frequently proposed use of biometrics for the health sector involves access to electronic health records; biometrics can be used to ensure that only authorized people have access to sensitive medical information. This application draws on many of the advantages of biometrics: A biometric cannot be lost or forgotten, and it cannot be lent to an unauthorized person. Zoe mentions in her e-mail that people need many different passwords for the different systems they have to access: patient records, appointment schedules, financial records. They commonly use the same password for all systems or write passwords down. The solution is a single sign-on system where one biometric is used as a password for all systems. This application offers convenience and leads to greater security as people use the system correctly [9].

Choice of Biometric Technology

The choice of biometric technology always depends on the context within which it will be deployed. In the medical sector, there are additional factors to take into account: Fingerprints will not work in environments where users wear latex gloves, face recognition will not work with surgical masks, voice recognition will not work in noisy environments. On the other hand, in the case of network access, if a doctor is accessing files with a laptop from remote locations, iris recognition will be unsuitable because the scanners are both expensive and bulky. Cross-contamination through contact readers is an issue of particular importance within a hospital environment, and the scenario mentions some ways of minimizing this risk [9].

A Failed Application

The third situation describes an example of a failed application. The specific details are not the issue, but the scenario tries to emphasize the point that biometrics are not a panacea for all ills. They are a tool with certain benefits and drawbacks, which may be used as part of a wider application in answer to a specific problem. Applications need careful design to fit in with working practices and other practical considerations [9].

Maternity Ward

Maternity wards are a field where biometrics have already been tried out for security reasons in order to prevent people taking someone else's infant. Once again, it is a small-scale, closed system (limited users and no issues of interoperability). Biometrics are a natural solution for confirming and linking the identities of mothers and their children, and there has been public support in areas where this has been implemented as people perceive the benefits [9].

The Health Card

The health card, described next, is a complex issue. Both private health insurance companies and public authorities have a vested interest in ensuring that only those eligible for treatment receive it. Biometrics could be instrumental in tackling fraud in the healthcare sector and in fact there are several instances where biometrics have already been introduced in order to cut down on health insurance fraud. There are two ways a biometric health card could be implemented: with or without a centralized database [9].

Tele-Care or Home Healthcare

A great benefit of biometrics is the ability for remote verification. This potential is mentioned in passing in the script, but is worth reflecting upon. So far security worries as well as technological limitations have stopped the widespread adoption of eHealth applications. For home healthcare in particular, it is important to be able to remotely identify patients. Biometrics offer the power to do this and could therefore enable many interesting applications that would otherwise not be able to make it off the drawing board [9].

Scenario on Biometrics at the Border

As part of the international drive for greater security at border control, the ICAO has recommended the introduction of biometric identifiers on machine-readable travel documents (MRTD). The European Parliament has voted in favor of proposals for biometrics on passports and visas, in accordance with ICAO recommendations. Taking the introduction of biometrics on MRTD as a given, the aim of the fourth scenario is to highlight issues raised by the implementation of biometrics at the borders. The story presents a father, daughter, and grandfather, making a trip around the world, with stops in Dubai, Beijing, and Bangkok. By focusing on three destinations and three family members, the scenario illustrates the use of biometrics in different countries, by different age groups. You'll follow the family through the process of obtaining visas to the journey itself. The analysis presented here briefly discusses the topics raised [9].

Visa Applications: Closed Versus Open Systems

Visa applications are a closed system and therefore each country (or group of countries in the case of the Schengen states) can choose a proprietary technology and store only biometric templates rather than full images. In contrast, passports are an open system, as they have to be readable by foreign border control authorities. In open systems, interoperability is an issue of particular importance and for this reason the ICAO has recommended storage of the full biometric image on passports [9].

A Few Dominant Technologies

Some countries may choose not to have visas (in this example, this is the UAE) while others may implement whichever biometric technology they see fit. If different countries use different technologies, it will lead to inconvenience for citizens, as they will have to go in person to enroll their biometrics at the embassy of the country for which they are obtaining a visa. Sovereign states will want to select the biometric technology that best fits their needs, but at the same time they may want to avoid costly enrollment procedures at local embassies by using a biometric available on the passport. It is likely that these two factors will lead to a few dominant technologies being used for all border control applications [9].

Correct Enrollment

The importance of correct enrollment is emphasized for the visa application, but the point is equally valid for any type of enrollment (passport, ID card, driver's license, etc.). An application is only as secure as its weakest point; if it is possible to make a fraudulent enrollment, the application quickly loses its value. For this reason the ICAO has suggested using biometrics in order to verify the identities of supervising staff and to confirm they have the authority to carry out the tasks they perform [9].

Schengen Zone

Although biometric controls will be introduced at external borders, the scenario shows that the Schengen Agreement continues to apply within the EU. The Schengen acquis is going to be further developed within the institutional and legal framework of the EU, including the use of biometric data for checks at external borders [9].

Confirmation of Presence

An article from the in-flight magazine draws attention to a different benefit of biometrics—the ability to confirm an individual's presence. Biometrics, in fact, are the only automatic tool that can verify the presence of a particular individual. Passwords and security cards can be shared or lost, but biometrics are an integral part of the individual. This unique property could have many applications. In the story, Schiphol Airport has introduced biometrics in order to verify the presence of airport control tower staff—something which would have been very useful during the recent crash of ComAir Flight 5169 in Lexington, Kentucky, which killed 49 people. The pilots took off on the wrong runway, which ended up being too short for takeoff.

It was pointed out that the airport control tower staff was composed of only one person at the time. According to FAA rules, there should have been at least two air traffic controllers at the time in the tower. If there would have been, perhaps the other controller would have noticed that the plane was on the wrong runway, and possibly aborted the flight in time.

Iris Scanner at Dubai and the Watchlist

The scenario imagines that at Dubai, a watchlist is used instead of visas (a database where the biometric data of certain individuals is stored). In this example, the watchlist contains the details of people who have been banned from the country and therefore should not be allowed entry. Passengers are checked against this database and if they do not match a record, they are allowed to enter the country [9].

Advanced Passenger Information (API)

API is used to carry out a type of watchlist operation in advance of travel. Data on each passenger (as contained in the machine-readable zone of the passport) is captured by the airline during the check-in process overseas, formatted by the airline's reservation-control system, and transmitted to the centralized Customs system, where it is checked against interagency databases and watchlists. The results of these checks are then downloaded to the airport of arrival, where they are distributed to both Immigration and Customs. The accomplishment of this part of the process prior to arrival of the flight substantially reduces or eliminates the time-consuming data entry and computer processing required during the examination of each passenger from a flight on which API data was not transmitted [9].

Revocation of Biometrics

An important question that has not yet been answered is whether biometrics can be revoked (If a person needs to change identity or finds that his or her biometric data has been compromised, what can be done to revoke that person's biometrics?). This question will assume even greater importance as biometrics diffuse into everyday life [9].

An Example of DNA Tests

There may be reluctance on the part of citizens to share biometric data, particularly of a sensitive nature such as DNA, with countries outside the EU. For those who travel for leisure, there will always be the option to avoid countries where they do not feel comfortable with visa application procedures. Business travelers, however, may not have the luxury of choice. Decisions taken unilaterally by one country may therefore affect a large portion of citizens [9].

Face Recognition: Controlling Conditions

The success of biometrics at border control will depend largely on the method of implementation. The face has been chosen by the ICAO and EU as the primary biometric identifier. But face recognition is currently one of the less accurate biometric technologies. It suffers from technical difficulties with uncontrolled lighting and it therefore may be necessary to install the face recognition readers in booths where lighting conditions are carefully controlled. Measures such as this one may lead to improvements in accuracy but also to an increase in costs [9].

Difficulties at Bangkok Airport

Biometric applications can and do go wrong sometimes and therefore secondary or back-up procedures are required to deal with these cases. The scenario shows just one such example. Iris recognition systems are believed to be able to match any person to their record by the third attempt. This may be true for regular users, but Gerard the grandfather suffers from glaucoma. It has been shown that glaucoma can cause iris recognition to fail, as it creates spots on the person's iris. When the machine rejects Gerard for the third time, officials take him aside for secondary procedures. This situation draws attention to several potential pitfalls for biometrics. Current border control staff are skilled employees who use personal judgment in deciding who needs further questioning. There is a danger that these skills could be sidelined if border control starts relying heavily on automated biometric checks. Furthermore, there has to be a recognition that biometric tests are statistical by nature, which means that there will always be a possibility, however small, that innocent individuals fail the verification. Secondary procedures must take this into account [9].

Lines

Biometrics at border control may be suggested as a way of automating the procedure, thus scaling back staffing requirements. The reality is that for the foreseeable future, border control staff will have an important role to play in supervising biometric checks, particularly early on in the implementation when travelers are still getting acquainted with the technology. Secondary procedures will always have to exist to deal with cases where the biometric check fails. Frequent traveler programs are sometimes cited as an example, where biometrics can improve passenger turnaround times, but they work with a limited user base of passengers who travel often and are therefore adept at using biometric readers. Furthermore, the travelers who may most need assistance (children, elderly people, disabled people, people without biometrics, etc.), are unlikely to be part of current frequent traveller schemes. Thus, existing performance data may not accurately reflect the difficulties that may arise when biometrics are implemented on a large scale [9].

Concluding Remarks on Scenario Exercises

The scenarios naturally place biometric applications at the center of attention, but it should be noted that in a future digital society, biometrics will be part of a larger IST (or ambient intelligence) environment that includes RFIDs and other digital technologies. As the cost of biometric technologies comes down and people grow accustomed to using them through border control and other government applications, it is likely there will be a diffusion of biometrics into everyday life. Tomorrow's diffusion effect provokes today's need for discussion. The critical issues raised by the scenarios can be categorized under three headings: privacy, security, and usability [9].

Privacy

The final e-mail of the medical scenario makes the assertion that biometrics can undermine or protect privacy, depending on the application and the implementation. The medical scenario demonstrates how biometrics can enhance privacy of medical records by replacing an easily compromised system of passwords with a theoretically more secure biometric and smart card combination. Similar situations occur in the everyday scenario with the use of biometrics to protect the parent's car and each family member's file space. The medical scenario suggests that a biometric template might be used as a key in a database of medical data so that a medical record can only be retrieved with someone's biometric. These applications show the positive side of biometrics [9].

On the other hand, biometrics can threaten privacy. The business scenario alludes to the potential for profiling with biometrics. Biometrics such as face, gait, or voice recognition that may in the future allow humans to be identified passively (without requiring their consent) have provoked surveillance fears in some privacy campaigners. A policy question for the future will be deciding on the appropriate safeguards (legislative or other) to deal with such issues [9].

The business scenario also shows the use of biometrics for auditing working hours. In this case, employees may resent or even obstruct the use of biometrics. In general, the principle of proportionality should apply when designing applications. The question to be answered is whether the use of biometrics is justified in the context or whether some other means of verification could fulfill the requirements equally well [9].

Security

The fundamental question from a security point of view is: How secure do systems need to be? For a particular application, is it more important to prevent impostors (low false-accept rate) or to let through the right people (low false-reject rate)? This question is broached by the everyday scenario, when comparing the access system at the daycare center to the senior pass for the bus. At what cost are we willing to achieve high security? The cafeteria system at the school installs cheap iris readers to save on costs, resulting in a system that can be spoofed. Arguably for a school cafeteria, the additional security provided by better readers does not justify the cost. In contrast, for the medical sector, it will be crucial to ensure that it is not possible to spoof access systems. If spoofing is possible, then a biometric system loses much of its security value and cannot guarantee privacy [9].

Security is not just determined by technical factors such as thresholds, hardware, and prevention of spoofing. All parts of the procedure have to be equally secure, including enrollment, storage of the biometric template (if using distributed storage), maintaining and updating the database (if using central storage), and secondary procedures for when biometric tests fail. Secondary procedures are shown in three of the scenarios, at the daycare center (the grandmother is checked against paper records), in the business (the employee has to go to a different gate when trying to gain access), and at border control (the customs officer has to receive confirmation of Gerard's visa from the embassy). Human factors have to also be taken into account; if biometric applications secure all other means of fraud, insider attacks may become more prominent [9].

Usability

The usability of biometric systems will greatly influence their success and acceptance. For universal applications (such as the health card) where all citizens are obliged to enroll, biometric systems will need to consider the needs of everyone, in particular people with disabilities, elderly people, children, and so on. This is a very different proposition to a frequent flyer program, for example, where users fit a fairly specific socio-demographic and socioeconomic profile [9].

In both the public and private sector, biometric applications will have to take into account working practices. The medical scenario shows an example of an application that fails because it disregards the practicalities of the environment in which it is being implemented [9].

Finally, secondary procedures also come under the category of usability. A person who fails a biometric test may either be an impostor or an honest person falsely rejected. For security purposes it is important that the secondary procedures are rigorous, but at the same time, the border control and everyday scenarios show the embarrassment and agitation that this rejection may cause in a law-abiding person. With current performance levels, the number of people falsely rejected may be 1 in 100 or even 1 in 10 depending on the application and the implementation of the technology. This stresses the need for user-friendly secondary procedures [9].

Summary/Conclusion

The biggest benefit of using biometric devices is that they are extremely secure, thanks to the combination of low false-accept rates at moderate sensitivity settings, combined with a short user keypad code. At the same time, biometrics are extremely convenient and error-free, thanks to low false-reject rates. So, with the preceding in mind, it is now time to summarize, and bring this book to a conclusion.

Summary

As previously explained throughout the book, biometrics are best defined as measurable physiological and/or behavioral characteristics that can be utilized to verify the identity of an individual. They include fingerprints, retinal and iris scanning, hand geometry, voice patterns, facial recognition, and other techniques. They are of interest in any area where it is important to verify the true identity of an individual. Initially, these techniques were employed primarily in specialist high-security applications; however, we are now seeing their use and proposed use in a much broader range of public-facing situations [8].

So What Was Wrong with Cards and PINs?

Personal identification numbers (PINs) were one of the first identifiers to offer automated recognition. However, it should be understood that this means recognition of the PIN, not necessarily recognition of the person who has provided it. The same applies with cards and other tokens. You may easily recognize the token, but it could be presented by anybody. Using the two together provides a slightly higher confidence level, but this is still easily compromised if one is determined to do so [8].

A biometric, however, cannot be easily transferred between individuals and represents as unique an identifier as you are likely to see. If you can automate the verification procedure in a user-friendly manner, there is considerable scope for integrating biometrics into a variety of processes [8].

What Does This Mean in Practice?

It means that verifying an individual's identity can become both more streamlined (by the user interacting with the biometric reader) and considerably more accurate (as biometric devices are not easily fooled). In the context of travel and tourism, for example, one immediately thinks of immigration control, boarding gate identity verification, and other security-related functions. However, there may be a raft of other potential applications in areas such as marketing, premium passenger services, online booking, alliance programs, and so on where a biometric may be usefully integrated into a given process at some stage. In addition, there are organization-related applications such as workstation/LAN access, physical access control, and other potential applications [8].

This does not mean that biometrics are a panacea for all of your personal identification–related issues—far from it! But, they do represent an interesting new tool in your technology toolbox, which you might usefully consider as you march forward through the millennium [8].

But surely, this is all science fiction? Right? You don't see them working in everyday applications? Eighteen years ago, this was an often-heard response and, frankly, a justified one, as many of the early biometric devices were rather cumbersome in use and priced at a point that prohibited their implementation in all but a few very-high-security applications where they were considered viable [8].

These days, things are different; not only has considerable technical progress been made, providing more accurate, more refined products, but unit cost has dropped to a point that makes them suitable for broader scale deployment where appropriate. In addition, the knowledge base concerning their use and integration into other processes has increased dramatically. This is no longer a black art practiced by a few high priests (who charged accordingly), but an everyday piece of relevant technology that the average five-year-old will soon be able to tell you all about [8].

Conclusion

Governments need to provide their citizens and consumers with a trusted online environment. Identification systems are key interfaces between the real world and the digital world, though often they are invisible to users. Biometric technologies provide a strong mechanism for verification and therefore can promote the development of a trusted information society. Therefore, deploying biometric technologies comes at the right moment, as it will supply the increasing need for identification in modern societies that are becoming more mobile, flexible, and networked [9].

However, biometric technologies are still under development. Although some applications (in particular for law enforcement) have been around for a long time and have been developed on a large scale, it is only recently that advances in technology have enlarged the field of possible applications. They have also lowered their cost to a point where it now seems plausible that biometrics may be used for many more purposes. Fingerprint, iris, face, and DNA have different strengths and weaknesses, making each one more suitable for certain applications than for others; however, they all can be expected to spread in the foreseeable future [9].

The diffusion of biometrics is currently led by government applications with the aim of improving public security, such as the inclusion of biometric data in passports, but it will go far beyond these specific uses. As citizens get used to biometric identification in their dealings with border control and customs officials, the association with criminal behavior will diminish and people may be more prepared to accept the use of biometrics for other purposes. This could include physical access control to private property and logical access control (online identity), and even simply to enhance their convenience or for fun [9].

Of course, the main reason for introducing biometrics is to increase security and the sense of security. Although increased efficiency in law enforcement does not directly improve security, it can be argued that the use of biometrics acts as a deterrent to criminal, illegal, or antisocial activities. In this respect, overblown claims about the performance of biometrics may actually prove helpful [9].

Nevertheless, since biometric identification is not perfect, neither is biometric security. There will be many false rejections (travelers with valid documents rejected by the system) depending on the threshold, which will create irritation. More importantly, there will be cases of false acceptance (allowing intruders access to the system by accident, and an allowable scope for circumventing the checks ["spoofing"]). As the sense of security increases, the scope for fraud once inside the system will increase, too. Besides, criminals are likely to respond by changing tactics: If the only way to receive cash is with a live finger, using violence to get someone's fingerprint could replace stealing a credit card [9].

Beyond the use of biometrics for physical or logical access control, one other important attribute of biometrics is that they can allow confirmation of presence (by asking a person to provide a biometric sample, it means that person is physically present). This can be useful for places such as airport control towers, medical operating rooms, or drug dispensaries [9].

Biometrics could also deliver improved convenience for the citizen in their everyday life based on the principle that they are always with you and can therefore be effortlessly used at any time. For this purpose, it is necessary that they be intuitive to use and be nonintrusive during enrollment and data acquisition, regardless of which biometric is used. Such applications could range from fancy e-toys for children to a rapid supermarket checkout for their parents [9].

Then again, if biometrics are established as the only means of access, they have a great potential for inconvenience, too. If biometric access is faster than traditional means during the introduction period (but once established resumes the same speed as previous techniques, because now everybody uses it, or because the increased efficiency is used to cut back on staff), people will end up with an obligation to use biometrics without any corresponding advantage. They will perceive biometrics as an inconvenience. This will be particularly true for those whose biometric samples are prone to problems, which can be a significant percentage of the population. In addition, the more biometrics are used for everyday convenience, the more data or samples may be diffused and become compromised, thus making life more difficult [9].

Whether secure and convenient or not, the implementation of biometrics raises great privacy-related fears, such as fears of a "surveillance society" or "function creep." The worry from this perspective is that biometrics will become the common mode of identity recognition; biometric data will be linked to all other personal data; it may be subsequently shared with third parties for all kinds of other purposes; and sensitive information will be prone to abuse. In order to allay these fears, a reinforced legal framework for privacy and data protection may be needed, one that adequately addresses the new technological possibilities of biometrics, thus preventing biometrics from becoming a tool in the service of surveillance. The particularly strong need for effective privacy and data protection provisions regarding biometrics reflects the fact that biometric data are an inseparable part of you, while any document is merely an item at your disposal. Thus, there is nothing separating the individual and his or her biometrics [9].

On the other hand, a key feature of biometrics is that they have the potential to enhance privacy. This is because biometrics, if properly used, can establish identity without connecting this identity to other data sets such as a Social Security number or driver's license. Moreover, in verification mode, biometric systems are able to verify a person's access rights without revealing his or her identity. Better protection against identity theft [10] also protects the privacy of those who avoid becoming victims. In other words, since you carry all of your biometrics with you all of the time, it is easier to use multiple biometrics to compartmentalize your personal information. You might not be able to remember 10 secret codes, but you are able to provide 10 different biometric samples to separately access 10 different systems [9].

Other Key Aspects (SELT)

Security and privacy are the obvious challenges presented by the deployment of biometrics. In addition, this chapter has provided some insights on the social, economic, legal, and technical (SELT) implications of biometrics for society. From this contribution, the following subjects emerge as the key characteristics of the transition to a biometric society:

- Social
- Economic
- Legal
- Technical [9]

Social

The spread of biometrics and therefore the replacement of weak or no identification by strong identification may reduce the scope for privacy and anonymity of citizens. Implicitly, this may challenge the existing trust model between citizen and state. Currently, the technical limits to government efficiency provide an important pillar of citizen's freedom and autonomy. If governments become more efficient at identifying citizens in all kinds of situations, that trust model is likely to change [9].

Therefore, it is important to be clear on the purposes of introducing biometrics and realistic about their performance. Concerning the former, one has to consider the possibility that "function creep" will set in over time (that biometrics will be used for purposes other than those envisaged and agreed to at the time of introduction). For example, currently separate biometric databases could be connected at some later stage. If biometrics are sold as a magic wand against all threats to society, expectations are bound to be disappointing and citizens might come to feel cheated. In that case, the automated decision making (the delegation of control from human to machine) may be resented even more than it would otherwise [9].

Another crucial point to keep in mind is that biometrics cannot work alone, but needs a fall-back procedure. For various reasons, including disabilities, age, or sickness, a significant number of individuals might not be able to participate in an automated biometric identity verification process. Clear and equivalent procedures (with comparable security and ease of use), and without stigma, need to be foreseen for these people. If your fingerprint is not easily legible, that should not make you a second-class citizen [9].

Economic

Biometrics provide strong identification. However, economic theory reveals that the strongest available identification is not always the optimal solution, as identification imposes a cost, which will only be compensated by the benefits of identity if these benefits are large enough. Moreover, an assessment of costs of biometrics should not only look at the cost of technologies, but also encompass the complete identification process, including for instance, the costs of (human) backup procedures [9].

Strong identification changes the risk profile of circumventing the system. A stronger wall against illegal entry into an area or system will make additional inside measures less efficient, thus leading to their disappearance, which means that once the outer wall is breached, all doors are open to the intruder. As a result, identity theft for example, may simultaneously become less likely and more serious [9].

In terms of the market development, the biometrics market has a number of characteristics that make a competitive market equilibrium unlikely. It is a network industry with a strong complementary, a tendency to "tipping," a few large launch projects establishing considerable first-mover advantage, and ample scope to use intellectual property rights to reduce or even prevent competition. Therefore governments, as launch customers with strong bargaining power, should use their public procurement policy to ensure that the market does develop into a competitive one. A competitive market is therefore attained by using intellectual property in the public domain, such as open source software, or by spreading their procurement among several competitors, thus forcing interoperable solutions to emerge [9].

Legal

The current legal environment in Europe and the United States is flexible and does not hinder the introduction of biometrics. However, it contains very few specific provisions with regard to the impact of biometrics on privacy and data protection. Existing data protection legislation does influence the implementation of biometrics, but it lacks normative content and some interpretation problems remain. Hence, new legislation will be needed when new applications become mandatory or biometrics become widely used [9].

Such legislation should be based on two pillars: opacity and transparency. On the one hand, opacity rules (privacy rules prohibiting use) should prevent inappropriate collection of biometric data and lay down the conditions under which the use of biometrics should be allowed. On the other hand, if use is allowed, transparency rules (data protection rules regulating use) should set out how the data can be processed and how the processing can be traced. Currently, users are not encouraged to consider the repercussions of the enrollment process, even if strong identity is not required. An evaluation of whether a biometric application is appropriate and how it will operate should always consider local storage (for instance on a smart card), proportionality, whether a less intrusive method exists, reliability, and consent. In this context, data encryption should be mandatory [9].

There is one further consideration for the increasing use of biometrics in law enforcement. In judicial processes, parties should have the right to meet the expert and be heard, an automatic right to counterexpertise is needed, and the likelihood of errors must always be contemplated [9].

Technical

Biometrics are different from paper documents or secret codes. They cannot be lost or stolen (though they can be copied) and they cannot be revoked. Many (face, voice) are in the public domain. A biometric match is never 100% certain; the match depends as much on the threshold of acceptance as it does on the two sets of data to be compared. Individuals making verifications and those being verified need to be aware of the variability of the threshold and how that may vary according to the application. They should also be aware that the biometric technology itself is merely a part of the whole security system, which will work well only if the acquisition environment is properly set up, the storage is secure, and the enrollment process is sufficiently controlled [9].

Recommendations

The overall message from this chapter is very clear: The introduction of biometrics is not just a technological issue; it poses challenges to the way society is organized. Thus, these challenges need to be addressed in the near future if policy is to shape the use of biometrics rather than be overrun by it. To address these challenges, many issues have been identified in this chapter that may require action. The following five major recommendations are proposed here as the most urgent ones to be dealt with:

1. Ensure clarity of purpose;
2. Promote privacy-enhancing use of biometrics;
3. Allow for the emergence of a vibrant biometrics industry;
4. Provide for flexibility;
5. Conduct large-scale trials.

Ensure Clarity of Purpose

The purpose and the limitations of any application must be clearly set out in order for biometrics to become acceptable to citizens. Legislators can allay citizens' fears by providing appropriate safeguards for privacy and data protection, in particular preventing the so-called "function creep." Since there is more potential for abuse in biometrics than in traditional identification systems, especially if their use becomes widespread, the existing safeguards may need to be adapted in order to guarantee that the accepted principles of privacy, human rights, and data protection maintain their effective force. This means in particular that it should be considered whether the legal framework will need specific provisions on biometrics [9].

Promote Privacy-Enhancing Use of Biometrics

While biometrics certainly raise fears related to the erosion of privacy, they also have the opposite potential to enhance privacy, because they are able to verify a person's access rights without revealing his or her identity. In addition, by using multiple biometric features, it is possible to keep various sets of personal data separate from each other. The more policy encourages such privacy-enhancing uses of biometrics, the more biometrics will become acceptable to the public at large [9].

Allow for the Emergence of a Vibrant Biometric Industry

The large-scale introduction of biometric passports in Europe and the United States provides a great opportunity to ensure that these have a positive impact. As the launch customer of the largest-scale implementation by far in Europe and the United States, they can ensure the emergence of a vibrant European and U.S. industry by insisting on interoperability and open standards. Avoiding automatic market dominance by the passport supplier and a concentration of key intellectual property rights in a few hands will not only lower barriers for entry, but also ensure that the forthcoming competition will provide improved products and thus the creation of stronger global industrial actors [9].

Provide for Flexibility

A biometric identification system must be able to deal with all kinds of implementation problems. This involves setting up appropriate fall-back procedures for those with difficulties in providing biometric samples; developing the necessary ease of use for all involved groups, including elderly people, children, overweight, very tall, disabled, ill, ethnic minorities, etc.; and ensuring appropriate supervision and procedures to deal quickly and efficiently with the non-negligible numbers of false rejections. All these elements will have to be included in calculating the cost of an application [9].

Conduct Large-Scale Trials

Finally, large information technology projects always have substantial infancy problems, whether implemented by the public or the private sector. The large-scale deployment of biometrics for identification will not be any different. Law enforcement use of large-scale biometric databases cannot contribute sufficiently to enhancing your expertise, since the number of operations is limited, they are not time-constrained, and they work with significant human involvement. Thus, at this stage, there is a need for more field trials with a heterogeneous sample population (not just frequent flyers). On the basis of such field trials, the actual running costs would also become much clearer and thus could provide sufficient data to allow a realistic cost-benefit analysis [9].

References

1. Tracy V. Wilson, "How Biometrics Works," HowStuffWorks.com, c/o The Convex Group, One Capital City Plaza, 3350 Peachtree Road, Suite 1500, Atlanta, GA 30326, 2005. HowStuffWorks, Inc., Copyright © 1998–2006.

2. Hitachi Engineering & Services Co., Ltd., Head Office, 2-2, Saiwai-cho 3-chome, Hitachi-shi, Ibaraki, Japan, 2006. Copyright © Hitachi Engineering & Ltd. 2006. All Rights Reserved.

3. Iridian Technologies, Inc., Iridian Headquarters, 1245 Church Street, Suite 3, Moorestown, New Jersey, 08057, 2006. Copyright © 2003 Iridian Technologies, Reserved.

4. SOFTPRO GmbH, Wilhelmstrasse 34, 71034 Boeblingen, Germany, 2006.

5. Ingersoll Rand Recognition Systems, 1520 Dell Avenue, Campbell, CA 95008. Copyright 2006 Ingersoll-Rand Company.

6. "Convenience vs Security: How Well Do Biometrics Work," TopickZ Inc., Toronto, Ontario Canada, 2006.

7. Kym Gilhooly, "Biometrics: Getting Back to Business," Computerworld, One Speen Street, Framingham, MA 01701, May 09, 2005. Copyright © 2006 Computerworld Inc. All Rights Reserved.

8. Julian Ashbourn, "The Biometric White Paper" [© 2005 Applied Engineering, Inc., 7999 Knue Road, Suite 300, Indianapolis, Indiana 46250], 1999. Copyright Ashbourn 1999.

9. "Biometrics at the Frontiers, Assessing the Impact on Society." For the European Parliament Committee on Citizens' Freedoms and Rights, Justice and Home Affairs (LIBE), European Commission, Joint Research Center (DG JRC), Institute for Prospective Technological Studies, 2005. Copyright © 2005 European Communities.

10. John R. Vacca, *Identity Theft*, Prentice Hall, Professional Technical Reference, Pearson Education (2002).

11. John R. Vacca, *Net Privacy: A Guide to Developing and Implementing an Ironclad eBusiness Privacy Plan*, McGraw-Hill (2001).

12. John R. Vacca, *Satellite Encryption*, Academic Press (1999).

13. John R. Vacca, *The Essentials Guide to Storage Area Networks*, Prentice Hall, Professional Technical Reference, Pearson Education (2001).

14. John R. Vacca, *Wireless Data Demystified*, McGraw-Hill (January 2003).

15. John R. Vacca, *Practical Internet Security*, Springer (2006).

Glossary

Accuracy: A catch-all phrase for describing how well a biometric system performs. The actual statistic for performance will vary by task (verification, open-set identification (watchlist), and closed-set identification).

AFIS: Automated fingerprint identification system. Associated with criminal systems rather than civil fingerprint systems.

Algorithm: A limited sequence of instructions or steps that tells a computer system how to solve a particular problem. A biometric system will have multiple algorithms, for example, image processing, template generation, comparisons, etc.

ANSI (American National Standards Institute): A private, nonprofit organization that administers and coordinates the U.S. voluntary standardization and conformity assessment system. The mission of ANSI is to enhance both the global competitiveness of U.S. business and the U.S. quality of life by promoting and facilitating voluntary consensus standards and conformity assessment systems, and safeguarding their integrity.

Application Programming Interface (API): Formatting instructions or tools used by an application developer to link and build hardware or software applications.

Arch: A fingerprint pattern in which the friction ridges enter from one side, make a rise in the center, and exit on the opposite side. The pattern will contain no true delta point. See also delta point, loop, whorl.

Attempt: The submission of a single set of biometric sample to a biometric system for identification or verification. Some biometric systems permit more than one attempt to identify or verify an individual. See also biometric sample, identification, verification.

Authentication: The process of comparing a biometric sample against an existing biometric template already on file in an automated system. If a match is determined at the level of template comparison, the person is considered "authenticated."

Auto ID: A term used to identify an automated biometric system as compared to a manual identification system. The term applies to both one-to-one (1:1) verification and one-to-many (1:N) identification.

Automated Biometric Identification System (ABIS): (1) Department of Defense (DoD) system implemented to improve the U.S. government's ability to track and identify national security threats. The system includes mandatory collection of ten rolled fingerprints, a minimum of five mug shots from varying angles, and an oral swab to collect DNA. (2) Generic term sometimes used in the biometric community to discuss a biometric system. See also AFIS.

Automated Fingerprint Identification System (AFIS): A highly specialized biometric system that compares a submitted fingerprint record (usually of multiple fingers) to a database of records to determine the identity of an individual. AFIS is predominantly used for law enforcement, but is also being used for civil applications (background checks for soccer coaches, etc). See also IAFIS.

Behavioral Biometric: Not a physical characteristic. Behavioral biometrics are traits that are learned or acquired over time as differentiated from physical characteristics. Some examples are voice authentication, signature recognition, and keystroke recognition.

Behavioral Biometric Characteristic: A biometric characteristic that is learned and acquired over time rather than one based primarily on biology. All biometric characteristics depend somewhat upon both behavioral and biological characteristics. Examples of biometric modalities for which behavioral characteristics may dominate include signature recognition and keystroke dynamics. See also biological biometric characteristic.

Benchmarking: The process of setting objective performance criteria tests to evaluate biometric systems against each other. Benchmarking typically measures accuracy and system throughput along with functionality and operational simplicity.

Bifurcation: The point in a fingerprint where a friction ridge divides or splits to form two ridges. See also friction ridge, minutia(e) point, ridge ending.

BIN or Binning: The process of classifying biometric data. This term is primarily used in conjunction with AFIS systems to speed system searches. In AFIS systems, fingerprints are evaluated for type, such as whorl, loop, or arch. The technique of binning, however, can be used with any biometric classification system.

BioAPI (Biometric Application Programming Interface): Defines the application programming interface and service provider interface for a standard biometric technology interface. The BioAPI enables biometric devices to be easily installed, integrated, or swapped within the overall system architecture.

Biological Biometric Characteristic: A biometric characteristic based primarily on an anatomical or physiological characteristic, rather than a learned behavior. All biometric characteristics depend somewhat upon both behavioral and biological characteristics. Examples of biometric modalities for which biological characteristics may dominate include fingerprint and hand geometry. See also behavioral biometric characteristic.

Biometric: A measurable physical characteristic or personal behavioral trait that is used to recognize or authenticate the claimed identity of a person.

Biometric Consortium (BC): An open forum to share information throughout government, industry, and academia.

Biometric Data: A catch-all phrase for computer data created during a biometric process. It encompasses raw sensor observations, biometric samples, models, templates, and/or similarity scores. Biometric data is used to describe the information collected during an enrollment, verification, or identification process, but does not apply to end-user information such as user name, demographic information, and authorizations.

Biometric Sample: Information or computer data obtained from a biometric sensor device. Examples are images of a face or fingerprint.

Biometric System: Multiple individual components (such as sensor, matching algorithm, and result display) that combine to make a fully operational system. A biometric system is an automated system capable of (1) capturing a biometric sample from an end user, (2) extracting and processing the biometric data from that sample, (3) storing the extracted information in a database, (4) comparing the biometric data with data contained in one or more references, (5) deciding how well they match and indicating whether or not an identification or verification of identity has been achieved. A biometric system may be a component of a larger system.

Biometrics: The science of automatic identification or identity verification of individuals using physiological or behavioral characteristics.

Biometric Applications: Some examples are: banking, ATM access, safe deposit access, physical access control, time and attendance monitoring, benefit payment systems, border control/passports, and PC/network access control.

Biometric System: An automated system that is capable of capturing a biometric sample from an individual and extracting data to construct a reference template. The template can then be used in various matching scenarios that can be used to authenticate the identity of an individual.

Biometric Taxonomy: A method of classifying types of biometric systems. San Jose State University researchers created an industry-accepted classification system for use in describing various automated biometric systems. Those classifications are cooperative versus noncooperative user, overt versus covert biometric systems, habituated versus nonhabituated users, supervised versus unsupervised users, and standard environments versus nonstandard environments

Biometric Technologies: As of 2006, some of the major recognized automated biometric technologies include body odor, DNA, finger imaging (AFIS), facial recognition, facial thermogram recognition, hand geometry or recognition, iris recognition, gait recognition, live grip recognition, palm recognition, retinal recognition, signature verification, skin print recognition, vein recognition, and voice authentication.

Body Odor: See odor recognition.

Capture: The process of collecting a biometric sample from an end user. Biometric data is captured, digitized, and entered into a database.

CBEFF (Common Biometric Exchange File Format): A standard that provides the ability for a system to identify, and interface with, multiple biometric systems, and to exchange data between system components.

CCD (Charge Coupled Device): A CCD is a semiconductor device that records images electronically. Utilized by some biometric sensors.

Challenge Response: A method used to confirm the presence of a person by eliciting direct responses from the individual. Responses can be either voluntary or involuntary. In a voluntary response, the end user will consciously react to something that the system presents. In an involuntary response, the end user's body automatically responds to a stimulus. A challenge response can be used to protect the system against attacks. See also liveness detection.

Claim of Identity: A statement that a person is or is not the source of a reference in a database. Claims can be positive (I am in the database), negative (I am not in the database), or specific (I am end-user 123 in the database).

Closed-Set Identification: A biometric task where an unidentified individual is known to be in the database and the system attempts to determine his or her identity. Performance is measured by the frequency with which the individual appears in the system's top rank (or top 5, 10, etc.). See also identification, open-set identification.

CMOS (Complementary Metal Oxide Semiconductor): An integrated circuit used by some biometric sensors.

Comparison: The process of comparing a biometric sample with a previously stored reference template. The biometric template is compared with a new sample.

Cooperative User: An individual that willingly provides his or her biometric to the biometric system for capture. Example: A worker submits his or her biometric to clock in and out of work. See also indifferent user, noncooperative user, uncooperative user.

Core Point: The "center(s)" of a fingerprint. In a whorl pattern, the core point is found in the middle of the spiral/circles. In a loop pattern, the core point is found in the top region of the innermost loop. More technically, a core point is defined as the topmost point on the innermost upwardly curving friction ridgeline. A fingerprint may have multiple cores or no cores. See also arch, delta point, friction ridge, loop, whorl.

Covert: An instance in which biometric samples are being collected at a location that is not known to bystanders. An example of a covert environment might involve an airport checkpoint where face images of passengers are captured and compared to a watchlist without their knowledge. See also noncooperative user, overt.

Crossover Error Rate (CER): See equal error rate (EER).

Cumulative Match Characteristic (CMC): A method of showing measured accuracy performance of a biometric system operating in the closed-set identification task. Templates are compared and ranked based on their similarity. The CMC shows how often the individual's template appears in the ranks (1, 5, 10, 100, etc.), based on the match rate.

Database: A collection of one or more computer files. For biometric systems, these files could consist of biometric sensor readings, templates, match results, related end-user information, etc. See also gallery.

Decision: The resultant action taken (either automated or manual) based on a comparison of a similarity score (or similar measure) and the system's threshold. See also comparison, similarity score, threshold.

Degrees of Freedom: A statistical measure of how unique biometric data is. Technically, it is the number of statistically independent features (parameters) contained in biometric data.

Delta Point: Part of a fingerprint pattern that looks similar to the Greek letter delta. Technically, it is the point on a friction ridge at or nearest to the point of divergence of two type lines, and located at or directly in front of the point of divergence. See also core point, friction ridge.

Detection and Identification Rate: The rate at which individuals who are in a database are properly identified in an open-set identification (watchlist) application. See also open-set identification, watchlist.

Detection Error Trade-off (DET) Curve: A graphical plot of measured error rates. DET curves typically plot matching error rates (false nonmatch rate vs. false match rate) or decision error rates (false-reject rate vs. false-accept rate). See also receiver operating curves.

Difference Score: A value returned by a biometric algorithm that indicates the degree of difference between a biometric sample and a reference. See also hamming distance, similarity score.

DNA (Deoxyribonucleic Acid): The building blocks of life. A unique genetic map of an individual's characteristics that is contained in each cell in every living thing. While this technology is considered a biometric, there are no completely unattended automated systems in existence.

D Prime: A statistical measure of how well a biometric system can discriminate between different individuals. The larger the D prime value, the better a biometric system is at discriminating between individuals.

Earlobe Recognition: Similar to facial recognition in that the user presents an earlobe to a camera for automated evaluation and comparison. I am not aware of any commercially available systems. This technique has been utilized in forensics to identify individuals. A physiological biometric.

Eavesdropping: Surreptitiously obtaining data from an unknowing end-user who is performing a legitimate function. An example involves having a hidden sensor co-located with the legitimate sensor. See also skimming.

EFTS (Electronic Fingerprint Transmission Specification): A document that specifies requirements to which agencies must adhere to communicate electronically with the Federal Bureau of Investigation's Integrated Automated Fingerprint Identification System (IAFIS). This specification facilitates information sharing and eliminates the delays associated with fingerprint cards. See also Integrated Automated Fingerprint Identification System (IAFIS).

Encryption: The act of transforming data into an unintelligible form so that it cannot be read by unauthorized individuals. A key or a password is used to decrypt (decode) the encrypted data.

End User: A person who interacts with a biometric system to enroll or have his or her identity checked.

Enrollee: An individual who has a biometric reference template on file in an automated biometric system.

Enrollment: The process of collecting a biometric sample from an end user, converting it into a biometric reference, and storing it in the biometric system's database for later comparison.

Enrollment Time: The amount of time it takes for an enrollee to complete the process on enrolling in a biometric system. The larger the number of unique records in the system, the longer it may take to complete the enrollment (1:N).

Equal Error Rate: The error rate occurring when the threshold of a system is set so that the proportion of false rejections will be approximately equal to the proportion of false acceptances.

Extraction: The process of converting a captured biometric sample into biometric data so that it can be compared to a reference template. A biometric template is created from this measurable data unique to the individual.

Face Recognition: A biometric modality that uses an image of the visible physical structure of an individual's face for recognition purposes.

Facial Recognition: A physiological biometric that analyzes facial features. Facial recognition can detect a person using static digital photographs or live video feeds. Normally, the system locates and tracks a person's head first. Depending upon the technology used, the system will extract a facial image using local feature extraction, eigenface comparison, or other methods to isolate unique aspects. These unique aspects, measurements, or features are then digitized into a template representation for comparison purposes. Popular applications include banking/financial/ATMs, access control, time and attendance, surveillance, antifraud welfare systems, passport, and general law enforcement.

Facial Recognition, Infrared: A technology announced in 2002 that is reportedly capable of performing infrared facial recognition and "continuous condition monitoring of individuals" using passive infrared imaging that is totally noncontact, noninvasive and works under any lighting conditions, even total darkness. A physiological biometric, the technology is reported to include automatic and continual validation that a person is present, alive, awake, alert, and attentive.

Failure to Acquire (FTA): Failure of a biometric system to capture and/or extract usable information from a biometric sample.

Failure to Enroll (FTE): The inability to enroll in a biometric system. The sample provided at enrollment is inadequate. For example, a finger is not properly placed on the sensing device, the fingerprint is not readable due to the physical condition of the finger, or the enrollment parameters of the system reject the sample. Some systems may accept the best of N attempts.

False Acceptance: When a biometric system incorrectly identifies an individual or incorrectly verifies an imposter against a claimed identity.

False-Accept Rate (FAR): The probability that a biometric system will incorrectly identify an individual or will fail to detect an impostor. When a system's FAR is too high, the threshold for the FAR is set too low. The false-accept rate may be estimated as: FAR = NFA (number of false acceptances)/NIIA (number of impostor identification attempts). The same formula can be expressed as FAR = NFA/NIVA (number of impostor verification attempts).

False Alarm Rate: A statistic used to measure biometric performance when operating in the open-set identification (sometimes referred to as watchlist) task. This is the percentage of times an alarm is incorrectly sounded on an individual who is not in the biometric system's database (the system alarms on Frank when Frank isn't in the database), or an alarm is sounded but the wrong person is identified (the system alarms on John when John is in the database, but the system thinks John is Steve).

False Match Rate: A statistic used to measure biometric performance. Similar to the false-accept rate (FAR).

False Non-Match Rate: A statistic used to measure biometric performance. Similar to the false-reject rate (FRR), except the FRR includes the failure to acquire error rate and the false non-match rate does not.

False-Reject Rate (FRR): The probability that a biometric system will fail to identify an enrollee, or verify the legitimate claimed identity of an enrollee. When a system's FRR is too high, the threshold for the FRR is set too high. Note that FAR and FRR are inversely related. The false-reject rate may be estimated as FRR = NFR (number of false rejections)/NEIA (number of enrollee identification attempts). The same formula can be expressed as FRR = NFR/NEVA (number of enrollee verification attempts).

Feature(s): Distinctive mathematical characteristic(s) derived from a biometric sample; used to generate a reference. See also extraction, template.

Feature Extraction: See extraction.

FERET (FacE REcognition Technology program): A face recognition development and evaluation program sponsored by the U.S. government from 1993 through 1997. See also FRGC, FRVT.

Filtering: The process of classifying biometric data according to information that is unrelated to the actual biometric data itself. Examples of this are information about the enrollee such as sex, age, etc. This term is sometimes used in conjunction with AFIS systems.

Finger Geometry: This technology uses the first two fingers of either hand. Users first make the "V for Victory" sign as their two fingers are placed on a platen. A three-dimensional digital image of the two-finger geometry is captured via a CCD camera. A physiological biometric, its enrollment data (template) of only 20 bytes may be stored inside the camera module or can be transmitted to other units.

Finger Imaging: A technology largely associated with law enforcement. A physiological biometric. Fingerprints are evaluated for type (whorl, loop, or arch) and minutiae is extracted. Current popular applications include AFIS/police/FBI, computer network access, time and attendance systems, welfare ID systems, voter registration systems, and physical access control systems. Criminal systems usually rely on all 10 fingerprints. Most civil systems utilize two index fingers or thumbs.

Fingerprint Recognition: A biometric modality that uses the physical structure of an individual's fingerprint for recognition purposes. Important features used in most fingerprint recognition systems are minutiae points that include bifurcations and ridge endings. See also bifurcation, core point, delta point, minutia(e) point.

FpVTE (Fingerprint Vendor Technology Evaluation): An independently administered technology evaluation of commercial fingerprint-matching algorithms.

FRGC (Face Recognition Grand Challenge): A face recognition development program sponsored by the U.S. government from 2003–2006. See also FERET, FRVT.

Friction Ridge: The ridges present on the skin of the fingers and toes, and on the palms and soles of the feet, that make contact with an incident surface under normal touch. On the fingers, the distinctive patterns formed by the friction ridges make up the fingerprints. See also minutia(e) point.

FRVT (Face Recognition Vendor Test): A series of large-scale independent technology evaluations of face recognition systems. The evaluations have occurred in 2000, 2002, and 2006. See also FRGC, FERET.

Gait: An individual's manner of walking. This behavioral characteristic is in the research and development stage of automation.

Gait Recognition: The system that can recognize an individual by the way they walk. This is a behavioral biometric. The system "... computes optical flow for an image sequence of a person walking, and then characterizes the shape of the motion with a set of sinusoidally varying scalars. Feature vectors composed of the phases of the sinusoids are able to discriminate among people."

Gallery: The biometric system's database, or set of known individuals, for a specific implementation or evaluation experiment. See also database, probe.

Goats: In voice authentication applications, goats are speakers who are exceptionally unsuccessful at being accepted.

Hamming Distance: The number of disagreeing bits between two binary vectors. This is used as a measure of dissimilarity.

Hand Geometry/Hand Recognition: Technology that primarily uses the front part of the hand. This is a physiological biometric. The hand is placed upon a reflective platen surface. Pegs on the platen guide hand placement. Mirrors are used to establish a 3D view of the hand. An infrared LED is used as a light source to capture the image. An image is captured by CCD sensor. Ninety (90) different measurements are taken of the hand. Popular applications include physical access control systems, time and attendance systems, voter registration, student meal access programs, immigration control.

Human Services Biometric ID Systems: Any biometric system used by government entitlement agencies to prevent the provision of duplicate benefits to the same individual.

ICE (Iris Challenge Evaluation): A large-scale development and independent technology evaluation activity for iris recognition systems sponsored by the U.S. government in 2006.

Identification: The one-to-many (1:N) process of comparing a biometric sample against all biometric templates in a system to determine if there is a match with any of the samples on file. Answers the questions: "Who is this? Is this person already known to the system under a different identity?"

Identification Rate: The rate at which an individual in a database is correctly identified.

Impostor: A person who submits a biometric sample in either an intentional or inadvertent attempt to claim the identity of another person to a biometric system. See also attempt.

INCITS (International Committee for Information Technology Standards): Organization that promotes the effective use of information and communication technology through standardization in a way that balances the interests of all stakeholders and increases the global competitiveness of the member organizations. See also ANSI, ISO, NIST.

Indifferent User: An individual who knows his or her biometric sample is being collected and does not attempt to help or hinder the collection of the sample. For example, an individual, aware that a camera is being used for face recognition, looks in the general direction of the sensor, neither avoiding nor directly looking at it. See also cooperative user, noncooperative user, uncooperative user.

Infrared: Light that lies outside the human visible spectrum at its red (low-frequency) end.

Integrated Automated Fingerprint Identification System (IAFIS): The FBI's large-scale 10 fingerprint (open-set) identification system that is used for criminal history background checks and identification of latent prints discovered at crime scenes. This system provides automated and latent search capabilities, electronic image storage, and electronic exchange of fingerprints and responses. See also AFIS.

IrisCode$^{\copyright}$: A biometric feature format used in the Daugman iris recognition system.

Iris Recognition: This technology uses unique iris patterns such as radial furrows, crypts, collarettes, pigment frills, and pits. This is a physiological biometric. An image is captured using monochrome CCD cameras. Zones of analysis are established. Pupil dilation is accounted for. The captured image is translated into a biometric template and encrypted (512-byte iris code). It is said that iris patterns are quite unique and do not change with age. The technology vendor claims that the probability of two individuals having the same iris pattern is 1 in 10^{78}. The current population of Earth is approximately 10^{10}. The entire population that has ever lived is approximately 10^{11}. Popular applications include financial services (ATMs), access control, computer security, public safety and justice, time and attendance, and airport security.

ISO (International Organization for Standardization): A nongovernmental network of the national standards institutes from 151 countries. The ISO acts as a bridging organization in which a consensus can be reached on solutions that meet both the requirements of business and the broader needs of society, such as the needs of stakeholder groups like consumers and users. See also ANSI, INCITS, NIST.

Keystroke Dynamics: Technology that analyzes the characteristics of one's typing. Users enroll by typing the same word or words a number of times. Verification is based on the concept that the rhythm with which one types is distinctive. This is considered a behavioral biometric. This works best for users who can touch type. This biometric prevents unauthorized access to a computer by comparing the typing rhythm of an intruder to the typing rhythm of the computer's owner. With this protection, only the computer's owner will have access to the system, even if her password has been compromised by an intruder.

Lambs: In voice authentication applications, lambs are speakers who are exceptionally vulnerable to impersonation by others.

Latent Fingerprint: A fingerprint "image" left on a surface that was touched by an individual. The transferred impression is left by the surface contact with the friction ridges, usually caused by the oily residues produced by the sweat glands in the finger. See also friction ridge.

Live Capture: Typically refers to a fingerprint capture device that electronically captures fingerprint images using a sensor (rather than scanning ink-based fingerprint images on a card or lifting a latent fingerprint from a surface). See also sensor.

LiveGrip Recognition: A new technology reported to be significantly different from other commercially available systems, according to the developer. LiveGrip technology analyzes highly unique internal features of the human hand such as veins, arteries, and fatty tissues. LiveGrip uses infrared light to scan and read the patterns of tissue and blood vessels under the skin of the hand presented in the gripped pose. The technology completely maps the substructure of the person's hand, and 16 scans are then taken. This is a physiological biometric.

Liveness Detection: A technique used to ensure that the biometric sample submitted is from an end user. A liveness detection method can help protect the system against some types of spoofing attacks. See also challenge response, mimic, spoofing.

Loop: A fingerprint pattern in which the friction ridges enter from either side, curve sharply, and pass out near the same side they entered as. This pattern will contain one core and one delta. See also arch, core point, delta point, friction ridge, whorl.

Match: A decision that a biometric sample and a stored template comes from the same human source, based on their high level of similarity (difference or hamming distance). See also false match rate, false non-match rate.

Matching: The process of comparing a biometric sample against a previously stored template and scoring the level of similarity (difference or hamming distance). Systems then make decisions based on this score and its relationship (above or below) a predetermined threshold. See also comparison, difference score, threshold.

Match/Non-Match: The existing biometric template matches the new biometric sample or it does not.

Mimic: The presentation of a live biometric measure in an attempt to fraudulently impersonate someone other than the submitter. See also challenge response, liveness detection, spoofing.

Minutia(e) Point: Friction ridge characteristics that are used to individualize a fingerprint image. Minutiae are the points where friction ridges begin, terminate, or split into two or more ridges. In many fingerprint systems, the minutiae (as opposed to the images) are compared for recognition purposes. See also friction ridge, ridge ending.

Modality: A type or class of biometric system. For example: face recognition, fingerprint recognition, iris recognition, etc.

Model: A representation used to characterize an individual. Behavioral-based biometric systems, because of the inherently dynamic characteristics, use models rather than static templates. See also template.

Multimodal Biometric System: A biometric system in which two or more of the modality components (biometric characteristic, sensor type, or feature extraction algorithm) occurs in multiple.

Multiple Biometric System: An automated biometric system that uses more than one type of biometric technology for identification or recognition.

Neural Net/Neural Network: A type of algorithm that learns from past experience to make decisions. See also algorithm.

NIST (National Institute of Standards and Technology): A nonregulatory federal agency within the U.S. Department of Commerce that develops and promotes measurement, standards, and technology to enhance productivity, facilitate trade, and improve the quality of life. NIST's measurement and standards work promotes the well-being of the nation and helps improve, among many others things, the nation's homeland security. See also ANSI, INCITS, ISO.

Noise: Unwanted components in a signal that degrade the quality of data or interfere with the desired signals processed by a system.

Noncooperative User: An individual who is not aware that his or her biometric sample is being collected. Example: A traveler passing through a security line at an airport is unaware that a camera is capturing his or her face image. See also cooperative user, indifferent user, uncooperative user.

Odor or Smell Recognition: Technology under development. Currently systems can recognize up to 30 different chemical elements. Some think that our body odor may be unique to each individual.

One-to-Many (1:N): Synonym for identification.

One-to-One (1:1): Synonym for verification.

Open-Set Identification: Biometric task that more closely follows operational biometric system conditions to (1) determine if someone is in a database and (2) find the record of the individual in the database. This is sometimes referred to as the "watchlist" task to differentiate it from the more commonly referenced closed-set identification. See also closed-set identification, identification.

Operational Evaluation: One of the three types of performance evaluations. The primary goal of an operational evaluation is to determine the workflow impact seen by the addition of a biometric system. See also technology evaluation, scenario evaluation.

Overt: Biometric sample collection where end users know they are being collected and at what location. An example of an overt environment is the US-VISIT program where non-U.S. citizens entering the United States submit their fingerprint data. See also covert.

Palm Print Recognition: A biometric modality that uses the physical structure of an individual's palm print for recognition purposes.

Palm Recognition: Like fingerprints, all palm prints are unique. Automated palm recognition systems are relatively new. Palm prints are acquired in a manner similar to fingerprints, utilizing larger surface area scanners. This is a physiological biometric.

Performance: A catch-all phrase for describing a measurement of the characteristics, such as accuracy or speed, of a biometric algorithm or system. See also accuracy, crossover error rate, cumulative match characteristics, d-prime, detection error tradeoff, equal error rate, false accept rate, false alarm rate, false match rate, false reject rate, identification rate, operational evaluation, receiver operating characteristics, scenario evaluation, technology evaluation, true accept rate, true reject rate, verification rate.

Physiological Biometrics: Recognition based upon physical characteristics. Some examples are fingerprint, hand geometry, iris recognition, retinal scanning, and facial recognition.

PIN (Personal Identification Number): A security method used to show "what you know." Depending on the system, a PIN could be used to either claim or verify a claimed identity.

Pixel: A picture element. This is the smallest element of a display that can be assigned a color value. See also pixels per inch (PPI), resolution.

Pixels Per Inch (PPI): A measure of the resolution of a digital image. The higher the PPI, the more information is included in the image, and the larger the file size. See also pixel, resolution.

Population: The set of potential end users for an application.

Probe: The biometric sample that is submitted to the biometric system to compare against one or more references in the gallery. See also gallery.

Radio-Frequency Identification (RFID): Technology that uses low-powered radio transmitters to read data stored in a transponder (tag). RFID tags can be used to track assets, manage inventory, authorize payments, and serve as electronic keys. RFID is not a biometric.

Receiver Operating Characteristics (ROC): A method of showing measured accuracy performance of a biometric system. A verification ROC compares false accept rate vs. verification rate. An open-set identification (watchlist) ROC compares false alarm rates vs. detection and identification rate.

Receiver Operating Curves (ROC): A graph that shows how the false-rejection rate and false-acceptance rate vary according to threshold.

Recognition: A generic term used in the description of biometric systems (face recognition or iris recognition) relating to their fundamental function. The term "recognition" does not inherently imply the verification, closed-set identification or open-set identification (watchlist).

Record: The template and other information about the end user (name, access permissions).

Reference: The biometric data stored for an individual for use in future recognition. A reference can be one or more templates, models or raw images. See also template.

Resolution: The number of pixels per unit distance in the image. Describes the sharpness and clarity of an image. See also pixel, pixels per inch (PPI).

Retinal Scanning: A technology that uses the pattern established by blood vessels in the retina. A 360-degree circular scan is taken using a low-intensity light source such as a laser. Approximately 400 readings are taken that produce 192 reference points. These reference points are reduced to a 96-byte biometric template. Popular applications include military or high security. The Illinois Department of Social Services experimented briefly with this technology.

Ridge Ending: A minutiae point at the ending of a friction ridge. See also bifurcation, friction ridge.

Rolled Fingerprints: An image that includes fingerprint data from nail to nail, obtained by "rolling" the finger across a sensor.

Sample: Term relates to the act of an individual interacting with a biometric system for the purpose of enrollment, identification, or verification. The biometric data collected by the system is referred to as a sample.

Scan: Term relates to the process of utilizing a sensor to read a user's sample into a system, for enrollment, identification, or verification. The term "live scan" is often used to refer to when a user's sample is taken for the purposes of identification or verification.

Scenario Evaluation: One of the three types of performance evaluations. The primary goal of a scenario evaluation is to measure performance of a biometric system operating in a specific application. See also technology evaluation, operational evaluation.

Score: The level of similarity from comparing a biometric sample against a previously stored sample.

Segmentation: The process of parsing the biometric signal of interest from the entire acquired data system. For example, finding individual finger images from a slap impression.

Sensor: Hardware found on a biometric device that converts biometric input into a digital signal and conveys this information to the processing device.

Sensor Aging: The gradual degradation in performance of a sensor over time.

Sheep: In voice authentication applications, sheep are speakers who exhibit good true speaker acceptance.

Signature Dynamics: A behavioral biometric modality that analyzes dynamic characteristics of an individual's signature, such as shape of signature, speed of signing, pen pressure when signing, and pen-in-air movements for recognition.

Signature Verification: Considered a behavioral biometric rather than an anatomical biometric. From the user's perspective, this is a natural and familiar action. It analyzes both the appearance and the dynamics inherent in signing your name, for example. Pressure, writing speed, and variation of speed, time to complete are all used as part of the analysis.

Similarity Score: A value returned by a biometric algorithm that indicates the degree of similarity or correlation between a biometric sample and a reference. See also difference score, hamming distance.

Skimming: The act of obtaining data from an unknowing end user who is not willingly submitting the sample at that time. An example could be secretly reading data while in close proximity to a user on a bus. See also eavesdropping.

Slap Fingerprint: Fingerprints taken by simultaneously pressing the four fingers of one hand onto a scanner or a fingerprint card. Slaps are also known as four-finger simultaneous plain impressions.

Speaker Recognition: A biometric modality that uses an individual's speech, a feature influenced by both the physical structure of an individual's vocal tract and the behavioral characteristics of the individual, for recognition purposes. Sometimes referred to as voice recognition. Speech recognition recognizes the words being said, and is not a biometric technology. See also speech recognition, voice recognition.

Speaker Recognition Evaluations: An ongoing series of evaluations of speaker recognition systems.

Speaker Verification: See voice verification.

Speech Recognition: A technology that enables a machine to recognize spoken words. Speech recognition is not a biometric technology. See also speaker recognition, voice recognition.

Spoofing: The ability to fool or "spoof" a biometric sensor. Incorrectly verifying an impostor against a claimed identity on file in a biometric system.

Submission: The process whereby an end user provides a biometric sample to a biometric system. See also capture.

Technology Evaluation: One of the three types of performance evaluations. The primary goal of a technology evaluation is to measure performance of biometric systems, typically only the recognition algorithm component, in general tasks. See also operational evaluation, scenario evaluation.

Template: Usually a proprietary mathematical representation of biometric data that represents the biometric measurement of an enrollee. Any graphical representation is reduced to a numerical representation. The template is then used by the biometric system as an efficient method to make comparisons with other templates stored in the system. The trend in automated biometric systems is toward the use of templates rather than, for example, images of a fingerprint.

Threat: An intentional or unintentional potential event that could compromise the security and integrity of the system. See also vulnerability.

Threshold: An adjustable means by which biometric system operators can be more or less strict in how efficient a match score is used to accept or reject matches.

Throughput Rate: The number of enrollment records that a biometric system can process within a given time interval. This is a very important number to have a good handle on when considering enrolling a large population group.

Token: A physical object that indicates the identity of its owner. For example, a smart card.

True Accept Rate: A statistic used to measure biometric performance when operating in the verification task. The percentage of times a system (correctly) verifies a true claim of identity. For example, Frank claims to be Frank and the system verifies the claim.

True Reject Rate: A statistic used to measure biometric performance when operating in the verification task. The percentage of times a system (correctly) rejects a false claim of identity. For example, Frank claims to be John and the system rejects the claim.

Type I Error: An error that occurs in a statistical test when a true claim is (incorrectly) rejected. For example, John claims to be John, but the system incorrectly denies the claim. See also false-reject rate (FRR).

Type II Error: An error that occurs in a statistical test when a false claim is (incorrectly) not rejected. For example: Frank claims to be John and the system verifies the claim. See also false-accept rate (FAR).

Uncooperative User: An individual who actively tries to deny the capture of his or her biometric data. Example: A detainee mutilates his or her finger upon capture to prevent the recognition of his or her identity via fingerprint. See also cooperative user, indifferent user, noncooperative user.

User: A person, such as an administrator, who interacts with or controls end users' interactions with a biometric system. See also cooperative user, end user, indifferent user, noncooperative user, uncooperative user.

US-VISIT (U.S. Visitor and Immigrant Status Indicator Technology): A continuum of security measures that begins overseas at the Department of State's visa-issuing posts, and continues through arrival and departure from the United States. Using biometric, such as digital, inkless fingerscans and digital photographs, the identity of visitors requiring a visa is now matched at each step to ensure that the person crossing the U.S. border is the same person who received the visa. For visa-waiver travelers, the capture of biometrics first occurs at the port of entry to the United States. By checking the biometrics of a traveler against its databases, US-VISIT verifies whether the traveler has previously been determined inadmissible, is a known security risk (including having outstanding warrants), or has previously overstayed the terms of a visa. These entry and exit procedures address the U.S. critical need for tighter security and ongoing commitment to facilitate travel for the millions of legitimate visitors welcomed each year to conduct business, learn, see family, or tour the country.

Veincheck: This technology scans the vein patterns in the back of the hand or wrist. Not widely used at this point in time.

Verification: The process of comparing a submitted biometric sample against the biometric reference template of a single enrollee (1:1) whose identity is being claimed. Answers the question: "Is the person who they claim to be?"

Verification Rate: A statistic used to measure biometric performance when operating in the verification task. The rate at which legitimate end users are correctly verified.

Voice Authentication: Technology that uses spoken words or phrases. Systems typically measure the dynamics of user annunciation. The physical construction of an individual's vocal cords, palate, teeth, sinus cavities, and mouth all impact our speech characteristics. In automated voice verification systems, background noise is filtered out to focus on key features. Samples are typically acquired using a telephone handset or PC microphones. Popular applications include network access, password reset, financial services (phone banking, etc.), physical access control, and e-commerce.

Voice Recognition: See speaker recognition.

Vulnerability: The potential for the function of a biometric system to be compromised by intent (fraudulent activity); design flaw (including usage error); accident; hardware failure; or external environmental condition. See also threat.

Watchlist: A term sometimes referred to as open-set identification that describes one of the three tasks that biometric systems perform. Answers the questions: Is this person in the database? If so, who are they? The biometric system determines if the individual's biometric template matches a biometric template of someone on the watchlist. The individual does not make an identity claim, and in some cases does not personally interact with the system whatsoever. See also closed-set identification, identification, open-set identification, verification.

Wavelet Scalar Quantization (WSQ): An FBI-specified compression standard algorithm that is used for the exchange of fingerprints within the criminal justice community. It is used to reduce the data size of images.

Whorl: A fingerprint pattern in which the ridges are circular or nearly circular. The pattern will contain two or more deltas. See also arch, delta point, loop, minutia(e) point.

Wolves: In voice authentication applications, wolves are speakers who are exceptionally successful at impersonating others.

WSQ (Wavelet Transform/Scalar Quantization): A compression algorithm used to reduce the size of biometric data. An example is WSQ compression of a gray-scale graphic of a fingerprint. Compressed images are especially efficient in speeding up the transmission of data back to the server.

Index

Printed and bound by CPI Group (UK) Ltd, Croydon, CR0 4YY

03/10/2024

01040338-0007